EAI/Springer Innovations in Communication and Computing

Series Editor

Imrich Chlamtac, European Alliance for Innovation, Ghent, Belgium

Editor's Note

The impact of information technologies is creating a new world yet not fully understood. The extent and speed of economic, life style and social changes already perceived in everyday life is hard to estimate without understanding the technological driving forces behind it. This series presents contributed volumes featuring the latest research and development in the various information engineering technologies that play a key role in this process.

The range of topics, focusing primarily on communications and computing engineering include, but are not limited to, wireless networks; mobile communication; design and learning; gaming; interaction; e-health and pervasive healthcare; energy management; smart grids; internet of things; cognitive radio networks; computation; cloud computing; ubiquitous connectivity, and in mode general smart living, smart cities, Internet of Things and more. The series publishes a combination of expanded papers selected from hosted and sponsored European Alliance for Innovation (EAI) conferences that present cutting edge, global research as well as provide new perspectives on traditional related engineering fields. This content, complemented with open calls for contribution of book titles and individual chapters, together maintain Springer's and EAI's high standards of academic excellence. The audience for the books consists of researchers, industry professionals, advanced level students as well as practitioners in related fields of activity include information and communication specialists, security experts, economists, urban planners, doctors, and in general representatives in all those walks of life affected ad contributing to the information revolution.

Indexing: This series is indexed in Scopus, Ei Compendex, and zbMATH.

About EAI

EAI is a grassroots member organization initiated through cooperation between businesses, public, private and government organizations to address the global challenges of Europe's future competitiveness and link the European Research community with its counterparts around the globe. EAI reaches out to hundreds of thousands of individual subscribers on all continents and collaborates with an institutional member base including Fortune 500 companies, government organizations, and educational institutions, provide a free research and innovation platform.

Through its open free membership model EAI promotes a new research and innovation culture based on collaboration, connectivity and recognition of excellence by community.

More information about this series at http://www.springer.com/series/15427

Tanupriya Choudhury • Abhirup Khanna
Teoh Teik Toe • Madhu Khurana
Nguyen Gia Nhu
Editors

Blockchain Applications in IoT Ecosystem

Springer

EAI
RESEARCH MEETS INNOVATION

Editors
Tanupriya Choudhury
Department of Informatics
School of Computer Science
University of Petroleum
and Energy Studies (UPES)
Dehradun, Uttarakhand, India

Teoh Teik Toe
Nanyang Technological University
Singapore, Singapore

Nguyen Gia Nhu
Duy Tan University
Da Nang, Vietnam

Abhirup Khanna
University of Petroleum and Energy Studies
Dehradun, Uttarakhand, India

Madhu Khurana
Queen's University Belfast
England, UK

ISSN 2522-8595 ISSN 2522-8609 (electronic)
EAI/Springer Innovations in Communication and Computing
ISBN 978-3-030-65690-4 ISBN 978-3-030-65691-1 (eBook)
https://doi.org/10.1007/978-3-030-65691-1

This Springer imprint is published by the registered company Springer Nature Switzerland AG
The registered company address is: Gewerbestrasse 11, 6330 Cham, Switzerland

Dr. Tanupriya Choudhury would like to dedicate this book to the Indian Army, for their dedication, sacrifice, and excellent service to the motherland, INDIA, and he would also like to dedicate this book to his parents Sri Mrigendra Choudhury and Smt. Minakshi Choudhury, his beloved wife Rituparna Choudhury, and dear son Rajrup Choudhury for their immense support throughout this work. And also he would like to dedicate this book to his research guides Prof. (Dr.) Vivek Kumar, Prof. (Dr.) V. Cyril Raj, Prof. Sumathy Eswaran, and Dr. Darshika Nigam, who have always mentored him during his master's and doctoral research. He would like to thank his uncle, late Girindra Mohan Choudhury, for his all-time love, blessings, and support and dedicate the book to his him. He would also like to thank another of his uncle, Dr. Tapobrata Chowdhury (MBBS), and brothers Mr. Supriya Choudhury and Mr. Debopriya Choudhury, who supported wholeheartedly to complete this work.

Mr. Abhirup Khanna would like to dedicate this book to his late grandmother

Smt. Usha Tandon, her love and warmth shall always be missed. He would also like to dedicate this book to his parents Mr. Ravi Khanna and Mrs. Abha Khanna for all their love and support.

Dr. Teoh Teik Toe would like to dedicate this book to his family.

Mrs. Madhu Khurana would like to dedicate this book to her mentor Dr. Daisaku Ikeda for the wisdom and courage immensely needed for this task. She would like to thank her husband Mr. Sanjeev Khurana and her children Sanjana and Samarth who supported whole-heartedly to complete this work.

Dr. Nguyen Gia Nhu would like to dedicate this book to his wife Vu Thi Van Nhung and children, Nguyen Vu Minh Thu and Nguyen Gia Bao.

Foreword

Blockchain technology and the Internet of Things (IoT) are amongst the fastest emerging technologies in recent times. Both the technologies have a great potential in impacting the lives of millions and can result in socio-economic advancements of the society in general. The editors have done a wonderful job in collating a variety of chapters from different areas of blockchain and IoT. They have been successful in presenting a comprehensive view of blockchain technology and its integration with IoT. The book facilitates its reader in having a valuable understanding of various application areas pertaining to blockchain technology and IoT. Moreover, the book assists the reader in exploring new research areas wherein blockchain and IoT can find their applicability based upon their list of benefits. I truly believe that the book will be a good read for those looking forward to exploring areas of blockchain technology and IoT.

Arba Minch, Ethiopia

Bhupesh Kumar Singh

Preface

Blockchain is an emerging technology that has drawn widespread interest across various verticals such as supply chain management, financial institutions, government agencies, healthcare organisations and technology developers. A blockchain is a digital data structure that comprises a distributed ledger chronologically containing records of transactions. A set of transactions is aggregated into a single entity called *blocks*. Each block present in a blockchain is timestamped and cryptographically linked to its predecessor block. The primary objective of blockchain technology is to remove the role of intermediaries and establish a peer-to-peer decentralised mechanism for performing transactions. Blockchain technology has emerged on the promise of offering transparent, secure, tamper proof systems that can represent novel business logic through the use of smart contacts. The integration of blockchain technology and IoT gives rise to a new set of opportunities in aspects of data security, traceability, privacy preservation, enhanced scalability and cost reduction. This book focuses on the fundamentals of blockchain technology along with the means and methods of its integration with the Internet of Things (IoT). It allows the reader to have a deeper understanding of blockchain technology, IoT and various application areas wherein both the technologies can be implemented. The book serves the purpose of providing adequate knowledge about the fundamentals of blockchain and IoT to a common reader along with facilitating a research scholar in identifying some futuristic problem areas that emerge from the convergence of both technologies.

Chapter 1 presents an introduction to blockchain technology and emphasises on the basics of smart contacts and popular blockchain platforms. The chapter also discusses the few prominent use cases for blockchain technology. Chapter 2 talks through the amalgamation of blockchain and IoT and discusses the architectural aspects for the same. Chapter 3 deals with explaining the role of blockchain and IoT concerning industry 4.0. It discusses the benefits and challenges of blockchain IoT integration along with popular real-life projects dealing with the same. Chapter 4 presents a detailed illustration of all the consensus algorithms being implemented in the blockchain arena along with their competitive study. Chapter 5 deals with smart contracts and how they can be implemented across different problem areas. Blockchain technology takes existing, proven concepts and merges them together into a single solution. Chapter 6 explores the fundamentals of how these technologies work and the differences between blockchain approaches. This includes how the participants in the network come to agree on whether a transaction is valid and what happens when changes need to be made to an existing blockchain deployment. Chapter 7 talks through the role of IoT and blockchain technology in creating a secure and transparent drug supply chain. Chapter 8 discusses ways in which IoT and blockchain can contribute towards creating a secure, transparent, economically efficient system for collection and management of patient-centric data in a healthcare environment. Chapter 9 illustrates various blockchain consensus algorithms regarding IoT networks and presents a comparative study among them based on scalability, throughput and security. Chapter 10 discusses the current challenges of IoT networks and presents a solution to them in form of the blockchain SDN integration. Chapter 11 deals with the concept of biometric signatures using blockchain technology for IoT domains. Chapter 12 explores the concept of digital twins and presents a blockchain-enabled digital twin for IoT-assisted applications. Chapter 13 talks through the security concerns prevailing in current IoT infrastructures and ways in which blockchain can assist in mitigating those risks. Chapter 14 presents a blockchain-based IoT-enabled supply chain model for the agriculture sector. Chapter 15 discusses the area of smart farming in which it talks about IoT-enabled agricultural practices and blockchain-assisted food supply chain. Chapter 16 explores cyber risks associated with smart homes and presents a blockchain-enabled solution for the same. Chapter 17 talks through the fusion of blockchain and edge computing and discusses ways of ensuring secure and reliable data transmissions. Chapter 18 explores the area of retail supply chain and presents a blockchain-enabled retail supply chain management system for local *kirana* stores. Chapter 19 discusses various application areas of blockchain technology with reference to the travel and tourism industry.

We hope that our efforts are appreciated and the reader benefits from this book.

Dehradun, India Tanupriya Choudhury

Dehradun, India Abhirup Khanna

Singapore, Singapore Teoh Teik Toe

England, UK Madhu Khurana

Da Nang, Vietnam Nguyen Gia Nhu

Acknowledgment

We, Dr. Tanupriya Choudhury and Mr. Abhirup Khanna, would like to thank our place of work, the University of Petroleum and Energy Studies, Dehradun, India, for providing us a positive research environment to start this proposal. We would like to thank all contributors from 12 different countries and especially reviewers throughout the globe who helped with reviewing the chapters to maintain the quality of the book and their valuable suggestions always whenever required. We are also thankful to the senior leadership of UPES and administration for giving us the opportunity to hold BAIOT 2020 and providing with all possible support. We are thankful to Sh. Sharad Mehra, CEO, GUS-Asia, for his "all-time-go-ahead" blessings and freedom of work. Dr. S J Chopra, chancellor UPES, for his blessings and guidance as always. Honorable Vice Chancellor Dr. Sunil Rai has been a continuous support to us as torchbearer, big thanks to you Sir for your mentorship. Not to mention the instrumental personality of Prof. (Dr.) Priyadarsan Patra, dean of the School of Computer Science, UPES, and Prof. (Dr.) T P Singh, HoD Informatics, UPES, as they have been a rock solid support behind everything, thank you so very much sirs. We would also like to thank our colleagues and friends for all-time support, especially Dr. Ravi Tomar and Dr. Praveen Kumar for moral and technical support as always whenever required. We cannot thank everyone enough for their involvement and their willingness to take on the completion of tasks beyond their comfort zones. See you all in the next edition of the book.

Dr. Teoh Teik Toe would like to thank his family for all time support.

Mrs. Madhu Khurana is grateful to her workplace, Gloucestershire College, for supporting and providing her with a working environment that she could devote time to projects such as this.

Dr. Nguyen Gia Nhu wishes to thank various people for their contribution to this project.

Contents

Chapter 1
Introduction to Blockchain

Ayushi Sharma, Shashwat Tiwari, Nitin Arora, and Subhash C. Sharma

1.1 Introduction to Blockchain

The blockchain is undoubtfully a superb creation; it has developed into something more worthy, and the question everyone is asking is: What is blockchain? Is blockchain technology the new internet? Well, blockchain has made it possible to circulate digital data, not by duplicating, which has contributed to the foundation of merely another sort of web. Initially, blockchain found its roots in the Bitcoin, but this tech network has potential enough to have other uses as well [1].

A blockchain can be understood as a periodic management of unchangeable blocks of information that are monitored by a bunch of PCs and not by any particular individual. Cryptographic standards maintain each of these blocks of data. Blockchain is a decentralized system. It is a mutual and unchangeable record, which means the data stored in it is accessible to everyone [5]. Consequently, anything that is based on the blockchain is by nature very straightforward, and every associated member is responsible for their activities.

Though blockchain is straightforward, it is a highlighted method for passing data from A to B in a completely secured environment. The initial step of this process is to make a block for storing information. PCs cross-check this block of information around the net. Then this checked block of data is added to a chain, which is circulated all over the Internet [12], making an exceptional chain of blocks with a brief history of records. Duplicating a single file would mean that the whole chain is distorted. That is virtually unthinkable. The fundamental concept, namely, Bitcoin,

A. Sharma · S. Tiwari
School of Computer Science, University of Petroleum and Energy Studies, Dehradun, India

N. Arora (✉) · S. C. Sharma
Electronics & Computer Discipline, Indian Institute of Technology, Roorkee, India

T. Choudhury et al. (eds.), *Blockchain Applications in IoT Ecosystem*,
EAI/Springer Innovations in Communication and Computing,
https://doi.org/10.1007/978-3-030-65691-1_1

utilizes this model for money-related exchanges; however, there can be more than one point of view.

1.1.1 How Does a Blockchain Work?

The data stored in a blockchain exists as an interactive database. This is eventually an ideology for using the system, which has clear advantages. It is more like a decentralized system, which means that its records are available to all. A programmer can degenerate none of the data in a blockchain. This information is cross-checked and circulated by PCs all over the web.

Blockchain technology has three fundamental properties, which have helped it to evolve over time; its features are as follows:

- Decentralization
- Straightforwardness
- Unchanging nature

The blockchain is blessed with a distributed environment—the blockchain system is an association of nodes that are interconnected to each other. Centers are a group of PCs that accept an input and apply an individual capacity to them. The blockchain makes use of a system known as the "shared system," which sends or transmits its whole remaining burden between the members of the chain, who are, for the most part, similarly particular, called "peers." There is not a single centralized server but a distributed collection of decentralized companions [1].

1.2 Consensus Algorithms in Blockchain

1.2.1 Proof of Work

This consensus algorithm is used to pick an excavator for the following square age. Bitcoin utilizes this proof of work (PoW) consensus algorithm. The central idea behind this algorithm is to handle a complex numerical puzzle and viably give out an answer. This mathematical puzzle needs a ton of digital force, and, accordingly, the center who comprehends the mystery as quickly as time allows gets the opportunity to mine the following square [3].

1.2.2 Practical Byzantine Fault Tolerance

Barbara Liskov and Miguel Castro presented a consensus algorithm in the late 1990s called the Pragmatic Byzantine Fault Tolerance (PBFT). PBFT was intended to work effectively in asynchronous (no upper bound on when the reaction to the solicitation will be gotten) frameworks. It is enhanced for low overhead time. Its objective was to tackle numerous issues related to effectively accessible Byzantine Fault Tolerance arrangements. Application zones incorporate conveyed registering and blockchain.

1.2.3 Proof of Stake

As opposed to PoW, proof of stake (PoS) is the most generally perceived alternative. Ethereum has moved from the PoW to the PoS agreement. In the consensus algorithm of this type, instead of putting resources into costly equipment to fathom a perplexing riddle, validators put resources into the framework coins, and this is done by locking a portion of their currencies and marking them as stake [8]. The approving of the squares by the validators will begin after that point. A wager is then put on it, and that's how the validators will validate blocks if any square is found, which according to them, can be added in the chain. Concerning the genuine squares included in the blockchain, all validators are rewarded with a prize according to their wagers and their increment of stake likewise [18]. At long last, a validator is picked to produce another square dependent on their monetary stake in the system. In this way, PoS energizes validators through a motivation system to agree.

1.2.4 Proof of Burn

In proof of burn (PoB), validators "burn" coins by sending them to an address from where they are irretrievable, instead of making expensive hardware equipment investments. By doing so, validators gain an opportunity to mine on the system, and this mining is based on a random selection method. So what burning coins mean here is that validators are committed for the long term at the cost of their short-term loss.

Depending on the PoB implementation, miners may burn the Blockchain application's native currency or any alternative chain currency, e.g., Bitcoin. If they consume more coins, the chances of them being selected to mine the next block increase. Although PoB is an excellent alternative to PoW, the wastage of resources involved is quite needless. Some people also question the fact that mining power simply goes to people who are willing to burn more money.

1.2.5 *Proof of Capacity*

In this, hard drive space should be contributed by the validators instead of putting resources into expensive apparatus or consuming coins. The chances of validators getting chosen for mining the following square and gaining the square prize increase if they have more hardware space.

1.2.6 *Proof of Elapsed Time*

Proof of elapsed time (PoET) is among the most desired algorithms used for consensus, which picks the following square utilizing reasonable methods as it were. Permission blockchain systems generally use it. This algorithm provides a reasonable opportunity for each validator on the system to make their square. All hubs do the same by hanging tight for an arbitrary time measure, including their holdup's proof in the square. The made squares are communicated to the system for the thoughts of others. The winner is the validator in the proof section with the least timer value. The square from the triumphant validator hub gets added to the blockchain. Extra checks are present in the algorithm to prevent centers from winning the political decision continuously and prevent hubs from producing the most reduced clock esteem [2].

There additionally exist different consensus algorithms, such as leased proof of stake, proof of weight, proof of importance, proof of activity, and so forth. In this way, it is imperative to admirably pick one according to the business network prerequisite because blockchain systems can't work appropriately without the consensus algorithms to check every single exchange that is being committed.

1.3 Protocols

A protocol is a set of standards and rules used to accomplish a particular task. Protocols provide a guarantee that this information can be productively moved. Enterprise blockchain protocol or blockchain protocols are responsible for maintaining the various parts of a blockchain. This means that there are blockchain security protocols, organize protocols, and blockchain agreement protocols. Each one of these protocols joins and turns into a blockchain system [7].

The central idea behind blockchain is its decentralized nature, as there is no concentrated substance. Some protocols additionally work to approve exchanges into blocks. These blocks, once made, can't be modified [19]. The entirety of this is obtained by using protocols. Typically, blockchain supports the distributed peer to peer methodology. These protocols mainly concern the following:

3.1. A "how to" system for governing as well as validating transactions.
3.2. A sure shot set of steps that ensure the participation of each node interactively

To be more precise, some of the essential features of the blockchain protocol are transactions and smart contracts, consensus, and truthlessness.

1.4 Applications of Blockchain

1.4.1 Decentralized Cryptocurrencies

Traditional structures lease a mediator, including a banker or a settlement enterprise, to make some agreement with. They stored a mental document of who claimed what and referenced this allotted community file when debates arose. The blockchain is this community report on a much extensive, digital scale. Each transaction is a virtual "block" that needs to be approved before it's allowed to go into the machine. Every PC on the organization contends on unscrambling the arrangements, and the triumphant organization includes this "block" to the "blockchain" in the succession that the "block" showed up. The champ communicates his confirmation to the remainder of the organization, which tests that evidence and checks it before lining the "block" to finish the exchange. Parties worried are assured that participants have screened and okayed the transaction. The technique now not simplest cuts down on fraud, consisting of double spending or spams, but also transfers price range definitely, competently, and speedy.

1.4.2 Insurance: Claims Processing

The processing of claims can be a frustrating and thankless process. Insurance processors have to wade through false statements, fragmented sources of information, or abandoned user policies to state a few and manually process these types. The space for error is immense. A perfect mechanism for risk-free management and accountability is provided by the blockchain. Its encryption features allow insurers to capture their own property [5].

1.4.3 Payments: Cross-Border Payments

The worldwide payments industry is error-prone, expensive, and vulnerable to money laundering. It takes days, if not more, to cross the globe with money. The blockchain already provides solutions that provide end-to-end blockchain-powered remittance services to remittance companies such as Abra, Align Commerce, and

Bitspark. Santander became one of the first banks to incorporate the blockchain into an app payment in 2004.

1.4.4 *Your Vehicle/Cellphone*

Primitive intelligent property types exist. For example, your vehicle key can be equipped with an immobilizer, where you can only unlock the vehicle if you tap the correct protocol on the key. Your mobile, too, will only function until you enter the correct PIN code. To secure your ownership, both run on cryptography.

The problem with primitive kinds of smart property is that the key is commonly held in a physical container, for example, the auto key or SIM card, and can't be easily moved or duplicated. The blockchain method solves this problem by way of allowing blockchain miners to replace and mirror a lost protocol.

1.4.5 *Supply Chain Sensors*

Sensors provide corporations to offer up-to-stop perceivability of their supply chain via providing facts at the area and state of the materials as they're transported across the globe. As of 2016, a Deloitte and MHI record surveyed 99 top deliver chain companies and found that sensors were utilized by 44% of those respondents; 87% of these companies said they intend to apply the generation by 2020. In 2013, it was reported that approximately 20 million sensors were in use in supply chains. This number is projected to rise to around 1 trillion in 2022, and by 2030, the Deloitte and MHI study predicts the deployment of around 10 trillion sensors.

Blockchain has many more applications: in healthcare, personal identification, and passport designing.

1.5 Use of Blockchain in Healthcare

Blockchain has a vast extent of utilization in human services. No matter the distance between the patient and the provider, the patient can record data and send it to healthcare providers in minutes.

1.5.1 *Medical Supply Chain Management and Drug Traceability/Safety*

To what extent do we think about our medication? Do we have the option to be sure that it hasn't been meddled with? Is it coming from a legitimate source?

These inquiries are the essential worries of the clinical stock network or the connection between the laboratory and the commercial center. Blockchain has positive changes for the pharmaceutical stock network, the executives, and its decentralization practically ensures full straightforwardness in the delivery procedure.

When a record for medication is made, it will stamp the purpose of the source (e.g., a research center). At that point, the record will keep on continuously documenting details, taking care of it and where it has been, until it reaches the buyer. Job costs and waste discharges may also be screened by the process. Companies: CHRONICLED, BLOCKPHARMA.

1.5.2 *Breakthroughs in Genomics*

The capability of genomics to improve the fate of human well-being, though a dream, is as of now a logical and monetary reality. In 2001, it cost $1 billion to process a human genome. Today it costs $1000, and companies like 23andMe and Ancestry.com are bringing DNA tests that open insinuations to our well-being and past to a significant number of homes. Blockchain is a perfect fit for this upcoming industry as it can safely house billions of genetic data. It's even become a business where people can offer their unordered genetic data to make a more extensive database, giving researchers access to useful information faster than at any other time [4]. Companies: Nebula Genomics, Encrypgen.

1.6 Blockchain Platforms

Blockchain is by all accounts getting pace these days. The innovation that once advanced in the year 2009 as the hidden stage for Bitcoin trade has now developed into a standard change. It discovers applications in different fields [17].

1.6.1 *Ethereum*

At that point, something that created a buzz in the market, after Bitcoin, was Ethereum. Vitalk Buterin established it in the year 2014. Ethereum is an open-source and one of the exceptionally dynamic blockchains, which likewise shapes the base for the advancement of different applications [1].

Ethereum's key features:

- It is open source.
- Work-based framework verification.
- Github follows it emphatically.
- Numerous language applications, such as Python and C++.

1.6.2 Hyperledger Fabric

This is one of the most recently established platforms for Blockchain. In the year 2016, the world became more acquainted with hyperledger, created by Linux Foundation. Its goal is to support the blockchain innovations' utilization across various businesses [6].

Hyperledger's key features:

- Inquiries into real-only history
- 180+ partnering companies
- Enterprise-ready development

1.6.3 IBM's Blockchain

It is the first organization to wander into the blockchain to make a stage for specific business activities. IBM's use of blockchain is different in that it just focusses on making applications based on blockchain. IBM boasts about an increasingly effective assent instrument that has made it earn the consideration of many people.

IBM blockchain's key features:

- It provides permission to organize, which is being looked at by the more significant parts of the organization because of the security issues.
- Bolstered languages: Java and Go.

1.6.4 Multichain

Multichain is an open-source Blockchain framework used to build authorization networks. It discovers use within as well as across various undertakings.

Multichain's key features:

- Formation of a network with permission.
- Github follows it unequivocally.

- Evaluation is free and open source.
- Bolstered dialects: C, JavaScript, Python, and C++.

1.6.5 Hydrachain

Hydrachain is a community-oriented activity of brainbot advances and Ethereum. As it is an augmentation of the Ethereum stage, it is utilized to make a private record that is valuable to carry out the undertaking even though it is not much popular; however, Github usually refreshes it.

Hydrachain's key features:

- Ethereum protocol is followed in it.
- Authorization network.
- Language maintained/upheld: Python.

1.6.6 Ripple

When we talk about popularity, Ripple is comparably more in demand than Ethereum or Bitcoin. It came into existence in 2012 and is extraordinary compared to other performing money [16].

Ripple's key features:

- It associates banks, computerized resources trades, and installment suppliers.
- Ability to send cash internationally effortlessly.
- The speed of wave ensures five exchanges each second, along these lines, making it a quicker and increasingly proficient stage.

1.6.7 R3 Corda

If a person is quite familiar with the blockchain, then there is a high possibility that they would be aware of the R3 consortium. It is a coordinated effort of the magnificent budgetary bodies that are looking forward to increasing the employability in the blockchain sector and trying to incorporate it with the framework. It is an open-source DLT stage [7].

Some of the essential highlights of R3 Corda are:

- It was founded in 2013.
- concentrating on the commercial segment, application Corda also discusses, aside from of blockchain to healthcare, trade finance, supply chain, etc.

- It is a permissioned Blockchain.
- It focuses on the ease of integration with the legacy framework for interoperability.

1.6.8 BigchainDB

BigchainDB is another open-source blockchain stage.

Some of its essential highlights are as follows:

- Customized resources.
- Use of in-fabricated cash is not allowed.
- An open and private system.

1.6.9 OpenChain

OpenChain is yet again a well-known open-source stage. It is significantly valuable to organizations that are dealing with digital resources [5]. It has tweaked consent at various stages.

Some of the essential highlights of OpenChain are:

- Private Network
- Bolstered language: JavaScript

1.6.10 IoTA

IoTA is one of the most recent participants, or we can say it is the youngest in the field of the Blockchain stage. It's an extra application of Blockchain's open source innovation [4].

Some of the essential highlights of IoTA are:

- It is used for making both open as well as authorized arrangement.
- It allows nano-installments.
- It is also famous for its token-based pricing.

1.7 Blockchain for IoT-Enabled Healthcare

The Internet of things, as we know, is widely spread across various segments; the increasing demand for wearable and ingestible IoT gadgets and devices has been incrementally contributing toward social causes. As we know everything has advantages as well as disadvantages, so is the case with IoT, the rapid increase in the popularity and advancement are raising worries over the information these gadgets gather. Blockchain's trust-building highlights support the extensive appropriation of these gadgets. Thus blockchain IoT use cases incorporate more trust into the business [11].

Clinical engineering developments often suggest improved instruments and increasingly sensitive consideration. New technology is enabling medical care specialists to convey better outcomes to their patients in social insurance frameworks worldwide. Network of things gadgets link to the web, collect data, and talk to each other. These regularly scan essential wellness information for medicinal services. On the one side, IoT clinical gadgets deliver new association routes to patients and specialists and have saved lives. Be it as it might, yet again, these devices gather a lot of sensitive information. Top issues should be security and secrecy [20, 21], however, various organisations have proven unfit to guarantee it. Blockchain IoT use cases involving powerful insurance and uprightness of knowledge will help protect these devices, putting patients in charge of who uses this data and for what reason [14].

1.7.1 Secure Remote Patient Monitoring

Within the medicinal services industry, IoT falls into two classifications: clinical administrations and bolster tasks. On the clinical administration side [10], IoT assists with improving remote patient checking, or RPM. RPM highlights are perhaps the most convincing blockchain IoT use cases in human services. Wearable IoT gadgets likewise help clinical preliminaries by following vital signs among different pointers, for example, glucose, pulse, or weight. Swiss Medtech startup HIT establishment is as of now utilizing blockchain innovation to encourage clinical preliminaries, to ensure clinical information, and to ensure clients know how their data is being used. Although it is not an IoT application per se [9], one option may be to integrate wearable or ingestible IoT devices that transmit patient data continuously during a clinical trial [22, 23]. This would furnish specialists with certain information, yet it likewise gives members it also provides participants with a good understanding of how the data will be used by researchers and for how long [13].

With blockchain-empowered IoT gadgets, end clients can control access to the information the devices gather. By conceding and disavowing access to touchy clinical details, clients can be sure it's appropriately utilized. Blockchain innovation in human services makes clinical gadgets increasingly hard to hack. But they may also show a comprehensive record of when your data is accessed by another person [11]. Customers and organizations worried about securing their information will benefit from incorporating blockchain into IoT gadgets.

1.7.2 Supply Chain

Alongside progressively secure remote patient checking in clinical IoT gadgets, the medication inventory network is another of a few blockchain IoT use cases. Pharma extortion is a significant risk to the trustworthiness of the worldwide pharmaceutical [12] industry—and to understanding well-being. Similarly, as you would utilize a blockchain IoT framework to follow nourishment provenance, following pharmacies through the production network would be one opportunity for blockchain IoT. Gadgets that continually convey the area, root, and temperature states of pharmaceuticals could impact the business. This would diminish squander as well as give controllers and shoppers dependable security [12]. In 2018 Swiss Post propelled a program utilizing blockchain and IoT that tracks the temperature dependability of pharmaceuticals dispatched via the post office. The bearer guarantees temperature-touchy medications; for example, insulin on its way to patients stays within an appropriate temperature range. Blockchain innovation permits Swiss Post to follow the information and, furthermore, share it with guarantors and clients.

1.8 Future Scope of Blockchain

Blockchain is one of the fastest emerging technologies in terms of maintaining the history of transactions and other records safely. There is a broad scope for blockchain being used in various sectors. In the case of India, after the commencement of demonetization, most of the financial organizations were unable to catch up with the heavy workload efficiently; as a result, the Reserve Bank of India encouraged banks to change their centralized system and opt for digitization, and what else could be a better alternative than a blockchain system which eliminates the need of centralization and is something much more reliable in terms of maintaining security [8]. Different blockchain organizations have already started to plan the transition phase with the Indian clients. Apart from this, there are many fields where blockchain stands strong in the bigger picture; some of these are as follows.

1.8.1 Blockchain in Digital Advertising

Talking about digital advertising, looking at the present scenario, the first thing that comes to mind is bot traffic, lack of transparency, and domain frauds. Blockchain can solve all of this quite efficiently by allowing only authentic companies to succeed, thus eliminating the bad players in the market [15].

1.8.2 Blockchain in Cybersecurity

Security is a key feature of blockchain and what else could be a better player in the field of cybersecurity [16]. The cryptography feature of blockchain makes the data less likely to be attacked.

1.8.3 Blockchain in Cloud Storage

In today's scenario, anything and everything around us is nothing but data. Cloud storage is widely used to store tons of data, but due to its centralized nature, this data is prone to hacking, loss, etc., so the decentralized security feature of blockchain comes handy here as well.

These are fields where blockchain growth will not only provide enhanced ability but will also provide security.

1.9 Conclusion

As a conclusion to this chapter, we can say that blockchain technology has a significantly broader scope in today's era. The recognition and the boost that blockchain technology got in the market is remarkable. Blockchain is establishing itself in almost every sector because of its vibrant advantages, which are security, transparency, and immutability. Blockchain is benefitting businesses as well through easier traceability. Due to its decentralized nature and cryptographic algorithms, hacking a blockchain is close to impossible as its behavior and nature make it immune to outside attacks. It even assures better quality. So it is essential for this generation to be aware of this technology, how to use this technology, and the various protocols related to blockchain.

References

1. A.D. Dwivedi, G. Srivastava, S. Dhar, R. Singh, A decentralized privacy-preserving healthcare blockchain for IoT. Sensors **19**(2), 326 (2019)
2. E. Tijan, S. Aksentijević, K. Ivanić, M. Jardas, Blockchain technology implementation in logistics. Sustainability **11**(4), 1185 (2019)
3. L. M. Bach, B. Mihaljevic, M. Zagar, Comparative analysis of blockchain consensus algorithms, in *2018 41st International Convention on Information and Communication Technology, Electronics, and Microelectronics (MIPRO)*. (IEEE, 2018, May), pp. 1545–1550.
4. P. Tasatanattakool, C. Techapanupreeda, Blockchain: Challenges and applications, in *2018 International Conference on Information Networking (ICOIN)*. (IEEE, 2018, January), pp. 473–475

5. C. Ehmke, F. Wessling, C.M. Friedrich, Proof-of-property: A lightweight and scalable blockchain protocol, in *Proceedings of the 1st International Workshop on Emerging Trends in Software Engineering for Blockchain*, (2018, May), pp. 48–51
6. M. Nofer, P. Gomber, O. Hinz, D. Schiereck, Blockchain. Bus. Inf. Syst. Eng. **59**(3), 183–187 (2017). https://doi.org/10.1007/s12599-017-0467-3
7. B. Arati, Understanding blockchain consensus models. *Persistent* (2017)
8. Z. Zheng, S. Xie, H. Dai, X. Chen, H. Wang, An overview of blockchain technology: Architecture, consensus, and future trends, in *2017 IEEE International Congress on Big Data (BigData Congress)* (IEEE, 2017, June), pp. 557–564
9. M. Mettler, Blockchain technology in healthcare: The revolution starts here, in *2016 IEEE 18th International Conference on E-health Networking, Applications and Services (Healthcom)*. (IEEE, 2016, September), pp. 1–3
10. K. Biswas, V. Muthukkumarasamy, Securing smart cities using blockchain technology, in *2016 IEEE 18th International Conference on High Performance Computing and Communications; IEEE 14th International Conference on Smart City; IEEE 2nd International Conference on Data Science and Systems (HPCC/SmartCity/DSS*. (IEEE, 2016, December), pp. 1392–1393
11. A. Kiayias, A. Russell, B. David, R. Oliynykov, Ouroboros: A provably secure proof-of-stake blockchain protocol, in *Annual International Cryptology Conference*, (Springer, Cham, 2017, August), pp. 357–388
12. K. Francisco, D. Swanson, The supply chain has no clothes: Technology adoption of blockchain for supply chain transparency. Logistics **2**(1), 2 (2018)
13. R. Moro Visconti, D. Morea, Healthcare digitalization and pay-for-performance incentives in smart hospital project financing. Int. J. Environ. Res. Public Health **17**(7), 2318 (2020)
14. V. Merton, The exclusion of pregnant, pregnable, and once-pregnable people (aka women) from biomedical research. Am. J. Law. Med. **19**, 369 (1993)
15. N. Wang, X. Zhou, X. Lu, Z. Guan, L. Wu, X. Du, M. Guizani, When energy trading meets blockchain in electrical power system: The state of the art. Appl. Sci. **9**(8), 1561 (2019)
16. L.J. Trautman, Bitcoin, virtual currencies, and the struggle of law and regulation to keep peace. Marq. L. Rev. **102**, 447 (2018)
17. A. Walch, The bitcoin blockchain as financial market infrastructure: A consideration of operational risk. NYUJ Legis. & Pub. Pol'y **18**, 837 (2015)
18. T.M. Fernández-Caramés, P. Fraga-Lamas, A review on the use of Blockchain for the internet of things. IEEE Access **6**, 32979–33001 (2018)
19. Y. Sompolinsky, A. Zohar, Accelerating Bitcoin's transaction processing. Fast money grows on trees, not chains. *IACR Cryptol. ePrint Arch., 2013*, 881, (2013).
20. D.S.R. Krishnan, S.C. Gupta, T. Choudhury, An IoT based patient health monitoring system, in *2018 International Conference on Advances in Computing and Communication Engineering (ICACCE)*, (2018), pp. 1–7
21. M. Khurana, T. Choudhury, P. Malik, A review on network security challenges and the Internet of Things (IoT), in *2019 International Conference on Contemporary Computing and Informatics (IC3I)*, (2019), pp. 228–233
22. A. Khanna, A. Sah, T. Choudhury, Intelligent mobile edge computing: A deep learning based approach, in *Communications in Computer and Information Science*: Vol. 1244 CCIS. (2020). https://doi.org/10.1007/978-981-15-6634-9_11
23. A. Khanna, R. Goyal, M. Verma, D. Joshi, Intelligent traffic management system for smart cities. *Communications in Computer and Information Science* **958** (2019). https://doi.org/10.1007/978-981-13-3804-5_12

Chapter 2
Block Chain and IoT Architecture

Sonia Chhabra, Parveen Mor, Hussain Falih Mahdi, and Tanupriya Choudhury

2.1 Introduction

2.1.1 Block Chain Technology

In 2008, Blockchain technology was created by Satoshi Nakamoto for use with the Bitcoin cryptocurrency as its open exchange record. The initial goal was to decentralize Bitcoin records; subsequently, the Blockchain was used to allow clients to control their own cash without anyone (including law enforcement) being able to access it. Blockchain has allowed Bitcoin to be the only motorized cash to handle the two-fold spending problem without the need for a separate focal or focal site. Blockchain is a special development package to alter the possible pre-determination between taking care of and disrupting two or three interactions with continuous innovative strategies. It is transparent and used in the same way both internally and externally under different circumstances.

The advancement extended enormous in-tribune from the move of electronic sorts of money at any rate it sees applications in numerous different divisions other than sponsor. Blockchain can be deciphered as two or three cryptographically joined squares [1]. The square contains an information structure with three areas: metadata,

S. Chhabra (✉) · P. Mor
Sharda University, Greater Noida, Uttar Pradesh, India

H. F. Mahdi
College of Engineering, University of Diyala, Baquba, Diyala, Iraq
e-mail: hussain.mahdi@ieee.org

T. Choudhury
Department of Informatics, School of Computer Science, University of Petroleum and Energy Studies (UPES), Dehradun, Uttarakhand, India

T. Choudhury et al. (eds.), *Blockchain Applications in IoT Ecosystem*,
EAI/Springer Innovations in Communication and Computing,
https://doi.org/10.1007/978-3-030-65691-1_2

a hash of the last square, and a hash containing the history of previous hashes [2]. Similarly, there is a trust certificate that is used to ensure the reliability of the entire Blockchain [3]. The hash is also modified when details in any of the squares change. This can serve as a winding effect if the hashes of the next squares are null. Those confirmation interactions on the Blockchain are unchanged [4].

This platform can be remarkably efficient at offering automated security plans in risky areas, such as IoT gadgets, systems, and data storage and transmission. Blockchains ensure trust by operating as a typical database, streamed across enormous shared systems that have no single point of failure and no single source of truth, meaning that no individual part can create or change information on the Blockchain autonomously or anonymously [24]. New information only can be added to a Blockchain through recognition by different parts of the system, an approach known as dispersed comprehension. Each piece of the system stores a duplicate copy of the Blockchain's information and maintains a variety of authentic focus points; on the rare chance that one community member changes its duplicate copy, different members ignore it. Blockchains are digital ledgers that keep track of data over time and release up forward indefinitely. New information is incorporated beyond what many would believe to be conceivable—and once included, it is persevering. Continuously arranged information cannot be deleted or changed because of way that an outline is captured in the squares of information that come after it (Fig. 2.1).

Why Are Blockchain Innovations Required? Blockchain developments have the potential to bring everyone to the highest level of responsibility. With the assistance of this development, no more trades will be recollected affectionately, human will be restricted, from that point forward. Lately, even large banks are implementing this technology for financial transactions, record keeping, and other backend limits. Blockchain's unalterable records allow them to follow documentation and meet contractual obligations. Additionally, Indian IT companies such as Infosys and TCS are using this innovation to design budgetary platforms for banks.

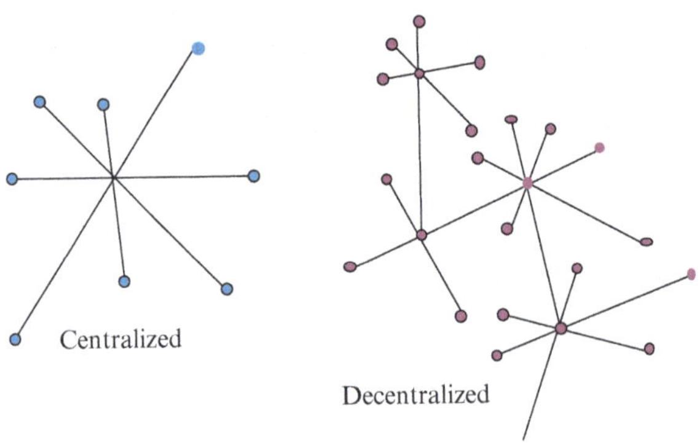

Fig. 2.1 Blockchain Structures

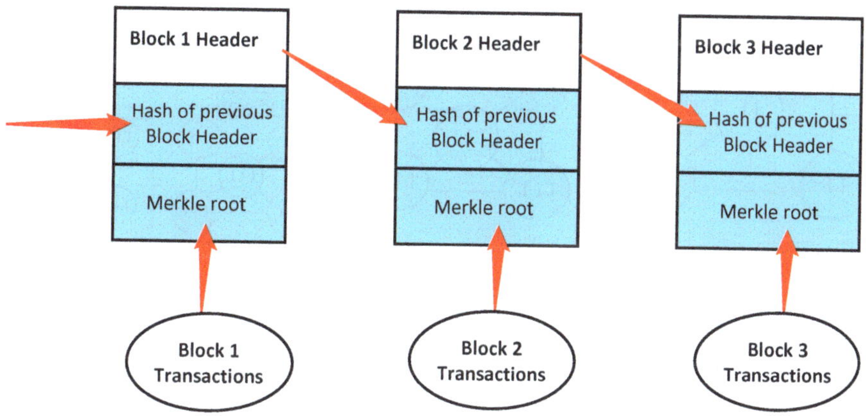

Fig. 2.2 Process of Blockchain transactions

The issue of trust in information platforms is extremely perplexing when no confirmation is given, especially for sensitive information such as financial transactions. In this regard, Satoshi Nakamoto [5] proposed two radical ideas in 2008 that have had extraordinary repercussions. The first of these is Bitcoin, which is virtual currency that operates without assistance from any bank or financial firm. The money is stored securely in a decentralized peer-to-peer arrangement comprising an auditable and transparent framework [6]. The second of idea is Blockchain, whose impact has been significantly greater than the cryptographic currency itself (Fig. 2.2).

2.1.2 Internet of Things

Customers' web-enabled devices and the arrangement of their "things" comprise the broad and progressively complex Internet of Things (IoT) [7]. More than 9 billion "things" have been identified thus far, and are predicted to exceed 20 billion in the not-so-distant future. These advancements include low-power embedded structures, distributed computing, big data, system learning, and framework organization. The specifications of web-associated machine-to-machinie accessibility frameworks extending beyond typical family devices.

IoT Network Configurations A portion of the IoT comprises devices restricted to local areas, closely resembling typical local and wide area networks, as shown in Fig. 2.3 [8]. The hubs are indicated by circles (LL), with neighborhood connection locations (LLL) and nearby connections that may be location-specific. Hubs have addresses that may correspond to a specific location. Similar locations might be rehashed within the different space. The entryway has a novel system prefix that can be used to recognize all hubs. This technique is very efficient, but the hubs need to convey to the web by means of the passage.

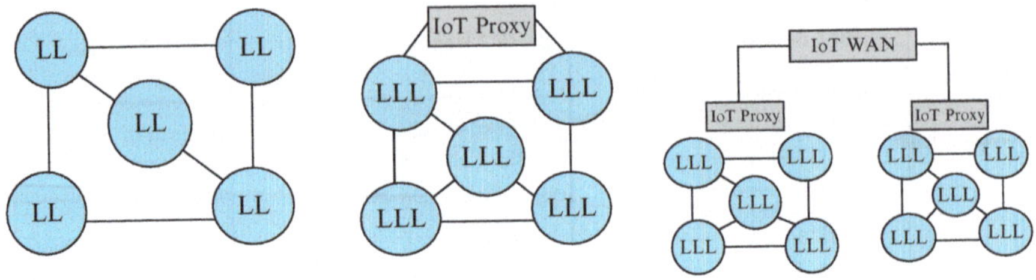

Fig. 2.3 IoT Network Configuration

2.1.3 *Hybrid Structure*

The role of Blockchain in IoT is to securely process records of information through IoT hubs. Blockchain could also be used for direct and transparent development. IoT requires such a development to allow secure correspondence among IoT hubs in heterogeneous conditions. The trades in Blockchain can be completed and confirmed by an approved individual inside the IoT. Using Blockchain in IoT may help to improve correspondence security [5]. Within proof-of-work (PoW) Blockchains, peers referred to as square diggers will use their hardware properties and the ability to repair a cryptographic puzzle as evidence of their work, so they have a valid option for a replacement square. PoW Blockchains use a relatively low throughput while scaling to a few centers to achieve understanding, whereas Byzantine fault tolerant (BFT) Blockchains can operate with a relatively disproportionate throughput at nearly any center.

Much knowledge is being gained for proof-of-stake (PoS) Blockchains [9]; as far as these arrangements are concerned, it is possible to produce and operate many performances, such as PoW and even BFT. PoW Blockchains have a relatively low throughput while scaling to a few centers, whereas BFT Blockchains have a relatively high throughput at hardly any center level. A hybrid version of IoT handles both PoW and BFT Blockchains. The PoW Blockchains are used to define the relationships between the numerous IoT devices. Using a hybrid PoW-IoT Blockchain allows acceptance of different Blockchain square sizes and time ranges, device locations, and increased compatibility.

PoW Blockchains without many geographically similar IoT gadgets have poor general execution (i.e., excessive trade throughputs) and small square multiplication delays; a hybrid IoT involves the development of only one PoW Blockchain. We are created with the agreement of extreme and rapid laws, called sweet spots, which agree to incorporate best practices in the arrangement of sub-block chains. Hybrid-IoT has an effect on a BFT between connector systems, including Polkadot and Cosmos, and is a good way to achieve interoperability between sub-block chains.

2.2 Architectures

2.2.1 *Blockchain Architecture*

All Blockchain structures can be classified as one of the three classes described here.

Public Blockchain A public Blockchain infers that the information and access to the system are available to anyone. Examples of such systems include Bitcoin and Ethereum Blockchain systems.

Private Blockchain A private Blockchain is available to customers with a particular affiliation. Members are confirmed customers who have credentials that allow them to access the system.

Consortium Blockchain A consortium Blockchain usually contains members with multiple affiliations. The methods in a consortium are defined and enforced by the initial users (Table 2.1).

Key segments The following are the key components of a Blockchain:

- *Node.* A customer or personal computer (PC) that exists inside the Blockchain structure. Each manages a free copy of the entire Blockchain record.
- *Transaction.* This is likely the most diminutive structure square of a Blockchain outline, containing records and information. However, it accomplishes essential work of the Blockchain.
- *Block.* A data structure that manages a large number of trades. This is later reliant upon movement to all centers in the framework.
- *Chain.* This is a game plan of squares in a specific solicitation.
- *Miners.* These are certain centers that confirm square activity before adding anything to the Blockchain structure.
- *Consensus.* A complex plan of choices and strategies that help accomplish distinctive Blockchain exercises.

The following list is a step-by-step procedure on how Blockchain configuration capacities work as a wallet:

Table 2.1 Detailed comparison of the three Blockchain systems

Property	Public Blockchain	Consortium Blockchain	Private Blockchain
Consensus determination	Miners	For selected nodes	One organization
Read permission	Accessible to all	Restricted or public	Public or restricted
Immutability	Cannot be modified	Can be modified	Can be modified
Efficiency	Low	High	High
Centralized	No	Partial	Yes
Consensus	Open	Authentication required	Authentication required

1. A request for a trade is made.
2. The creation of a square that reflects the trade occurs.
3. Every center point existing inside the framework will receive the new square.
4. The center points confirm the trade. As a result of their efforts, they receive a monetary reward.
5. The square will officially transform into another extension to the Blockchain.
6. The trade is successfully executed.

Building the Architecture The arrangement of a decentralized Blockchain application resembles other normal programming efforts. Requirements include Prerequisites for its progression have a realistic detail, UX/UI structures, and a solid structure plan. It is important to completely consider the system stream and the communication occurring between the customers and information [10–15].

Thusly, when creating a new Blockchain, it is essential to consider two things:

- *Block chain orchestrate.* This defines the application's structure inside a specific area for one or more affiliations.
- *Block chain code.* This defines the tasks and targets that the development of this Blockchain needs to perform.

System Parties At the point when an affiliation (or affiliations) makes the choice to complete a Blockchain, they are now starting to build a framework. It is possible to consider the associations with their work power from the point of view of the particular establishment that exists inside these associations.

Consider precious stones as an example. Risk related to valuable stones occurs during all parts of the transaction, from mining to sales. Customers generally need a guarantee that they are purchasing legitimate gems, not fakes. Likewise, governments wish to monitor for their own needs. Blockchain configurations can be a valuable instrument in overcoming these risks. Thus, this framework may include the following:

- Manufacturer of the precious stone
- Government establishments
- Transporters of the valuable stone
- Diamond shippers

Sharing the Information Blockchain courses of action consolidate all of these activities into a common framework. One that guides the removal taking everything into account and the development of a obvious structure. Everyone can access the synchronized information of an invariable record. They can monitor the valuable stone's activities, from the mine to the customer. The Blockchain record captures the progression of all activities, including mining, refining, and distribution.

Normally, each association within a system claims a duplicate that corresponds to the Blockchain's smart conventions and specialised layers. In order to track multiple methods occurring at the same time, an ordering service is used for all events. Specifically, parties select the trades inside the Blockchain structure by demand.

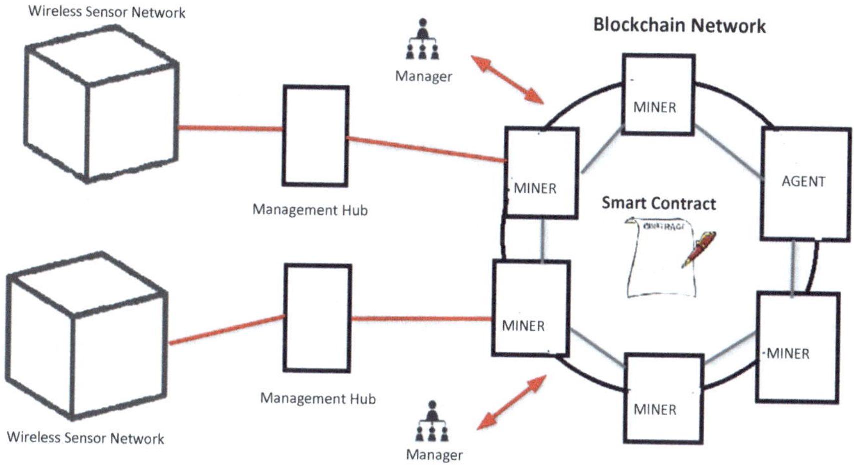

Fig. 2.4 Blockchain Architecture

Customers may have a membership services provider (MSP), which grants access for specific customers inside the framework.

Inevitably, all trades are documented inside a general record. Using the gem model, this includes photos, extraction territory, successive number, and where it was cut, among various other items. This information is used to prove authenticity (Fig. 2.4).

2.2.2 *IoT Architecture*

IoT includes more than just Internet-related customer devices. IoT has advanced to manufactured systems arranged to self-monitor and responding to events without human intervention. Thus, a strategy stream can be created for a chosen structure over which an IoT answer is collected.

IoT architecture has four stages. Stage 1 organizes the devices, such as wi-fi sensors and actuators. Stage 2 integrates sensor structures and modernized databases. In Stage 3, local IT systems perform preprocessing of activities before being launched in the community or cloud. Finally, in Stage 4, the information is analyzed, regulated, and verified on back-end community structures. The sensor/actuator domain is in the realm of activity innovation (OT) specialists, as is Stage 2. Stages 3 and 4 are ordinarily controlled through IT, regardless of the location. The vertical line in Fig. 2.5 is the customary limit among OT and IT obligations, regardless of the reality.

Stage 1. Sensors/Actuators A sensor recognizes changes inside its device or inside the state of another gadget or system, then responds to this information

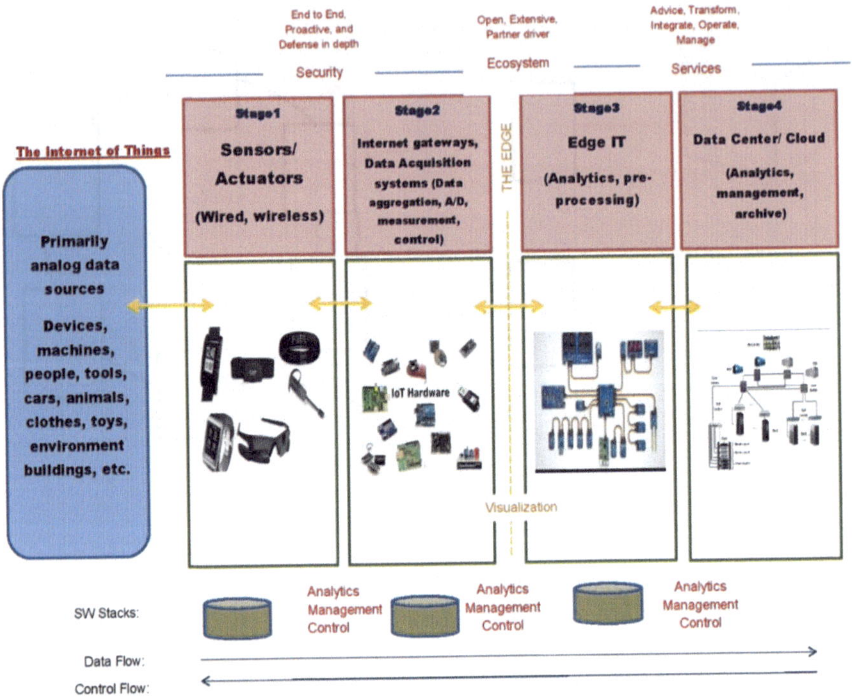

Fig. 2.5 The Four Stages of IoT Architecture

in a beneficial way. Sensors may distinguish or detect physical changes in the qualities of a device. The term "transducer" refers to two sensors that can be used to detect differences in nearby events, light, temperature, or pressure, among others. Furthermore, actuators may be used to change voltages or streams [16]. An actuator moves or controls the instrument or the structure. An actuator can form a framework followed by a device. An actuator requires a signal and a source of action. After a signal is received, the actuator responds by changing the power into a mechanical response. These controls are often essential (in a strictly mechanical or electronic gadget), software-based (e.g., a printer driver, robot control system), a human, or other information.

Stage 2. The Internet Gateway IoT gateways with or without mediators are used for internet accessibility and IoT LAN integration. An upstream course of action with prefixes is obtained using instruments such as DHCPv6 and assigned to the centers using stateless address autoconfiguration (SLAAC). LU (unique local link) addresses are maintained for routable areas, in cases where internal strength is of prime concern [18]. However, LU cannot communicate quickly with the Internet or the upper layers, which is improved by adding an item layer mediator. Application layer delegates also can be organized to process data, instead of simply passing data. In center points with no assistance for computationally focused tasks, IoT delegates aggregate information, which is dispatched to the closest association by multicast addresses and distributed universally.

Stage 3. Edge IT Once IoT records have been digitized and gathered, they are prepared to transfer into their locations. The estimations can similarly require equivalent preparation before entering the areas of knowledge focus. This is frequently located in edge IT systems, which complete extra analysis. Edge IT also can be arranged in remote work environments or other segments, but they typically are located inside the workplace or region where the sensors reside [17]. Because IoT records can challenge transmission limits and community resources, it is beneficial to have structures on the periphery to reduce the burden on the IT setup.

Stage 4. The Server Farm and Cloud Data that need progressively noticeable in-power getting ready, and where remarks don't have the opportunity to be immediate, are sent to physical or cloud-based systems with adequate IT platforms to process, monitor, and securely store the information. It takes more time to receive results if you wait until data arrive at Stage 4; however, you will be able to execute a more thorough assessment [17]. Stage 4 can occur on-premises, in the cloud, or in a crossbreed cloud structure.

2.2.3 *Hybrid Architecture*

Hybrid IoT involves more than one PoW sub-block chain that achieves a shared understanding between IoT gadgets that are friends in the block chain. In order to link the sub-block chains, hybrid IoT uses a BFT within the connector system that guarantees the inter-block chain exchange.

Framework Execution The trade stream in hybrid IoT is as follows: exchanges at the PoW sub-block chains are taken care of and associated with squares, which can be passed on to their different sub-block chain upon PoW accord. When a trade among specific sub-block chains happens, it is preferred by the BFT between the connector structures [18]. The BFT between the connector structures checks the consistency and validity of the trade. After an acceptable response, the BFT between the connector structures passes the trade to the target sub-block chain trading pools, which hold daily trades. Finally, the transaction is done in a newly created obstruct inside a different sub-blockchain, based on a PoW agreement. The outlook for the option of a BFT within the connector system falls outside the limits of BFT understanding, which indicates that a high throughput with a wide range of peers is obtained [19]. Therefore, the interface of some sub-block chains have acceptable throughput between block chain transactions. Similarly, by using low inaction in the transfer of interblock chain transactions, the BFT within the connector system allows for a new sub-block chain to be interlocked without giving rise to programming execution. An example of hybrid IoT architecture involving sub-block chains is shown in Fig. 2.6.

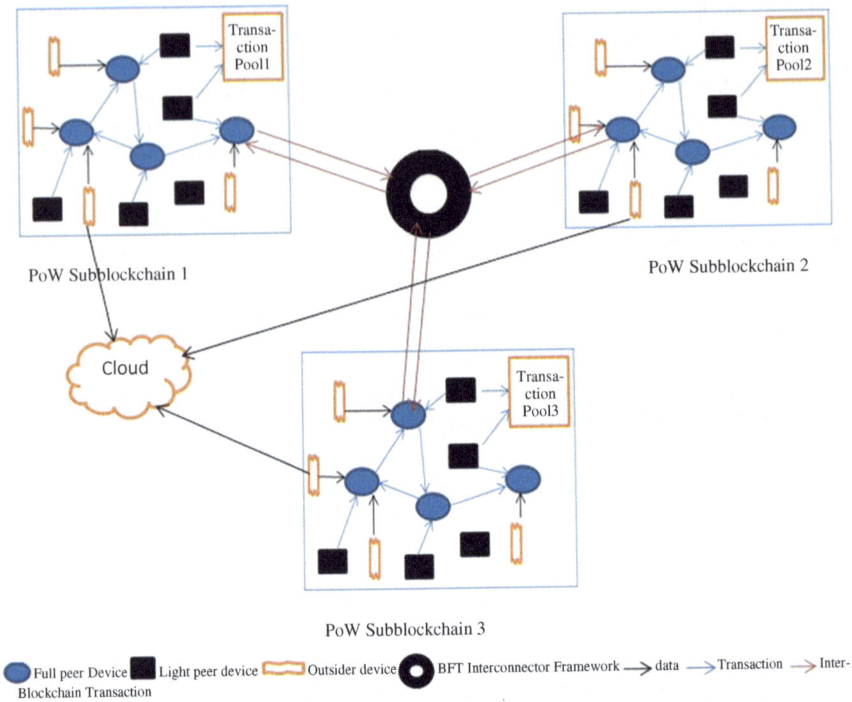

Fig. 2.6 The Blockchain in IoT Architecture (Hybrid Architecture)

Accord Interest Hybrid IoT is a permissioned Blockchain framework. This is especially significant for the explanation that sub-block chains depend on PoW; the character of IoT gadgets can become part of a larger attack [20]. In fact, some mining equipment may not address issues associated with an IoT gadget or have enough square mining power to control the PoW Blockchain.

Remediation Unlike specific PoW mining for cryptocurrency, IoT gadgets have hardware requirements and are mostly power-constrained devices [21]. IoT devices generally do not have enough gear or essential resources to solve complex PoW puzzles. In hybrid IoT, the PoW puzzle is set with respect to the hardware requirements of IoT devices. Thusly, IoT gadgets can currently total their application accurate endeavors, extensive of information dealing with, while at the same time keep up to mine squares [22].

Uses of IoT Devices IoT gadgets have heterogeneous limits, and their uses should reflect their capacities. In hybrid IoT, we discuss three uses for IoT gadgets as companions to the Blockchain: full companion work, light companion work, and outcast work.

Full Companion Work IoT gadgets with sufficient power can be used to perform complex exercises, similar to a Raspberry Pi 3, in full companion work. They have adequate resources and run stable platforms like Raspbian. Subsequently, as companions on the Blockchain, they mine squares and participate in the knowledge system inside the PoW sub-block chains. Despite that, full companion devices are entry gadgets serve as gateways to link a set of low-power companion devices to the blockchain network, referred to as the complete peer node subnet [23]. Thus, a square formed with the guide of a full companion merges its own trades and exchanges via its instrument subnet. The measure of light companion gadgets inside the full companion subnet works harmoniously with its mining ability to ensure genuine square period rates.

Light Companion Work IoT gadgets that have defined limits such as power, including Arduino Yun, perform light companion work. They have some good platforms such as Alpine Linux, and they can partner and participate inside the Blockchain by performing clear endeavors and sending trades. Light companion devices transport trades to the Blockchain trade pool and to the full companion as an entry. This allows the full companions to process the aggregate of the trades inside the sub-block chain.

Outcast Occupation IoT gadgets that have confined capacities are most useful as direct sensors (outcast occupation). They are not blockchain companions, but they can join full companions for additional reality combinations (such as realities accumulation). To avoid reality overloading, crude information generated by an outcast is no longer saved on the blockchain.

2.2.4 Performance

Through hybrid IoT, light-company systems submit trades to full-company apps, and they can aggregate them into late-delivery squares. Subsequently, there is a requirement for full companions to handle significant commercial loads. Appropriately, the first execution test for a hybrid IoT is a weight test, in which a set of light companions sends trades to and from full allies. In PoW sub-block chains, full companions focus on the methodology of knowledge. Sub-block chains are made by a particular number of complete partners, which has an effect on the time taken to gain comprehension and on how the full companions manage their advantages. In this way, the second form of execution check is often carried out by fluctuating sub-block chain scales, determining the time required to achieve comprehension and the maximum value of the companions.

Result We found that the knowledge cycle is longer with sub-block chains with even more full companions as the square and the spread of trade takes longer. Overall, the same circumstance with more full companions uses less resources. Therefore, for sub-block chains with even more complete sidekicks, resource usage is seen at the midpoint of longer comprehension intervals.

2.3 Conclusion

IoT network architecture poses a range of difficulties, including complexity, unstable interoperability, infrastructure impediments, and protection and safety vulnerabilities. The continuous development of Blockchains is basically a response to issues such as increased interoperability, protection, stability, accessibility, and secure consistency. This chapter introduced a comprehensive Blockchain map and provided an overview of Blockchain progressions. We also provided overview of IoT designs.

This chapter introduced a novel crossbreed IoT Blockchain program, referred to as hybrid-IoT. Through hybrid-IoT, subgroups of IoT contraptions are peers on PoW sub-block chains, connected to the BFT through the connector network. Within this discussion, we also examined the structure of the PoW sub-block chains. The display shows the authenticity of the PoW sub-block chain structure.

References

1. M. Swan, *Block Chain: Blueprint for a New Economy* (O'Reilly Media, Inc, Beijing, 2015)
2. M. Iansiti, K.R. Lakhani, The truth about block chain. Harv. Bus. Rev. **95**(1), 118–127 (2017)
3. M. Crosby et al., Block chain technology: Beyond bit coin. Appl. Innov. **2**(6–10), 71 (2016)
4. C. Cachin, Architecture of the Hyperledger block chain fabric. Workshop Distrib. Crypto Curr. Consens. Ledgers **310**(1), 4 (2016)
5. Madinah, Saudi Arabia, Block chain and its Role in the Internet of Things (IoT), Int. J. Sci. Res. Comput. Sci. Eng. Inf. Technol. ISSN: 2456–3307 UGC Journal No: 64718
6. S. Nakamoto, "Bit Coin: A Peer-to-Peer Electronic Cash System," 2008
7. ("The Internet of Things", ITU Internet Report 2005) The Internet of Things. Executive Summary [Online]
8. T. Savolainen, J. Soininen, B. Silverajan, IPv6 addressing strategies for IoT. IEEE Sensors J. **13**(10), 3511–3519 (2013)
9. F. Tschorsch et al., Bit coin and beyond: A technical survey on decentralized digital currencies. IEEE Commun. Surv. Tutorials **18**(3), 2084–2123 (2016)
10. A.S. Chhabra, T. Choudhury, A.V. Srivastava, A. Aggarwal, Prediction for big data and IoT in 2017, in *2017 International Conference on Infocom Technologies and Unmanned Systems: Trends and Future Directions, ICTUS 2017, 2018-January*, (IEEE, Piscataway, 2018). https://doi.org/10.1109/ICTUS.2017.8286001
11. K.S. Obheroi, A. Chaurasia, T. Choudhury, P. Kumar, Economical home monitoring system using IOT. Adv. Intell. Syst. Comput. **712**, 627–637 (2018). https://doi.org/10.1007/978-981-10-8228-3_58
12. K. Jaswal, N. Kashyap, M. Singla, T. Choudhury, A framework for security and protection in Internet of things (IoT) devices, in *Proceedings of the 2018 International Conference on Communication, Computing and Internet of Things, IC3IoT 2018*, (IEEE, Piscataway, 2019). https://doi.org/10.1109/IC3IoT.2018.8668121
13. A. Khanna, R. Anand, IoT based smart parking system, in *2016 International Conference on Internet of Things and Applications, IOTA 2016*, (IEEE, Piscataway, 2016). https://doi.org/10.1109/IOTA.2016.7562735
14. A. Khanna, R. Tomar, IoT based interactive shopping ecosystem, in *Proceedings on 2016 2nd International Conference on Next Generation Computing Technologies, NGCT 2016*, (IEEE, Piscataway, 2017). https://doi.org/10.1109/NGCT.2016.7877387

15. A. Sharma, T. Choudhury, P. Kumar, Health monitoring & management using IoT devices in a cloud based framework, in *Proceedings on 2018 International Conference on Advances in Computing and Communication Engineering, ICACCE 2018*, (IEEE, Piscataway, 2018). https://doi.org/10.1109/ICACCE.2018.8441752
16. http://www.electronics-tutorials.ws/io/io_1.html
17. https://techbeacon.com/enterprise-it/4-stages-iot-architecture
18. E. Buchman, Tendermint: Byzantine fault tolerance in the age of block chains, PhD dissertation, 2016
19. Y. Dai, D. Xu, S. Maharjan, G. Qiao, Y. Zhang, Artificial intelligence empowered edge computing and caching for internet of vehicles. IEEE Wirel. Commun. Mag. **26**(3), 12–18 (2019)
20. G.O. Karame et al., Misbehavior in bit coin: A study of double spending and accountability. ACM Trans. Inf. Syst. Secur. (TISSEC) **18**(1), 2 (2015)
21. L. Da Xu et al., Internet of things in industries: A survey. IEEE Trans. Ind. Inf. **10**(4), 2233–2243 (2014)
22. S.K. Singh, S. Rathore, J.H. Park, Block IoT intelligence: A block chain-enabled intelligent IoT architecture with artificial intelligence, in *Future Generation Computer Systems*, (Elsevier, Amsterdam, 2020)
23. Z. Li, J. Kang, R. Yu, D. Ye, Q. Deng, Y. Zhang, Consortium. IEEE Trans. Ind. Inf. **14**(8), 3690–3700 (2018)
24. L. Lao, Z. Li, S. Hou, B. Xiao, S. Guo, Y. Yang, A Survey of IoT Applications in Block chain Systems: Architecture, Consensus, and Traffic Modeling. ACM Computing Surveys ..., 2020. dl.acm.org

Chapter 3
Fusion of Blockchain and IoT: The Future of Industry 4.0

Ruchika Gupta, Shiv Ranjan, and Gagan Kukreja

3.1 Introduction

Industry 4.0 consists of the integration of manufacturing networks using cyber-physical structures, thereby allowing manufacturing facilities and development processes to transition into automated and complex networks. This autonomous system involves communication between the components of a "smart factory," both internally and with other factories that are built using IoT.

The number of attacks on IoT gadgets has greatly increased in recent years. The vast majority of these attacks attempt to assume responsibility for IoT gadgets with the goal of shaping their own botnet. Botnets are a gathering of PCs acting together to finish some specific assignment. A few botnets are devoted to dispatch only DDoS attacks on explicit administrations. A Distributed Denial of Service (DDoS) assault is an endeavor to make an online service inaccessible by overpowering it with traffic from different sources. Therefore, IoT gadgets that are not protected with appropriate username-password combinations can be effectively undermined and shaped into a botnet to dispatch attacks on different administrations. Examples of malware that misused this IoT vulnerability are listed below [1]:

- **Mirai**: There was a major DDoS attack directed at KrebsOnSecurity on September 20, 2016 that came from various areas across the globe. The malware that was responsible for this attack was the Mirai malware, whose source code was made

R. Gupta (✉) · S. Ranjan
Amity University, Greater Noida, Uttar Pradesh, India
e-mail: rgupta@gn.amity.edu; sranjan@gn.amity.edu

G. Kukreja
Ahlia University, Manama, Bahrain
e-mail: gkukreja@ahlia.edu.bh

T. Choudhury et al. (eds.), *Blockchain Applications in IoT Ecosystem*,
EAI/Springer Innovations in Communication and Computing,
https://doi.org/10.1007/978-3-030-65691-1_3

publicly available by the pseudonym Anna Senpai. This same malware was used on October 21, 2016 to dispatch a massive reach attack on Dyn Domain Name Servers (DNS), which culminated in a large number of downtime pages. The working guideline of Mirai malware is as follows:

  - **CnC server:** A single Command-and-Control (CnC) system monitors all Mirai-infected computers and places them under its control. The ransomware emerges like a worm and attacks primarily IoT users. Once a device has been compromised, it waits for CnC server commands to start its attack.
  - **Loader:** The loader system distributes the malware to computers until it identifies a computer with the default username-password combination.
  - **IoT devices:** In the most operating concept, the malware seeks a series of default username-password combinations on every network. Because many IoT device users do not adjust their default username-password combination, they are very easy victims of the malware and quickly fall under the CnC server power.

- **Hajime:** Hajime is a Mirai variant that adopted the same operating concept as Mirai. However, its CnC service is not a single node but rather a distributed collection of peer-to-peer (P2P) servers that make it more challenging to take down. The hackers are still in active development.
- **IoTWorm:** A few analysts from Israel discovered another novel assault that could begin a chain response to make smart lights unusable and can disable a smart light system of an entire community. Such lights use an encryption key AES-CCM (Advanced Encryption Standard- Counter with CBC-MAC (Cipher Block Chaining, Message Authentication Code) to validate the reset of the firmware, so all lights use an AES-CCM equivalent switch. The AES (Advanced Encryption Standard) key can be acquired by a side channel test. Using powerlessness in the ZLL (Zigbee Light Link networks) convention's touch link component, the lights can be ejected from their specific controllers and malevolent software updates can be subsequently sent by labelling them with the AES key. When lights use the same AES key, heightening this attack exponentially and resolving any single smart light in a region is incredibly easy [1].

A potential solution to these problems is to use encrypted public-key networks for all interactions between IoT devices and Blockchain to ensure the security of the public-keys. Using a public-key dependent authentication system with a shared key infrastructure (PKI) would have prevented both the Mirai and Hajime attacks because it would have become more difficult to determine the private keys of many IoT users. The IoTWorm would also not have spread rapidly because separate IoT devices would have private individual keys and not a common key for all devices. To obtain the key and send out the malicious updates, each light would need to be hacked manually.

Moreover, with the advancement of 5G technology and future developments up to 8G technology, IoT has been growing rapidly and has been implemented by almost 85% of companies. IoT endpoints are anticipated to expand at a compound yearly development pace of 32% from 2016 to 2021, reaching an introduced base of 25.1 billion units, as indicated by Gartner's forecast [2]. However, due to security and scalability issues, its benefits are unclear.

Therefore, it is imperative to understand what IoT is. What are the major concerns of users of IoT? How will we resolve these concerns? What is Blockchain? What are the challenges of the fusion of Blockchain with IoT? What is the future of the fusion of Blockchain with IoT? Will advanced features of IoT be developed to replace the benefits of Blockchain? This chapter describes one of the possible ways IoT and Blockchain technologies can be integrated to solve the given problems. The chapter also examines the feasibility of integrating Blockchain with IoT technology, prospective obstacles, and the advantages such convergences can bring.

3.2 IoT: Benefits and Challenges

IoT integrates humans, locations, and objects, rendering them 'smart' and able to interact with each other. IoT collects big data by continuously observing the real environment, processing, and taking a smart action centered on the same data. It makes the seamless convergence of the digital and physical realms feasible, thereby shifting the very meaning of our real-world experience [3].

Complex processors, cameras, and actuators are implanted into physical items, with each moving information to the IoT network. Thereafter, the analytical tools of IoT use this information to transform thoughts into training, impact advertising practices, and add creative methods of working. Thanks to the network of user-friendly devices, a person may access data irrespective of their role. Because communications are not fluent and transparent, inefficiencies are caused. However, in a network of interconnected devices, better connectivity is feasible because sending data packets over a wired network saves time and resources. Not only does IoT save time and money, it also encourages automation – the most critical aspect of today's tech-savvy existence, in which all tasks can be performed without human intervention with improved service quality [4].

The platform is gaining immense traction in the industry; as a result, it was projected that approximately 75.44 billion IoT-linked devices will be accessible worldwide by 2025 [5]. However, a variety of technological and health issues remain unaddressed. One of the issues with current IoT implementations is the need for a centralized entity (such as a cloud server) to communicate and interact through the Internet, which poses a major challenge to the privacy and security of the vast amounts of sensitive data that are produced. The original network concept requires a decentralized infrastructure, such as a peer-to-peer or distributed framework.

In contrast, the client-server model is very expensive in terms of high latency costs and low interoperability due to insufficient data exchange, device heterogeneity, coordination requirements with other distributed IoT networks, maintenance costs, and network equipment expenditures. The cloud service becomes a single point of failure, thereby undermining the entire network and making resources ineffective when evading the scenario.

Security has become a significant IoT worry that has upset its enormous scope organization. IoT gadgets often experience the ill effects of security vulnerabilities, which make them an obvious target for attacks, including Distributed Denial of Service (DDoS). A few DDoS assaults have caused disturbances for associations and individuals lately. Unsecure IoT applications provide digital lawbreakers access to hack them into propelling DDoS assaults [3].

Another issue with current IoT frameworks is that of flexibility. As the amount of devices connected through an IoT network increases, current fused structures to validate, support, and interface different centers in a framework will become a bottleneck. This would require huge endeavors to create servers that can manage the colossal amount of information exchange, and the entire framework can go down if the server becomes difficult to reach.

While talking about the issue of IoT security, some rush to accuse the clients. Shoppers, they state, do not think enough about the associated gadgets they are introducing in their homes and working environments. They do not comprehend that the gadgets need refreshing, like personal computers and cell phones. Also, they overlook that the microwave or refrigerator is online and requires updates like some other devices. However, governments around the globe are endeavoring to manage this test of consumer mindfulness with regards to IoT security, progressively coordinating efforts and exhortation towards end clients.

In such a situation, with the end goal for IoT to succeed, highly coordinated activities are required across the board. IDC (International Data Corporation) had anticipated that practically 90% of associations using IoT will experience an IoT-based penetration of their back-end IT frameworks in the near future [6].

3.3 Blockchain Technology: Functions and Usefulness

On the most fundamental level, a Blockchain is an extraordinary sort of database wherein 'blocks' of successive and unchanging information related to physical/virtual resources are connected through cryptographic hashes, which are also appropriated as ever-developing 'chains' among numerous shared hubs. Blockchain increases are rendered after ratification by multiple hubs that use an arrangement method; the two simplest methods are Proof of Stake (PoS) and Proof of Work (PoW), upon which the new squares are allotted to all hubs. PoW is the most well-known negotiation instrument right now, with Bitcoin mining by solving cryptographic riddles being the most common concept. In any case, PoS requires lower processing assets and power, and it can convey quicker throughput.

Each personal computer (PC) in the network organization has its own network replication, which means that there are thousands, or a massive amount of duplicates of the corresponding Blockchain due to Bitcoin. Although Blockchain replication is identical, distributing the data through a computer network makes it extremely challenging to monitor the data. For Blockchain, there is no singular, definitive record of controllable times. Alternatively, a programmer needs to monitor any Blockchain duplication on the network. Therefore, Blockchain is referred to as a "distributed database."

There are two types of Blockchain frameworks: permissionless (which anyone can join) and permissioned (in which individuals are confirmed by whoever is running the network). The latter can be further divided into 'private' and 'network' Blockchain frameworks – a lone endeavor versus associations connected by a particular business process, for example. In permissionless blockchains, such as those that underpin Bitcoin and Ethereum, consensus mechanisms are considered more reliant for affirming identities and verifying transactions [7].

As an appropriated record, Blockchain can be used to record any exchange, as well as monitor any advantage and related installments. Contrasted with conventional business forms, Blockchain can convey time and cost reserve funds, along with better security – particularly in a permissioned arrangement.

Operation of Blockchain The foundation for building a Blockchain is a P2P network that includes all the devices required to achieve the goals of the application. Asymmetric cryptography is used to assign two keys to each of the nodes: a public key to identify a machine on the network, and a private key to enable transactions on the network between themselves or other computers. When a device wants to make a transaction, it signs with its private key, transfers it to its neighbors for verification, and then disseminates it across the network. The private key offers confidentiality and integrity; miners bundle numerous such transactions into a block of timestamped transactions before they are verified by the network. Block validation may be achieved by several methods and then transmitted to the network, where all the nodes verify the block and its transactions, as well as the hash connection with the previous iteration. It is attached to and modified in the chain when tested, and otherwise discarded [6].

Blockchain uses four main concepts as its basis:

- **P2P network**: Eliminates the central Trusted Third Party and means that all network nodes have the same rights.
- **Accessible and distributed ledger**: A clear network where each node independently assesses the authenticity of a transaction.
- **Mining**: Network delays occur in a distributed system and not all nodes receive blocks of transactions at the same time. Therefore, any node must be prohibited from introducing a transaction to the chain, because the chain must have only one true and organized branch.
- **Synchronization of ledger copies**: Nodes come with a backup of the same ledger. The updating of ledgers by techniques is then used to transmit the new transactions to the network, validate the new transactions, and generally connect the authorized transactions to the ledgers [6].

Blockchain technology takes care of encryption and faith problems in many respects. Firstly, new blocks are still placed chronologically and linearly. Therefore, they are still connected to the Blockchain's " top." When you glance at the ledger in Bitcoin, you will find that each block has a place on the list, labeled as a "height." As of January 2020, the height of the block had reached 615,400 [8].

After connecting a square at the edge of the record, it is extremely difficult to reach back and change the block's substance. Each square has its own hash, and a square hash before it. A mathematical formula makes hash codes and changes advanced data into a series of numbers and letters. In the event that the data are changed at a certain point, the hash code consistently moves.

Blockchain Usefulness Within the ledger, the 'blocks' store details on cash-related trades. However, for unexplained purposes, Blockchain is still a fully secure way of transmitting knowledge among different forms of exchanges. Blockchain software can be used to store property trading records, stops in a scalable chain, and even applicant-friendly decisions.

Deloitte interviewed 1000 organizations in seven nations on integrating Blockchain in their activities. Their research showed that 34% currently had a Blockchain implementation, while another 41% have plans to implement a Blockchain program over the next year. Likewise, approximately 40% of the organizations surveyed expected to invest at least $5 million in Blockchain in the coming year [3]. Blockchain has the potential to improve performance in the following areas:

- **Banks:** Buyers will have their transactions managed in as little as 10 min when integrating cryptocurrency into banks; this is basically the period it takes to connect a square to the network, with no restriction on the time of day or day of the week. Furthermore, through Blockchain, banks have the ability to exchange assets more easily and anonymously between organizations. Capgemini, a French consultancy, assessed that buyers may have saved $16 billion in banking and security charges in 1 year via Blockchain-based software [20].
- **Health care:** Medical care providers may use Blockchain to store the health data of their patients in a safe manner. This can be built into the Blockchain exactly where a health report is made and signed, which provides patients with verification and confidence that the document cannot be altered. Such unique accounts of success will be stored and managed with a private key on the internet, and they would only be reached through clear individuals, thereby ensuring security.
- **Voting:** Casting a Blockchain vote may prevent misrepresentation of the electoral judgment and boost voting participation. Each vote on the Blockchain can be placed away as a square, making it impossible to alter. The Blockchain conference will similarly make the administrative process clear, reducing the workers needed to direct a government judgment and providing quick outcomes to authorities.
- **Monitor supply chains:** When it comes to tracking supply chains, blockchain is extremely advantageous. By discarding the paper-based method, companies can easily identify unnecessary elements within their supply chains, even by

gradually discovering items. Therefore, Blockchain would enable companies, and probably even customers, to see how products are handled from a quality assurance point of view from their source to the store.
- **Copyright and eminence assurance:** Copyright and distribution laws for music and entertainment have been unclear in an environment of evolving Internet access. With blockchain, copyright laws for advanced material downloads will be greatly extended, ensuring that the craftsman or manufacturer of the substance being purchased earns a substantial amount of money. Blockchain can likewise provide performers and creators with constant and clear ownership knowledge.
- **Smart contracts.** On top of a ledger, smart contracts may be designed and work as decentralized applications. These programs may provide functions that are becoming increasingly complex as the need for traditional legal contracts fades.
- **Property.** Property titles, sales, and value can be assembled onto the Blockchain, providing clarity and decreasing the time and cost related to property transactions.

These are only a portion of the conceivable outcomes that accompany the new innovation. Currently, there are only murmurs in the business world of how Blockchain innovation can disturb the current models. Blockchain must be implemented with a goal to drive operational efficiencies. If appropriately executed, Blockchain innovation has even more extensive ramifications – without a doubt, positive ones.

3.4 How Can Blockchain Solve IoT Safety Issues and Scalability?

An IoT network can maange data transfers through numerous devices controlled and operated by different entities, making it impossible to determine the cause of any data breach in the event of a cyber-criminal attack. Furthermore, the IoT produces a large array of data, and the control of the data is not necessarily transparent for many parties involved.

Blockchain has the ability to help solve some of the issues surrounding IoT insurance and versatility. Blockchain, because of its unique features, is a technology game changer. It provides a method to gather clients to archive and trade the subtleties. Selected agents of this gathering hold their duplicate of the record and will commonly check every single new exchange through an agreement component until they are allowed onto the record [3].

Blockchain can help reduce the security and adaptability concerns related to IoT in the following manner:

- The flowed record in a Blockchain system is deliberately planned, which provides the necessity for trust among the included parties. No single affiliation has order over the enormous proportion of data generated by IoT devices.

- Utilizing Blockchain to store IoT data would incorporate another layer of security that software engineers would need to avoid in order to pick up induction to the framework. Blockchain provides a more robust level of encryption that makes it very difficult to overwrite existing data records.
- Blockchain engineering can decentralize the DNS, move the content to numerous hubs, and make it nonsensical to hackers. Modification rights should be provided only to the persons who require them (space owners) and no other user may make modifications, which significantly minimizes the possibility of accessing or altering knowledge by unauthorized people. A system can guarantee that it is resistant to programmers by using Blockchain to protect the details, even if each and every hub is cleaned off at the same time.
- When an exchange is started, approval of the information square is done through agreement between the system partners. At that point, the information is confirmed for all time, except if any approved changes are made to the information. At that point, it is encoded with the strongest encryption conventions to ensure information security.
- Smart contracts, a two-party agreement that is held in the Blockchain, can additionally permit the execution of legally binding understandings between parties, subject to certain conditions being met. For instance, when the prerequisites for the conveyance of assistance have been satisfied, smart contracts will support installments naturally, without the necessity for human intervention.

3.5 Feasibility Considerations for Integrating Blockchain and IoT Technologies

Blockchain addresses IoT issues by decentralizing the dynamic to an agreement-based shared system of devices. Be that as it may, when structuring the design for IoT devices related to a Blockchain record, there are a few plausibility issues to consider [3]:

1. The protection of exchange history in the mutual record for a system of IoT gadgets cannot be effectively allowed on an open Blockchain. That is because an exchange design investigation can be applied to make inductions about the personalities of clients or gadgets behind open keys. Affiliations should examine their security requirements to determine whether cross variety or private Blockchains may better suit their requirements.
2. One of the most important problems still facing IoT is one of scale: how to handle the vast volumes of data produced by a massive array of sensors. Defining a sensible data model in advance can save time and thwart difficulties when bringing the course of action into creation.
3. IoT devices have limited capabilities, and they lack virus and malware protection software, making them easy targets for hackers.
4. High exchange costs are likewise an inhibitor for collaboration.

5. Similarly, the resolute nature of IoT sensors may be undermined by interfering with the correct estimation of the measures that ought to be met to execute a trade. Measures should ensure the trustworthiness of IoT devices, with the ultimate objective that they cannot be changed by outside interventions; this ensures the protection of data recording and trades.

3.6 Examples and Use Cases of IoT Blockchain Technologies

There are numerous examples and use cases of IoT-enabled Blockchains, such as the following [3, 9, 10]:

1. **Smart appliances:** A smart appliance is a web-enabled device that offers additional data and power compared with traditional appliances. For example, when your medications are scheduled or your laundry cycle has ended, a code linked to your device can be attached to the web and alert you. Such alerts keep your devices in perfect order, save the user money in terms of productivity output, and allow the user to monitor their equipment while away from home, among other benefits. Encoding these Blockchain devices guarantees your ownership and allows transferability.
2. Helium: Helium is the world's first decentralized network of machines. The organization uses Blockchain to associate low-control IoT machines (e.g., switches, microchips) to the Web. Helium's Blockchain-based remote web foundation uses radio innovations to fortify web associations and radically decrease the resources needed to operate smart machines.
3. **Supply chain sensors:** Sensors include items of measurable consistency in the supply chain to collect data on the region and state of the structures when they are sent for distribution. In a 2016 survey of 900 leading supply chain companies, Deloitte and MHI Research discovered that 44% of respondents used sensors, with 87 percent planning to use the technology by 2020. These advancements would results in up to 1 trillion sensors by 2022 and up to 10 trillion sensors by 2030. The Blockchain holds the data, then controls, guarantees, and transfers it [21].
4. **Riddle & Code:** It integrates smart card authentication by incorporating blockchain technologies with cryptography to establish a hardware-based digital identity for all physical objects connected together. These connected objects communicate securely using highly secured crypto chip which enables them to become individual blockchain node. Communication and transactions between devices are autonomously and securely performed using a highly secure crypto chip made up of an adhesive nonremovable tag which enables each device to become a blockchain node. In addition, an Android program is used to perform a blockchain transaction in order to register the chip's special, tamper-proof identity. It will communicate with other devices once it has been tested in the network.

5. ArcTouch: This service creates and fabricates Blockchain-based programming for a scope of smart, associated things, including voice aides, wearable devices, and smart televisions. The service has assembled customized, decentralized apps (DApps) for many organizations that connect to IoT gadgets. ArcTouch's DApps provide an additional degree of IoT security and can process requests quicker through smart agreements. The organization has assembled a few Blockchain DApps that can be associated with IoT gadgets, such as Amazon Alexa.
6. **Modum.io:** This service solidifies IoT sensors with Blockchain advancements, providing data integrity to trades, including physical items. The sensors record normal conditions, such as temperature, that product is reliant upon while in movement. When the product appears at a transit point or end customer, the sensor data is verified against fated conditions in a smart agreement on the Blockchain. The agreement ensures that the conditions meet all requirements set out by the sender, their clients, or a controller and triggers various actions, such as notifications to the sender and beneficiary, amount, or appearance of product.
7. HYPR: This service utilizes decentralized systems for associated ATMs, vehicles, and homes. One of the fundamental reasons cyberattacks are so obliterating and across the board is that unified databases store a huge number of passwords. HYPR stores biometric logins on its Blockchain, verifying and decentralizing significant data. The organization's biometric security conventions incorporate remarkable facial, eye, voice, and palm recognition instruments for IoT gadgets.

3.7 Benefits of Integrating Advanced IoT for Blockchain Networks

Blockchain enables IoT gadgets to upgrade security and acquire straightforwardness in IoT systems. As indicated by IDC, 20% of all IoT arrangements were using Blockchain-based arrangements by 2019. Banks and financial institutions, such as HSBC, are using PoC to affirm the Blockchain development. In addition to financial establishments, a wide range of associations are using the capacity of the Blockchain [6].

IoT provides boundless open entryways for associations to run smart assignments. Many devices are equipped with sensors, sending data to the cloud. Thus, consolidating the two innovations of IoT and Blockchain can make the frameworks more productive. Some advantages offered by this coordination are as follows:

- Increased security and trust in shared multi-party exchanges and information
- Increased business proficiency and reduced expenses.
- Increased income and business opportunities
- Improved constituent or member experiences

However, the advantages provided by this combination vary by industry. Some of the advantages associated with this reconciliation from various ventures are described in the following sections [11].

Logistics and Supply Chain A worldwide supply chain includes numerous partners, such as specialists and crude material suppliers. Additionally, the supply chain can extend over long stretches of time and comprise a huge number of installments and solicitations. Because of the contribution of different partners, conveyance delays become the greatest test. In this way, organizations are attempting to make the vehicles IoT-empowered to follow the development all throughout the shipment process. Because of the absence of transparency and intricacies in the current supply chain and logistics, a Blockchain and IoT combination can help to upgrade the quality and discernibility of the system [12].

Both the organization and the purchaser can follow the item's entire life cycle throughout the supply chain utilizing Blockchain and IoT features. Blockchain is a complete information record where all the interchanges among IoT gadgets are captured in the history. It gives instant access to all data associated with the items, such as the dates a fish was harvested, processed, and sold – a total record of its excursion from sea to fork. Once information is saved on the Blockchain, collaborators that have signed Agreements eventually obtain access to the details. Supply chain members can similarly prepare for shipment and run cross-border exchanges.

Construction Industry The construction process includes many experts who need to exchange data to configure and actualize projects effectively. There are many intermediaries who are used to verify the entire procedure, such as controllers, investors, backup options, legal advisors, and so on. There is a need to build trust among all the partners. The progress of tasks from conventional strategies to advanced structures is suited for a computerized approach, such as Blockchain, which can encourage and empower trust among players. For the individuals who need open, reliable IoT correspondences without depending on go-betweens, a private Blockchain could provide the arrangement and empower information security between IoT gadgets.

Blockchain additionally makes a dependable chain of events, exchanges, resources, and basic project details. This permits messages, project management frameworks, and bookkeeping frameworks to meet up in one spot, creating a validated record of all exchanges on the Blockchain and guaranteeing that information cannot be lost. It ensures that with each project, there is a single mutual adaptation of fact, and this is what the organization wants to prevent replication, minimize errors, and maintain data integrity. Some of the advantages it can offer include the following: Improve the transparency and trustworthiness of construction logbooks, works completed, and material supplies reported, providing a consistent infrastructure for information management at all phases of the building life cycle [13].

Automotive Industry Today's vehicles are advancing to be much more than just a transportation tool. The cars of the twenty-first century are moving server farms with local sensors and computers that collect vehicle information. With progressively stable, identifiable transfers and improved data access and integrity, Blockchain may reinforce trust and unify efforts between organizations, consumers, and even vehicles. The automotive sector is a fascinating use of IoT with Blockchain, in which centralized engineering interacts with computerized fuel deployment, self-sufficient cars, smart departures, and mechanized traffic management.

Blockchain innovations have assisted with the assembling of self-governing vehicles. Smart vehicles are outfitted with autopilot modes, which permit the vehicle to independently depart or perform different undertakings by training the vehicle's advanced computer utilizing voice commands. Because of the RFID labels on Blockchain-produced vehicle parts, every segment can be easily confirmed for credibility. The blockchain platform offers the highest degree of confidentiality and speeds up the vehicle ownership process. Blockchain additionally uses a smart contract, in which the dealer and purchaser can exchange merchandise without the requirement for a broker. Blockchain likewise triggers machine-to-machine (M2M) exchanges with the use of smart contracts [14].

Pharmacy Industry Blockchain-coordinated IoT can accelerate the pace and reliability of clinical trials and improve pharmaceutical supply chains. Additionally, as one survey from BIS Research showed, healthcare companies worldwide lose approximately $200 billion annually because of counterfeit or contaminated medications [15]. The transparent nature of Blockchain allows the distribution of medications to be monitored from manufacturing to shipment.

Agriculture From ranchers to makers and merchants, Blockchain combined with IoT is reinventing the food manufacturing industry. Blockchain can improve agricultural management practices using a simpler approach to maximize farming services such as water, labor, and fertilizers.

In IoT-empowered smart agriculture, the crop field is monitored with sensors for measurements such as temperature, pH, soil dampness, moistness, and light. IoT sensors and gadgets capture information that can assist farmers with making educated choices in the development of the crops. Artificial intelligence (AI) is applied to the information captured from the sensors to provide valuable bits of knowledge with respect to crop identification and crop yield forecast, among others [16].

The high-quality information assembled by applying AI is stored in an interplanetary file system, an appropriated stockpiling stage where addresses are hashed and put away on the Blockchain. The data captured in the Blockchain trigger smart agreements to process rules characterized inside them. Smart contracts encourage the trading of information stored on the Blockchain inside particular partners in the framework. Because data are available to each agricultural business member, they will be able to effectively produce crops or food.

3.8 Prospective Barriers to the Convergence of Blockchain and IoT

The combination of Blockchain with IoT is not simple. Blockchain was intended for an online environment with powerful PCs, and this is a long way from the IoT reality. Blockchain exchanges are carefully marked; thus, devices suitable for working with cash must be equipped with this ability. Fusing Blockchain into the IoT requires further testing. The following are some of the recognized difficulties:

- **Certification security:** Given the way that Blockchain is recognized for its high-security standards, a network built on Blockchain is equally as safe as the direction of the application. Other authentication protocols, such as TLS (Transport Layer Security), are currently used by IoT application protocols to provide safe communications. Such reliable protocols are complicated and resource-intensive, and they necessitate centralized maintenance and governance of key infrastructure, which is usually accomplished by Private Key [17]. Also in such a scenario, loss of a record's private keys can provoke absolute loss of advantages, or data, obliged by this record. Therefore, this needs to be reviewed further.
- **Capacity limits and versatility:** As discussed, stockpiling limits and the adaptability of Blockchain are still under investigation. However, with regard to IoT applications, the inalienable limit and adaptability constraints make these difficulties a lot more noteworthy. In this sense, Blockchain may give off an impression of being incompatible with IoT applications, but there are methods by which these impediments can be mitigated or avoided. In the IoT, where gadgets can produce gigabytes of information progressively, this constraint is an incredible hindrance in its combination with Blockchain. Some current Blockchain executions can easily process several trades per second, so this could be a potential bottleneck for the IoT. Moreover, Blockchain is not intended to store a lot of information like those created in the IoT. A reconciliation of these advancements should manage these difficulties.
- **Processing Speed:** Concerns have been raised about the computing power needed to encrypt all of the objects in a blockchain-based ecosystem. The IoT networks, unlike traditional computer networks, are very complex and made up of computers with a wide range of computing capacities, and not all of them would be able to execute the same encryption algorithms at the same speed.
- **Unwavering quality:** Blockchain may be a key development to provide important security improvements in the IoT. One of the essential challenges in the combination of the IoT with Blockchain is the steadfast nature of the data generated by the IoT. Blockchain should ensure the data in the chain is unchanging and that it can recognise updates, regardless of whether data is somewhat compromised in the blockchain.

3.9 Conclusion

We are at the beginning of another period of Industry 4.0. The advancements in sensors and smart chips is growing rapidly, making them dynamically minimal and appropriate for a continuing relationship with Blockchain records. The blend of Blockchain and IoT has vast potential for the creation of a business focal point of organizations among devices, as well as providing organizations with the ability to create a driving force from accumulated data [18, 19]. The increase in Blockchain shows, affiliations, and IoT device providers demonstrates that there is a place for Blockchain in the IoT division.

To achieve an ideal, secure model of IoT, security must be included in the foundation of the IoT's natural framework, with intensive authenticity checks, approval, and data validation. All data must be encoded at all levels. Without a solid foundation, more risks will be created with every device added to the IoT. A protected and safe IoT with guaranteed security is needed. If we can overcome the disadvantages of Blockchain technology, it is an outrageous trade-off. It's important to note the Blockchain isn't a guarantee of stable IoT. Its suitability is determined by the way it is applied.

References

1. A. Kumar, T.J. Lim, A secure contained testbed for analyzing IoT botnets. https://arxiv.org/pdf/1906.07175.pdf
2. M. Arnott, P. Middleton, K. Sharpington, Internet of Things forecast database. Gartner Research, 05 November 2019
3. Driving Into Digital: Journey to the future, Deloitte. https://www2.deloitte.com/in/en.html
4. S.A. Wright, Privacy in IoT Blockchains: With big data comes big responsibility, in *IEEE Int'l Workshop on IoT Big Data and Blockchain (IoTBB'2019)*
5. Internet of Things (IOT) Connected Devices installed base worldwide from 2018 to 2025 (in Billions), Statista Research Department, May 2020. https://www.statista.com/statistics/1101442/iot-number-of-connected-devices-worldwide/
6. R. Thakur, R. Vaghasiya, C. Patel, N. Doshi, Blockchain based IoT – A survey. Proc. Comput. Sci. **155**, 704–709 (2019) https://www.sciencedirect.com/science/article/pii/S1877050919310178
7. T.K. Sharma, Permissioned and permission less Blockchain – A comprehensive guide, Blockchain Council, Insights and Resources, November 2019. https://www.blockchain-council.org/blockchain/permissioned-and-permissionless-blockchains-a-comprehensive-guide/
8. A. Chatterjee, What is bitcoin and Blockchain? Learn Blockchain and bitcoin quickly. Techtravelhub, May 2020. https://www.techtravelhub.com/blockchain/
9. Blockchain Infographics: Blockgeeks. https://blockgeeks.com/blockchain-infographics/
10. Blockchain: What is Blockchain technology? How does Blockchain work? Built-in Report. https://builtin.com/blockchain
11. IBM Blockchain is changing business, industries- and even the world, IBM Cloud Forum 2020. https://www.ibm.com/in-en/cloud/blockchain-platform/developer
12. Oracle Report: Blockchain in manufacturing – Answering the Clarion call for better traceability, January 2018

13. T. Ziga, Klinc Robert: potentials of Blockchain technology for construction management. Proc. Eng. **196**, 638–645 (2017)
14. Deloitte Report: Accelerating technology disruption in the Automotive Market. https://www2.deloitte.com/content/dam/Deloitte/cn/Documents/consumer-business/deloitte-cn-consumer-blockchain-in-the-automotive-industry-en-180809.pdf
15. BIS Research: Global Blockchain in healthcare market 2018. https://bisresearch.com/industry-report/global-blockchain-in-healthcare-market-2025.html
16. Leewayhertz Report: Blockchain in agriculture-improving agricultural techniques. https://www.leewayhertz.com/blockchain-in-agriculture/
17. G. Chandra, R. Gupta, N. Agarwal, Role of artificial intelligence in transforming the justice delivery system in COVID 19 pandemic. Int. J. Emerg. Technol. Learn. **11**(3), 344–350 (2020)
18. P. Srivastava et al., Fuzzy methodology approach for prioritizing maintenance 4.0 attributes, in *2020 International Conference on Computation, Automation and Knowledge Management (ICCAKM)*, (IEEE, Dubai, 2020), pp. 308–311. https://doi.org/10.1109/ICCAKM46823.2020.9051483
19. A. Gupta, A.O. Salau, P. Chaturvedi, S.A. Akinola, N.I. Nwulu, Artificial neural networks: Its techniques and applications to forecasting, in *IEEE International Conference on Automation, Computational and Technology Management (ICACTM)*, (IEEE, London, 2019), pp. 320–324. https://doi.org/10.1109/ICACTM.2019.8776701
20. Smart Contracts in Financial Services: Getting from Hype to Reality, Capegemini Report, 2016. https://www.capgemini.com/news/consumers-set-to-save-up-to-sixteen-billion-dollars-on-banking-and-insurance-fees-thanks-to/
21. Accelerating change: How innovation is driving digital, always-on supply chains, The 2016 MHI Annual Industry Report, http://cpbucket.fiu.edu/1168-geb6368x81168_emba-97075%2F2016-industry-report-2016-(1).pdf

Chapter 4
Blockchain Consensus Algorithms: Study and Challenges

Avita Katal, Vitesh Sethi, and Saksham Lamba

4.1 Introduction to Blockchain Technology

Before blockchain technology, when customers used to make transactions, they had to trust a third party that ensured the transaction is valid, so that it can be put into action. But sometimes the middle party cannot be trusted blindly as it could be exploited. The root cause for this problem is formal centralization, in which there is a single centralized authority on which everything depends. This problem was avoided by the use of blockchain which is based on decentralization. Blockchain was developed by Haber and Stornetta [1] and uses multiple independent organizations to validate the transactions. In more formal words, blockchain is defined as:

The distributed and decentralized database of public and private ledger of all transactions or digital assets that is unalterable and visible to each and every one who is associated with that particular network with the use of cryptographic hashing.

This technology gained popularity after the introduction of Bitcoin by Satoshi Nakamoto and group [2] in the year 2008. The main reason to introduce Bitcoin was to overcome the problems of traditional transactional methods, i.e. trust in third parties. The ledger in Bitcoin contains the records of all the transactions. It is established containing many blocks with transactions inside them. That is how the name Blockchain was introduced. The previous block could be referred by the block using a hash value. The first block added to the chain is known as Genesis block that contains the first transaction. The blocks are arranged according to a particular chronology. The transactions are verified by the nodes which are a part of the network. Once the transaction is considered to be valid and digitally

A. Katal (✉) · V. Sethi · S. Lamba
Department of Virtualization, University of Petroleum and Energy Studies, Dehradun, India

T. Choudhury et al. (eds.), *Blockchain Applications in IoT Ecosystem*, EAI/Springer Innovations in Communication and Computing,
https://doi.org/10.1007/978-3-030-65691-1_4

signed by the sender, the block containing the transaction details is added to the chain and thereafter cannot be changed. This method of verification of transactions can be confusing since each node would broadcast their block information. To find a solution to this problem, a consensus is made among all the nodes on what all blocks must be mined and which all nodes would have the authorization to make changes to their proposed blocks. Blockchain has different deployment strategies: Public Blockchain and Private Blockchain [3]. A Public blockchain or Permissionless or Unpermissioned blockchain is the one in which any node can enter and leave the chain at any point of time. Private blockchain also known as permissioned blockchain is different from public blockchain as it allows only trusted or authorized nodes to participate, thus ensuring privacy of the chain data. They can further be categorized as Consortium blockchain and Fully Private blockchain. In the Consortium blockchain, few nodes validate a transaction [3]. In the next section of the chapter, consensus algorithms will be discussed in detail.

4.2 Introduction to Consensus Algorithms

Consensus algorithms are considered to be an important component of blockchain technology which helps in achieving a tamper free environment. It accepts only one form of truth and is accepted by all the miners present in the network. This technology uses consensus mechanisms among nodes to verify the information, thus preventing the need of intermediaries [4]. All the nodes present in this network should come up with a consensus about the present state of the blockchain. This improves the security of the system as it becomes difficult for the attacker to create a tampered block and introduce it in the decentralized network [5]. Thus, choosing the correct consensus algorithm is very important to implement a blockchain solution. In addition to this, consensus algorithms help to solve two major problems in blockchain, Double Spending in which the same currency is used in two transactions at the same point of time and Byzantine Generals Problem which occurs in the distributed systems.

Double spending occurs when the cryptocurrency is stolen from the blockchain network. The attacker creates and sends a copy of the currency transaction so that it appears to be legitimate. The most common method used by the attacker is to send multiple packets to the network, and reverse the transactions so that it appears that the transactions never happened. Double Spending is solved by validating the transactions by many nodes in the distributed network. Since the data can be communicated among various nodes, some of the nodes may be attacked leading to alteration in communication which is also known as Byzantine Generals Problem. To distinguish the data that has been changed and to get efficient results with the help of other miners, consensus algorithms may be used.

In certain architectures, consensus can be easily achieved under the following scenarios when:

- There is no fault in the system and each node can receive and transfer the messages correctly.
- The system is synchronized.

Three properties that ensure the correctness of the distributed consensus protocol are [6]:

- *Safety/Consistency*: It makes sure that nodes involved in the process of consensus will never converge to an incorrect state.
- *Liveliness/Availability*: The acceptance of each and every correct value occurs eventually or in other words, all non-faulty nodes participate in the consensus process and produce an output.
- *Fault Tolerance*: The blockchain network performance goes unchanged even amidst node failures.

The traditional distributed consensus algorithms are based on certain approaches. They are:

- A message passing system that requires a cloud environment where each and every entity should know about other entities that are a part of the network.
- The shared memory approach in which a common memory space is created so that shared variables can be read and written by everyone present in the network. This approach cannot be implemented for internet grade computing as a readable and writable memory space needs to be created, so that it can be accessed by every user which is a part of the network. Similarly, message passing is not suitable for an open environment like the Bitcoin system where anyone can be a part of the network at any point of time.

4.3 Consensus Algorithms Characteristics

This section describes the different characteristics of the consensus algorithms. The consensus algorithm properties can be categorized into: Structural, Block and reward, Security and Performance [7].

4.3.1 Structural Properties

These properties tell how various nodes/miners in the decentralized network are arranged to take part in a consensus algorithm. The structural properties have the following categories:

(a) Node type: It helps in achieving consensus by engaging different types of nodes in the consensus algorithms. The node types depend upon the consensus algorithm like some of the algorithms may have clients, miners, minters, validators, electors and stakeholders, etc. Node types will be discussed in detail in the coming sections.

(b) Structure type: It defines different ways of structuring nodes within the consensus algorithm by utilizing the committee mechanism. The committee can be subdivided in two types: single and multiple committees.

- Single committee: It defines a particular set of nodes among the nodes that take part in the consensus mechanism by the production of blocks and the extension of the blockchain network.
- Multiple committee: The time required to achieve consensus in a single committee increases as the number of nodes in a decentralized network increases. This reduces the efficiency. To maintain network performance, the concept of multiple committees is introduced in which each committee has different validators.

(c) Underlying mechanism: It is a mechanism of selecting a particular node that is deployed by a consensus algorithm. This mechanism utilizes lottery, coin age or a voting mechanism. A lottery uses a probabilistic mechanism which is based on cryptography or the other mechanisms that are randomized [7]. Voting can take place in single rounds or multiple rounds. The property used by coin age depends on the time for which owner owns a coin.

4.3.2 Block and Reward properties

These properties are used to differentiate the cryptocurrencies on the basis of quantitative metrics. Some of the properties are genesis date, block reward, total supply, formula and block creation time [7]. Even though these properties do not represent different consensus algorithms in a direct manner, they do have an impact on how consensus is achieved in cryptocurrency based blockchain networks. Some of the properties are mentioned below:

(a) Genesis Date: shows the timestamp of the first block created for a specific cryptocurrency.
(b) Block Reward: refers to the incentive that a miner achieves when he creates a new block.
(c) Total Supply: shows the total amount of cryptocurrency that is supplied.
(d) Block time: it refers to the average time taken for block creation of cryptocurrency.

4.3.3 Performance Properties

The performance properties are used to calculate the efficiency of a consensus algorithm. Some of the properties are described below:

(a) Fault tolerance: represents the number of defective nodes that a consensus algorithm can handle.
(b) Throughput: refers to the rate at which the transactions are processed.
(c) Scalability: means increasing the size and functionality of the system without affecting the throughput of the original system.
(d) Latency: defined as the total duration taken by the consensus to reach out and process the proposed transaction.
(e) Energy consumption: addresses whether the mechanism or the using system has high energy consumption.

4.3.4 Security Properties

The consensus protocols must follow the following security properties:

(a) Authentication: Ensures that the nodes participating in a consensus protocol are verified and authenticated.
(b) Non-repudiation: Checks if a consensus protocol satisfies non-repudiation i.e. it cannot deny the validation of transaction.
(c) Censorship resistance: Implies whether the corresponding algorithm is able to withstand any censorship resistance.
(d) Attack vectors: A combination of attack vectors that are relevant to consensus algorithms are presented.
(e) Adversary tolerance: Represents the maximum Byzantine nodes that can be accepted by the protocol.
(f) Sybil protection: In this type of security breach, the thief duplicates its identity to achieve the advantages that are against rules and protocols. In a blockchain network, the Sybil attack can be implicated when an adversary creates many nodes according to the need in the underlying peer to peer network so as to influence the consensus algorithm and gain advantage from it.
(g) Denial of Service (DoS) resistance: Ensures whether the consensus algorithm has mechanisms that are against the DoS attacks.

4.4 Consensus Algorithms

In this section, various algorithms that are used to achieve consensus in blockchain will be discussed in detail. These algorithms are required to provide equality

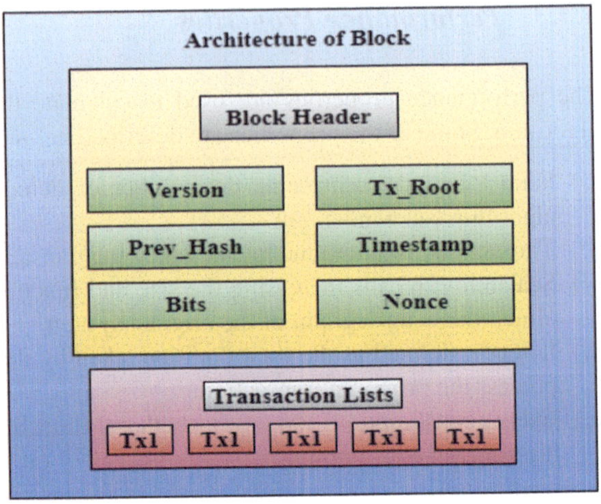

Fig. 4.1 Architecture/various fields of a block in a blockchain

and fairness to the whole system. Consensus algorithms in blockchain can be categorized as Proof based consensus algorithms and Voting based consensus algorithms. Before getting into the details of consensus algorithms, the block structure in the blockchain needs to be discussed and is shown in Fig. 4.1.

- *Prev Hash*: It can be defined as the connection of the block to its previous one and can also be considered as the reference to the parents.
- *Timestamp*: It refers to the time duration at which the block was obtained.
- *Tx Root*: Tx Root is also called the Merkle root. This field consists of all the hash values of the verified transactions contained inside a block. All the transactions that are present in the block are hashed by SHA 256 algorithm into a hash value. After hashing is done, these transactions are combined with each other, pair by pair and again put into some other hash function. The above process continues till only a particular value is obtained which is called as the Merkle root.
- *Version*: It refers to the protocol version which is used by each node in order to put the block into the chain.
- *Nonce*: Nonce also called as 'number only used once' is a pseudo random number that acts as a counter during a process of mining. It also describes the efforts made by node to append a block.
- *Bits*: This field generally shows the complexity of Proof of Work.

4.4.1 Proof Based Consensus Algorithm

Many types of Proof based consensus algorithms have been implemented, that are based on Proof of Work (PoW), Proof of Stake (PoS) or a combination of both and various other types that are independently made from the two important ones that are listed [8]. Proof based algorithms concept works around the fact that among numerous nodes, part of the network, only the node with suitable proof will have the permission to append the node to the chain.

4.4.1.1 Proof of Work (PoW)

If each and every node in a blockchain network tries to present their blocks holding validated transactions, it will result in a lot of confusion. To get a solution to this issue, Proof of Work algorithm is introduced.

PoW was introduced in 1992 by Dwork and Naor in order to stop junk/spam mails in which the user had to do some work so that he or she can send and receive a valid email [9]. Proof of Work allowed only trusted users to calculate the result of the puzzle, thus preventing the attacker from sending junk mails. The receiver received the mail only when the result was correct. As discussed in section 3, Node types are one of the properties of consensus algorithms. In PoW, two types of nodes exist: Requestors and Verifiers.

One of the most important features of PoW is asymmetry [6], which ensures that even if the task is relatively complex it should be feasible for the service requestor.

In Bitcoin, PoW is used to extend the hashcash based PoW system so as to come up with the methodology in protecting the blockchain through distributed consensus mechanism. Hash cash system is based on the puzzle friendliness property which is a part of cryptographic hash function. In blockchain, before the puzzle is solved, the verifying nodes must put their validated transactions including information like hash of previous block (Prev Hash) and Timestamp into a block. Each puzzle is solved by guessing a secret value known as nonce field which should be present in the block.

All of this information present in the block header is combined and then added in a SHA 256 hash function. The secret value is accepted only if the result of the hash function is less than a given threshold which represents the complexity. Otherwise, node keeps guessing the other secret value until the correct answer is obtained. To ensure that the average speed for the addition of the block in the chain is 1 block per 10 min, the difficulty of the puzzle is adjusted and managed after appending every 2016 blocks [8]. The threshold value is based on the complexity of the puzzle. The more the difficulty level, the lesser the threshold value. The work for predicting the correct value is called Proof of Work (PoW). The node which joins the network using the PoW is known a miner and the work for getting a correct nonce is called as mining.

The proposed block is broadcasted to the other nodes by a particular node when the secret value has been found. This process is followed to notify the other nodes. After getting the signal the nodes who still have not got the correct mysterious value of their puzzles will stop guessing and would start checking whether all the transactions in the broadcasted block are valid or not. The proposed block will be added to the current chain if all the verifications are validated. Figure 4.2 explains the various steps in PoW.

Fig. 4.2 Flow chart representing proof of work algorithm

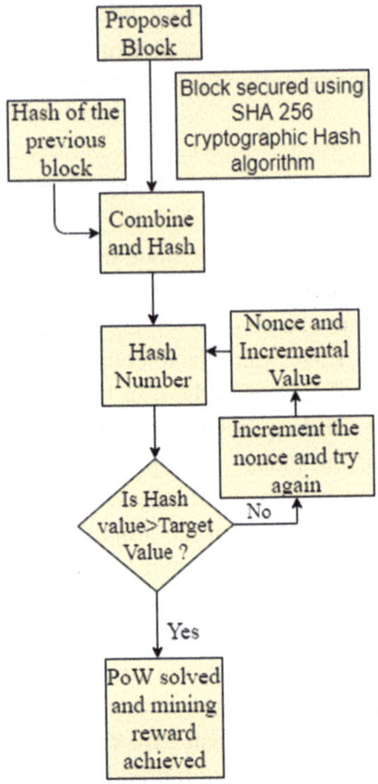

Sometimes, a case may occur when one miner gets the suitable answer for the puzzle before it is informed by some other miner. In this case, the block would still be presented by the miners with the discovered nonce. The other miners when receiving the first incoming block would neglect other blocks coming thereafter. This will lead to the Forking Problem in which there are various chains of blocks in the validated network. Satoshi proposed that the nodes who have got the correct answers will continue appending a fresh block on their respective forks, till one fork will be greater than the other. Therefore, the longest fork has to be followed by all the nodes at this time. After performing all the tasks, when the verified block is put into the chain, the node which appends this block gets rewarded in the form of bitcoins. The solution to forking problem is shown in Fig. 4.3.

4.4.1.2 Proof of Stake (PoS)

PoS provides equal opportunity to all the miners. The PoS consensus algorithm decides which node would get the permission to append the consecutive block on the basis of the stake it owns. This method can be very advantageous as the miner with more stake would be more trustful. PoS also provides the required security to

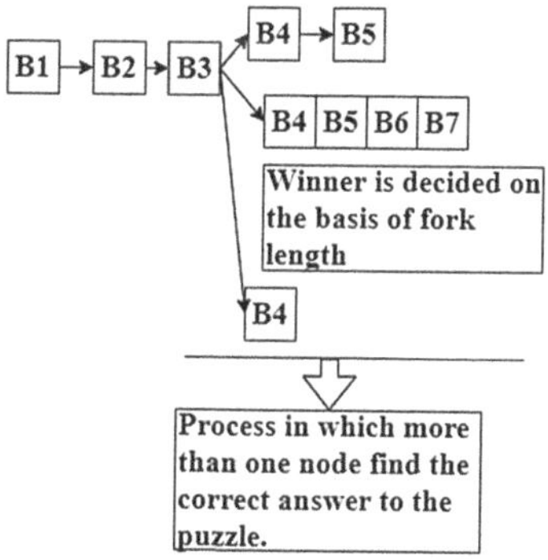

Fig. 4.3 Solution to the forking problem

the network because any attacker cannot perform a double spending attack until or unless he or she owns at least 51% [8] of the total stakes in the network.

The implementation of PoS can be seen in Nextcoin [10] which ensures that the miner having more stake will get the opportunity to mine a recent block.

Bentov et al.[11] gave a mechanism similar to Nextcoin which stated that the chance of a miner to append a block lies on the amount of stake he owns. The more the stake the node owns, the more would be the chance of him appending a block. Bentov also proposed the procedure called the Satoshi procedure.

Kiayias et al.[12] implemented the same technique as Bentov, to execute PoS consensus by following the procedure of Satoshi. They proposed that the leader selection should be done randomly by some calculation that should be secure enough. This would prevent the protocol from predicting the calculation easily. Leader selection is the process of selecting a particular miner to issue the next block. The steps for PoS are shown in Fig. 4.4.

4.4.1.3 Implicit Consensus

In the implicit consensus model, individual blockchain is processed by each node. Unbound throughput is one of the major advantages of this consensus model. Here special types of blocks known as check point blocks are considered instead of the transaction blocks. Since consensus cannot be implemented in each and every transaction, the implicit consensus algorithm leads to the scalability of the scheme in addition to linear message complexity. Some of the main features of this consensus model are the following [13]:

- With the help of self-interest formula this algorithm replaces the termination property of Byzantine Fault Tolerance schemes.
- Agreement and correctness for each transaction will hold on till the transaction is validated by the validation scheme locally.

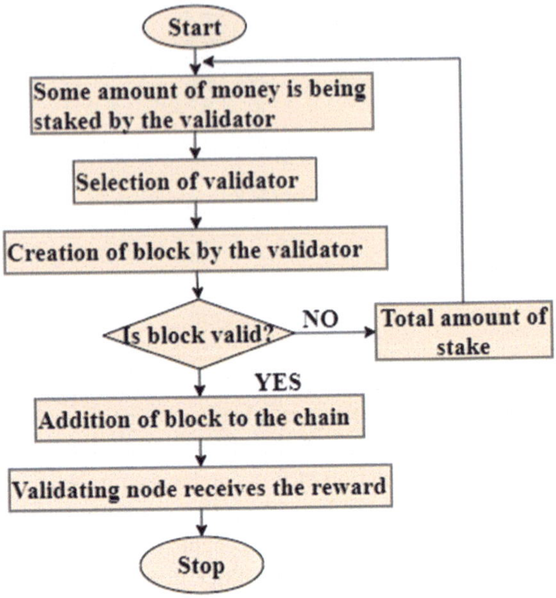

Fig. 4.4 Flow chart representing proof of stake algorithm

4.4.1.4 Secure Sharding Algorithm (ELASTICO) [14]

ELASTICO consensus algorithm is a great asset for permissionless blockchain. It is a scalable agreement algorithm in which the transaction rate varies proportionally to the processing present in the mining process i.e. more transactions can be processed in the blocks with more computational power. The idea upon which this algorithm works is to segregate the network into smaller parts which are known as committees. Each committee is responsible for the processing of a disjoint arrangement of exchanges and the entire technique is parallelized simultaneously.

4.4.1.5 Hybrid Algorithms (PoW/PoS)

The hybrid consensus algorithms use the combined features of PoW and PoS so as to overcome the weakness of each of them. PoW suffers from intensive energy consumption as it requires plenty of computational power. Also, PoW is vulnerable to a '51% attack', meaning that if the node earns 51% of the entire computational energy in blockchain network, that miner/node can make changes in the blockchain network. It can also introduce double spending which can be a serious problem. Similarly, in PoS only the rich stakeholders are allowed to control the consensus in the blockchain. And only those stake owners are allowed to have a control of consensus in blockchain.

The first variant that implemented a hybrid consensus algorithm was PPcoin proposed by King and Nadal [15]. A term called 'coin age' was defined for each node in the network which is computed by multiplying the stake by the time, till the time the node has earned that stake.

Vasin [16] did not associate coin age with his Blackcoin. The concept behind his idea was that with coin age there is an increased chance of attackers to collect sufficient amounts. Also, some nodes would keep a hold on their own money till they receive sufficient amounts of coin age, when they are not online in the validation system. To overcome this problem, Vasin [16] proposed that instead of using coin age, raw stake would provide the miners to append a new block. Thus, ensuring that more nodes are online in order to gain rewards.

If the nodes owned greater than 51% of the computational power, there would be a high security risk. Duong et al. [17] thought of mitigating the double spending attack using the Hybrid consensus algorithm. Duong et al. [17] addressed that each of the Proof of Stake blocks is connected to another Proof of Work block and each Proof of Work block is connected to a preceding Proof of Stake block. Thus, it becomes tough to make a double spending attack. This attack can only happen when the attacker owns 51% of the mining power but also he should own greater than 50% of all the stakeholders.

The main problem associated with the work proposed by Duong et al. [17] was that initially the executing environment was not dynamic because the physical hardware that was invested and stake remained unchanged. To improve this proposed method, Chepurnoy et al. [18] proposed a method to adjust the difficulty on the basis of the environment the rate at which the block is created.

4.4.1.6 Proof of Stake Velocity (PoSV)

To enhance security in the blockchain network, another consensus called Proof of Stake Velocity algorithm was implemented in 2014 [19] as a substitute to PoS and PoW algorithm. The PoSV was first used in Reddcoin's inception and was based on the traditional PoS algorithm. The idea behind using PoSV was to validate the transactions of the Reddcoin, which was the cryptocurrency developed mainly for social interactions in the digital age. PoSV ensured ownership (stake) and activity (velocity), the two major functions of Reddcoin as a real currency. The formula to calculate the velocity of money in a particular time is: [19]

$$V_T = nT/M \tag{4.1}$$

where V_T is the velocity with which the money flows, nT stands for the aggregate notional of transactions and M is the total amount of stake which is in flow. One of the disadvantages of PoSV is that it is particularly designed for the digital social currency; Reddcoin cannot be implemented for other cryptocurrencies. PoSV is evaluated as a part of the Reddcoin system and not as a standalone.

4.4.1.7 Proof of Activity (PoA)

Bentov et al. [20] came up with the idea and implemented a combination of Proof of Work and Proof of Stake called Proof of Activity consensus algorithm. This algorithm stopped the double spending attack as well as examined the tragedies made by Proof of Work known as tragedies of common. The first tragedy is that only miners who solve the puzzle get the reward, whereas the ones who preserve the ledger, update it and validate the new block do not get any reward. Another tragedy is that the nodes can cooperate with other miners in order to increase the transaction fee to charge from the user. This would lead to the less usage of blockchain technology. So as to overcome these tragedies, the researchers came up with the idea to create a vacant block by Proof of Work where all the nodes would try to crack the nonce associated with the block with no transactions. The block is broadcasted to other nodes for authentication when the nodes find the correct nonce. On the basis of the received block, they would also check if they have won another lottery. This fortunate chance is similar to that of followed in Satoshi procedure (this method receives index as an input of a Satoshi (smallest value of cryptocurrency) between 0 and the entire Satoshis in distribution. It gets the block from the ledger data into which Satoshi is appended and keeps a check on the transactions which had carried this Satoshi to the consecutive addresses. This process continues till it gets the stakeholder who could presently spent this Satoshi) in [21].

4.4.1.8 Proof of Burn (PoB)

Ian Stewart [22] proposed another consensus algorithm called Proof of Burn. In PoB, the miners have to first burn their coins in order to take part in the mining activity. Burning of coins refers to the sending of the coins to an address without the private key so that coins are never usable. This means that burning of coins is similar to the idea of investing for the building of the mining rig. The value of the coins burned has a specific relation with the probability of being chosen to mine the following block. This process is more or less like proof of work in which the nodes, in order to uphold the hash power invest in modern equipment.

Cryptocurrency implements the idea of Proof of Burn in combination with Proof of Work and Proof of Stake is Slimcoin [23]. The concept is very much identical to the Proof of Stake algorithm with additional PoB mechanism which is sandwiched between PoW and PoS mechanism. For the generation of initial coin supply using the bitcoin mechanism, PoW is used. It plans to transfer to the hybrid mechanism which consists of PoW and PoS when enough amount of stake is supplied to the system similar to Peercoin where miner is selected by the PoB. For the participation in the PoS minting process, the miners have to burn their accumulated coins. The PoB mechanism is generally used in the selection of minter without affecting the security of the network.

4.4.2 Voting Based Consensus

For the implementation of the consensus mechanism based on voting, miners in the blockchain network must be recognized and should be adjusted, so that message passing between the nodes becomes easier. Before deciding to mine the proposed block they would have to first communicate with other nodes in the network.

The implementation of the voting based consensus algorithm is very much identical to the conventional methods in similarity to tolerate the faults in the network. The Voting based consensus is proposed to solve some of the problems that arise in the blockchain network:

- Crashing of nodes.
- Damage of the well established system by the nodes.

4.4.2.1 Proof of Trust (PoT) [24]

The Proof of Trust algorithm enhances the efficiency of crowdsourcing services. The Shamirs secret sharing algorithm and RAFT leader election algorithm are used for the election of these nodes. PoT is implemented through four phases. Phase 1 uses the Raft leader algorithm for leader selection. Phase 2 is used for the selection of validators of the transactions. The next phase focuses on the validation of transactions by those which were selected in the previous phase and the fourth phase links the verified transactions with the blockchain network. This algorithm implements fault tolerance in the network when [24]

$$p => 3q + 1 \tag{4.2}$$

where p represents the nodes which participate in the blockchain network and q stands for the number of Byzantine nodes.

4.4.2.2 Proof of Vote (PoV) [25]

The PoV algorithm is responsible for the validation of blocks by using the voting process. The Proof of Vote algorithm is superior from all algorithms in terms of power usage or power consumption. There are 4 roles that are described in a consortium network model: commissioner, butler candidate, butler, and ordinary user.

- *Commissioner*: Various organizations around the globe form a league committee so that the consortium blockchain is being maintained. One of the members is called the commissioner who is selected by the alliance law. The machine working is responsible to represent the commissioner in a consortium decentralized network. The commissioner evaluates, recommends and votes for the butler. They also validate and process blocks and transactions. The block is considered to be verified when it gets at least 51% of the votes.

- *Butler*: The production of blocks is done by the butler. Butler is specially designed for segregating voting and execution right. The butler collects the data related to a transaction from a network and puts all the information into a block. After putting all the data, butlers need to put their signature on the block.
- *Butler Candidate*: Since there are only a few number of butlers, the election of the butler should be done by the butler candidates, and the candidates will be voted by all the commissioners. If any candidate is not selected for the butler, it can remain active and can wait for the upcoming elections.
- *Ordinary user*: Ordinary users need not to authorize their identity and they have the permission to enter and exit the network anytime they want. They do not have the right to take part in the procedure of creation of blocks and can only participate in block distribution and message passing. The whole consensus mechanism can be seen by them while utilizing the services of the network.

4.4.2.3 Ripple Algorithm [26]

The Ripple consensus algorithm makes use of subnetworks that are considered to be trusted within the decentralized network. So as to ensure the correctness of the network system, the protocol is deployed after a few seconds. The ledger is considered to be closed after the consensus is achieved. The last closed ledger that is executed by each node in the network should be the same. This protocol runs in different rounds. At the beginning, a candidate list is being announced publicly by each of the nodes in a network which contains all the verified transactions. After this process, voting is done by every node so as to check the correctness of all the transactions. It then combines the candidate list prepared by all the nodes. In order to consider a transaction as verified the number of votes received is compared with the threshold value and the decision is made on the basis of it. The last round requires at least 80% of the nodes participating in the network must reach consensus on the transaction. When the transaction meets all of the above mentioned criteria, it is considered to be validated and is then applied to a ledger. Thus, creating a fresh closed ledger.

4.4.2.4 Practical Byzantine Fault Tolerance (PBFT)

Castro and Liskov [27] implemented the Practical Byzantine Fault Tolerance consensus algorithm. PBFT consists of two categories of nodes: A leader node and a few verifying nodes. For the addition of a block in the chain, the miners need to execute some rounds of mining. After this process, the execution of three phases of PBFT occurs. The first phase is called the Prepare phase, in which the proposed block is being presented to the other nodes by the leader. This block is stored locally. To validate the authenticity of the block that is being received, the nodes re-evaluate through presenting it in the prepare phase and commit phase. When the node gets the same block which is stored locally after the prepare phase or more than two third of the entire nodes present in the blockchain system, the commit phase is executed. The similar process is followed even after the commit phase that is the major need of the node for the processing of the transaction in a particular block and then mining it to the current chain.

4.4.2.5 Delegated Byzantine Fault Tolerance (DBFT)

The DBFT consensus algorithm was initially implemented in NEO blockchain and was proposed by Da HongFei and Erik Zhang [28]. DBFT consensus can be achieved in a public network in a very fast manner.

Since DBFT can also work with few miners and the system can also handle up to [28]

$$(p - 1)/3 \tag{4.3}$$

faulty nodes where p stands for a group of consensus nodes and not like PBFT where p represents the nodes that take part in the blockchain network.

Delegated Byzantine consensus algorithm works on the voting mechanism. All the nodes that are a part of a network and have NEO token are called ordinary nodes. These nodes have the power to implement transactions in the network and voting procedure for consensus nodes in real time. The consensus nodes contain the speaker and delegates. The responsibility of the speaker is to fetch the transactions from the memory, verify them and then put them into a new block whereas delegates are responsible for verifying the blocks by the voting mechanism.

4.5 Comparison of Different Consensus Algorithms

This section compares the various Consensus algorithms based on generic and performance parameters as shown in Table 4.1

4.6 Research Challenges and Future Scope

This section discusses the main issues in consensus algorithms in blockchain technology. Some of them are as:

(a) Security Problems:

Security issues in consensus algorithms lead to various security attacks. This can lead to unauthorized access of a blockchain network by the individuals who

Table 4.1 Comparison between different consensus algorithms

Algorithms	Blockchain type	Mining	Category	Scalability	Latency
ELASTICO (2016)	Permissionless	On the basis of computational power	Proof based	Scalable	Low
Implicit consensus (2017)	Permissioned	On the basis of proof based	Proof based	Not Scalable	High
Proof of trust (2018)	Permission based consortium	Based on probabilistic and voting mining	Vote based	Scalable	Low
DBFT Consensus Algorithm (2018)	Permissioned	Non-proof of based mining(Random selection of miner)	Vote based	Not Scalable	Very low
Ripple (2014)	Permissioned	On the basis of voting mining	Vote based	Scalable	Low
Proof of vote (2017)	Consortium	On the basis of voting mining	Vote based		Very low
Proof of work (2008)	Permissionless	On the basis of computational power	Proof Based	Not Scalable	Very high
Proof of Stake (2011)	Permissioned and permis-sionless	On the basis of nodes wealth and staking age	Proof based	Scalable	High
Proof of stake velocity (2014)		On the basis of stake and amount (velocity)	Proof Based Hybrid (PoW/PoS)	Scalable	Low
Proof of activity (2014)	Permissionless	On the basis of effectiveness Of work by the miner	Proof Based Hybrid (PoW/PoS)	Scalable	Low
Proof of burn (2014)	Permissioned and Permis-sionless	On the basis of coin burn-ing(Probabilistic lottery)	Proof Based Hybrid (PoW/PoS)	Scalable	Average
Practical Byzantine Consensus algorithm (1999)	Permissioned	On the basis of round of mining	Vote Based	Not Scalable	Very low

act as miners but in real sense they are not miners but attackers. Various forms of attacks which can harm the blockchain system are: Distributed Denial of Service Attacks, Double Spending Attack, Denial of Service Attacks, etc.

(b) Performance Problem:

Performance of the blockchain network is dependent on the consensus algorithm implemented. Since performance problems can decrease the efficiency of the blockchain network by reducing scalability and increasing the latency in the network, choosing the right consensus algorithm becomes a major challenge. Some of the failures affecting performance are: Temporal failure, Omission failure, Transient failure and Software failure [29].

- Temporal failure: This failure occurs due to the latency in the network, though it generates correct results but takes more time to be processed.
- Omission failure: It arises due to transfer problems like buffer overflow and improper functioning of the transmitter.
- Transient failure: This failure is permanent in nature. It occurs in hardware due to the issues in batteries and a sudden spike in power whereas in case of software these issues can be in the form of bugs in the codes which cannot be detected even when the testing face is going on.
- Software failure: These occur because of the flaws in designing and modelling. Software failures can further trigger other failures such as omission.

(c) Consensus mechanisms:

Each consensus algorithm discussed in the above sections performs a particular task for the validation of the transactions. Thus, it becomes necessary to implement the right algorithm for the verification of any transaction. For example, when a stakeholder owns 51% of the total computational power, PoW should not be implemented as it is vulnerable to the double spending attack. Instead a hybrid and PoS consensus algorithm should be implemented.

(d) Energy Management:

The computational power which is being utilized by the consensus algorithms is very high due to their complex mechanisms. Thus, proper management of energy resources is required to prevent its unnecessary usage.

(e) Byzantine failure:

It is a fault that shows various symptoms in the network. It prevents the nodes from reaching a consensus because the system gets confused and cannot handle faults in the network. For example, a particular server operating in the network might appear to be functioning improperly to one of the nodes and operating properly to the other, this server cannot be called as failed as both the nodes would not come to an agreement because both the servers do not have the same information.

Over the time, the evolution of blockchain has developed decentralized applications beyond financial transactions in different areas. Nowadays the need for an open, energy efficient and scalable blockchain becomes important. This is because of the services provided by the blockchain network in a large scale collective ecosystem like social networking, smart healthcare, smart cities and social networking with the aim of cost reduction and green computing. Initially, blockchain network was developed using a public network and it was open allowing anyone to participate disabling access control to data. Security, scalability and consumption of energy were the major threats in the system [30–33]. Afterwards, blockchain architecture and consensus protocols evolved in order to tackle the above mentioned issues. However, one of the major issues which is still a topic of research in these architectures is the high utilization of energy so as to maintain the security of the system. More the number of participants, higher is the consumption of energy in the consensus protocols which also puts a negative impact on the environment.

The Proof and Voting based Consensus protocols were developed in order to overcome the problem of high energy utilization but this led to architectures which were not scalable. A possible solution should be developed that should be less computationally complex and more energy efficient.

Also, inflexible and non-adaptive behaviour of present consensus protocols and architectures act as an obstacle for a growing collaborative digital ecosystem because they target a particular field. These architectures should be modified to provide a suitable environment for the applications.

4.7 Conclusion

With the increasing popularity of blockchain technology, according to International Data Corporation (IDC) the transactions made in the blockchain system could reach a very high value of 12.4 billion dollars by 2022 [34]. Blockchain can be implemented in many sectors like business activities, IoT, Medical informatics, etc. It is expected that this technology would overcome the other technological domains. A suitable consensus algorithm must be used to implement the blockchain technology as it is one of the major components of the decentralized network and dictates the efficiency of the system. In this chapter, we have discussed the consensus algorithms, their categorization, their implementation and usefulness in the blockchain network. We have compared the various mentioned consensus algorithms on different parameters and how the implementation of each consensus algorithm differs from other. Apart from the advantages, we have listed the various research challenges being faced in this subdomain of blockchain.

References

1. S. Haber, W.S. Stornetta, How to time-stamp a digital document. J. Cryptogr. **3**(2), 99–111 (1991)
2. S. Nakamoto, *Bitcoin: A Peer-to-Peer Electronic Cash System* (2008). https://bitcoin.org/bitcoin.pdf
3. S.M.H. Bamakana, A. Motavali, A.B. Bondarti, A survey of blockchain consensus algorithms performance evaluation criteria. Expert. Syst. Appl. **154**, 113385 (2020)
4. H. Watanabe, S. Fujimura, A. Nakadaira, Y. Miyazaki, A. Akutsu, J.J. Kishigami, Blockchain contract: a complete consensus using blockchain, in *Proceedings of the IEEE 4th Global Conference on Consumer Electronics* (2015), pp. 577–578

5. Ashok Kumar Yadav, Karan Singh.: Comparative analysis of consensus algorithms of blockchain technology, ambient communications and computer systems, in *Advances in Intelligent Systems and Computing*, vol 1097 (Singapor, Springer, 2020), pp. 205–218
6. S.S. Panda1, B.K. Mohanta, U. Satapathy, D. Jena, D. Gountia, T.K. Patra, Study of blockchain based decentralized consensus algorithms, in *IEEE Region 10 Conference* (2019)
7. Md.S. Ferdous, M.J.M. Chowdhury, M.A. Hoque, *Blockchain Consensus Algorithms: A Survey* (2020). shortcomarxiv: 2001.07091v2 [cs.DC]
8. G.-T. Nguyen, K. Kim, A survey about consensus algorithms used in blockchain. J. Inf. Process. Syst. **14**(1), 101–128 (2018). https://doi.org/10.3745/JIPS.01.0024
9. C. Dwork, M. Naor, Pricing via processing or combatting junk mail, in *Proceeding of the Annual International Cryptology Conference* (Springer, Berlin, 1992), pp. 139–147
10. Nxt Whitepaper (2016). Last accessed 28 March, 2020. https://nxtwiki.org/wiki/Whitepaper: Nxt
11. I. Bentov, A. Gabizon, A. Mizrahi, Cryptocurrencies without proof of work, in *Financial Cryptography and Data Security* (Springer, Heidelberg, 2016), pp. 142–157
12. A. Kiayias, A. Russell, B. David, R. Oliynykov, *Ouroboros: A Provably Secure Proofof-Stake Blockchain Protocol* (2016). https://eprint.iacr.org/2016/889.pdf
13. Z. Ren1, K. Cong, J. Pouwelse, Z. Erkin, *Implicit Consensus: Blockchain with Unbounded Throughput* (2017). arXiv:1705.11046v3 [cs.DC]
14. L. Luu, V. Narayanan, C. Zheng, K. Baweja, S. Gilbert, P. Saxena, A secure sharding protocol for open blockchains, in *Proceedings of the 2016 ACM SIGSAC Conference on Computer and Communications Security* (ACM, New York, 2016), pp. 17–30
15. S. King, S. Nadal, *PPcoin: Peer-to-Peer Crypto-Currency with Proof of Stake* (2012). https://decred.org/research/king2012.pdf
16. P. Vasin, *Blackcoin's Proof-of-Stake v2, The BLK Community* (2014). https://blackcoin.co/blackcoin-pos-protocol-v2-whitepaper.pdf
17. T. Duong, L. Fan, H.S. Zhou. *2-hop Blockchain: Combining via Proof-of-work and Proof-of-Stake Securely* (2016). https://eprint.iacr.org/2016/716.pdf
18. A. Chepurnoy, T. Duong, L. Fan, H.S. Zhou, *TwinsCoin: A Cryptocurrency via Proof of-work and Proof-of-stake* (2017). https://eprint.iacr.org/2017/232.pdf
19. L. Ren, *Proof of Stake Velocity: Building the Social Currency of the Digital Age* (2014). https://www.reddcoin.com/papers/PoSV.pdf
20. L. Bentov, C. Lee, A. Mizrahi, M. Rosenfeld, Proof of Activity: extending bitcoin's proof of work via proof of stake. ACM SIGMETRICS Perform. Eval. Rev. **42**(3), 34–37 (2014)
21. I. Bentov, A. Gabizon, A. Mizrahi.: Cryptocurrencies without proof of work, in *Financial Cryptography and Data Security* (Springer, Heidelberg, 2016), pp. 142–157
22. *Proof of Burn*. Last accessed 22 March, 2020. https://en.bitcoin.it/wiki/Proofofburn
23. *Slimcoin* Last accessed 22 March, 2020. http://slimco.in/
24. J. Zou, B. Ye, L. Qu, Y. Wang, M.A. Orgun, L. Li, Proof-of-trust consensus protocol for enhancing accountability in crowdsourcing services. IEEE Trans. Serv. Comput. **12**(3), 429–445 (2018)
25. K. Li, H. Li, H. Hou, K. Li, Y. Chen, Proof of vote: a high performance consensus protocol based on vote mechanism and consortium blockchain, in *Proceedings of the 2017 IEEE 19th International Conference on High Performance Computing and Communications; IEEE 15th International Conference on Smart City; IEEE 3rd International Conference on Data Science and Systems (HPCC/SmartCity/DSS)* (IEEE, New York, 2017), pp. 466–473
26. D. Schwartz, N. Youngs, A. Britto et al. The ripple protocol consensus algorithm, in *Ripple Labs Inc White Paper*, vol. 5 (2014)
27. M. Castro, B. Liskov, Practical Byzantine fault tolerance, in *Proceedings of the Third Symposium on Operating Systems Design and Implementation* (1999), pp. 173–186
28. A distributed network for Smart Economy, in *Neo, White Paper* (2019). https://docs.neo.org/docs/en-us/basic/whitepaper.html

29. N. Chaudhry, M.M.Y. Punjab, Consensus algorithms in blockchain: comparative analysis, challenges and opportunities, in *International Conference on Open Source Systems and Technologies (ICOSST)* (2018)
30. A. Khanna, R. Anand, IoT based smart parking system, in *Proceedings of the 2016 International Conference on Internet of Things and Applications (IOTA)* (2016)
31. S. Purri et al. Specialization of IoT applications in health care industries, in *Proceedings of the 2017 International Conference on Big Data Analytics and Computational Intelligence (ICBDAC)* (2017)
32. A. Khanna, R. Tomar, IoT based interactive shopping ecosystem. *Proceedings of the 2016 2nd International Conference on Next Generation Computing Technologies (NGCT)* (2016)
33. A. Khanna et al. Intelligent mobile edge computing: a deep learning based approach. Commun. Comput. Inf. Sci. **1244**, 107–116 (2020)
34. https://www.idc.com/getdoc.jsp?containerId=prUS44898819. Last accessed 30 May, 2020

Chapter 5
Block Chain Platforms and Smart Contracts

Dakshita Negi, Anushree Sah, Saurabh Rawat, Tanupriya Choudhury, and Abhirup Khanna

5.1 Introduction

The need for modernization has driven rapid development in technologies during the past decade. For example, blockchain has now become an integral part of daily life. Blockchain has received considerable hype, starting with "cryptomania" in the trading markets and then expanding its impact to private and public sectors of society. Blockchain is one of the most exalted technologies with a pervading impact on almost all industries, including banking and financial services, supply chains, agriculture, healthcare, and government.

Blockchain is derived from the principles of cryptography, peer-to-peer networks, and game theory. It evolved as a formal name for tracking the databases underlying cryptocurrency such as Bitcoin, but now it has become a distributed ledger with software algorithms to record all the transactions in form of chain of blocks with trustworthiness and anonymity [1]. Blockchain also uses the concept of smart contracts, in which business rules are implied by agreements that are embedded in the blockchain and executed with transactions.

Blockchain has reevaluated the interoperability of databases. It has pushed reliability, verifications, interaction, and data security to different actors in the arrangement. Despite this, data immutability, digital scarcity deficiency, and the

D. Negi · A. Sah (✉) · A. Khanna
University of Petroleum and Energy Studies, Dehradun, India

S. Rawat
Graphic Era University, Dehradun, India

T. Choudhury
Department of Informatics, School of Computer Science, University of Petroleum and Energy Studies (UPES), Dehradun, Uttarakhand, India

T. Choudhury et al. (eds.), *Blockchain Applications in IoT Ecosystem*,
EAI/Springer Innovations in Communication and Computing,
https://doi.org/10.1007/978-3-030-65691-1_5

solution to duplication of digital data. Blockchain technology is integrated with applications from different domains:

Faster settlements: it provides faster settlements for transactions than in traditional banking, which uses a very slow and time-consuming process.

Security: Cryptography functions and consensus provide secure transactions.

Transparent: Because blockchain is a decentralized platform, third-party inventions are not required at all. All stakeholders of the network can participate in the network, thereby providing transparency [2].

Inexpensive: Blockchain does not require the expensive brick-and-mortar model that facilitates traditional financial transactions, nor is it required to pay large commissions to avail financial services [3].

Blockchain technology provides a better sharing platform, where the suppliers and buyers trade on a trusted network. It does not involve any intermediates or any third parties. Traders can obtain lower prices and earn higher profits. Thus, the technology creates a trusted and transparent marketplace and also a better economy [4].

Today, numerous variations of smart contracts with blockchain innovations have been developed. However, there are differences between an open-system blockchain, such as Bitcoin and Ethereum, and private system types.

5.2 Literature Review

Another paper [5] applied mean peril speculation to methodically research how the risk attitude of purchasers impacts the ideal assessment of the on-demand platform, customer surplus (CS), expected profit (EP) and profit risk (PR) of the platform, and higher help administrators. When the customers were more risk-seeking, then the ideal assist cost with willing drop (increase). When examining the changed risk viewpoints of the customers, it was found that the customers were risk seeking. Ultimately, the customer's risk-seeking can be increased by blockchain advancement.

Another study [6] examined BitFund, which is a worldwide investment platform. Blockchain innovation helped to create a decentralized system where the exchanges are recorded in an open appropriated record, consequently making it transparent and without cost. Thus for the requirement of a good and effective "crowd funding platform" in order to build up a smart country or nation and having the features of blockchain, BitFund was introduced or proposed. Here, speculators can request a particular job along with an offer regarding the time, cost, and required maintenance; the developers can offer proposals with the same parameters to establish a venture proprietorship. A smart contract is created between the speculator and developer to arrive at a solution for the speculator.

The effects of blockchain technology and the Internet of Things (IoT) on the modern healthcare market have been investigated [7]. Medical data needs to be handled in a secure manner. These innovations have improved a number of factors in this regard, including data security, effectiveness, protection, analysis accuracy, and transparency, among others. By applying blockchain to healthcare services, data security can be improved across the board, ensuring the protection and security of patients' medical information.

Another study [8] reviewed how blockchain technology characterizes the concerns and the functionality identifies the blockchain-based solution for food traceability concerns, includes the advantages and challenges of the blockchain-based traceability systems, helps the researchers and scientists to apply blockchain-based food system by proposing a suitable framework, and analyzing flowchart for the blockchain-based food traceability systems. This paper examines food acceptability and provides critical information to researchers and professionals on the most ideal approach to use blockchain-based food structures.

Researchers [9] also investigated the mechanisms used in blockchain technology and analyzed the security protection and lack of transparency in digi-cash. Blockchain has become a hot issue in the market due to its qualities of decentralization, verification ability, and protection against alterations. Blockchain is a key strategy that can maintain Bitcoin and achieve data encryption while remaining anonymous and secure.

A review [10] of blockchai focused on the existing public key infrastructure (PKI) innovations and the key administration of a blockchain wallet. The PKI is used in blockchain technology to guarantee the verification of the entities and the integrity of the blockchain. To maintain privacy of the records in the blockchain, a group key management scheme for batch correspondence was proposed.

The potential of blockchain technology in the energy sector has also been discussed [11]. The energy sector has expanded the use of renewable energy sources beyond decentralization in the market. Blockchain technology plays a key role in this ongoing change by offering decentralized interfaces and frameworks as a possible way to deal with the current associations in the energy market. Blockchain incorporates all socio-specialized and authoritative setups in the energy sector, dependent on the usage of the blockchain technology for energy exchange, data stockpiling, or expanded direct energy services.

Another paper [12] examined the connection between blockchain innovations and social networking sites. In recent years, the decentralization of social networking administration has been viewed as a major opportunity to tackle numerous issues on online informal communities. Blockchain is a decentralized strategy that has been considered to establish a new age of social networking platforms. This paper examined the platforms in detail by describing the administrations they offer, their primary advantages, and their disadvantages.

Another paper [13] described how blockchain can be used to support economically important systems. The execution of blockchain is isolated form the external world and hence requires the blockchain operators or agents to get the data from the outside world. Despite the fact that blockchain is considered to be very dependable, agents and operators are off-chain segments that could be a point of failure in the whole blockchain-based network. Accordingly, the paper examined the dependability of the instruments, as well as upgrades for the weak connections in a blockchain-based framework.

The incorporation of blockchain with 5G systems has also been explored [14]. 5G wireless systems still face some security challenges, including decentralization, transparency, information interoperability, and system security. Here, blockchain becomes an integral factor because it is very important for 5G to have decentralized and secure provisions. Blockchain may engage significant 5G administrations, including execution, information sharing, and virtualization.

The possibility of overcoming the disadvantages of distributed cloud storage by blockchain technology have been discussed [15]. The expanding number of cloud storage customer leads to issues with the integrity of the cloud. The traditional examining scheme excluding the third-party auditor or evaluator which is not always available in the outside world and also tends to increase the price of the service. Consequently, a blockchain-based smart contract may increase capacity.

An information-sharing model based on a transformative game hypothesis using blockchain with smart contracts has been proposed [16]. Information-sharing methods have been considered to fundamentally reduce recurring work. However, there are difficulties with respect to the arrangement of common trusted connections and the expanding degree of client support. Blockchain 2.0 with smart agreements has the option to automate trusted transaction between users.

Smart contracts have been examined in another paper [17]. They could allow engineers and developers to send decentralized and secure blockchain applications for IoT. To allow smart contracts to obtain off-chain information, the paper proposed an information exchange that is savvy and flexible for a blockchain-empowered IoT environment.

Elsewhere, researchers [18] proposed a convention for contract marking that is dependent on blockchain innovation. Electronically signed agreements are fundamental for web-based business exchanges, so contract signing conventions should facilitate trade. The method proposed in this paper does not need third-party verification. The proposed convention fulfills the essential security requirements.

An overview [19] of the various uses of blockchain technology in financial transactions is provided elsewhere. Blockchain offers unprecedented opportunities for innovation in financial transactions in banking, money transfer, insurance, and lending. This paper examined the risks, security requirements, and challenges of such innovations.

5.3 Survey of Existing Blockchain Platforms

This section discusses the various blockchain-based platforms that have emerged in recent years.

5.3.1 Waltonchain

One of the platforms for the IoT sector is Waltonchain. Waltonchain can integrate the transparency, responsibility, and provenance properties of blockchain technology with radiofrequency identification (RFID), which is used to capture and read information that is stored on a tag attached to an object equipped with IoT hardware.

The Waltonchain open blockchain platform and the software stage or platform that uses or interferes the hardware with the blockchain. The general aim of Waltonchain is to assist with Value IoT (VIoT), which is appropriate for a broad range of IoT applications, such as supply chain racking, authentication, and identification and so on.

Waltonchain consists of two types of blockchain: the parent blockchain and the child blockchain. Waltonchain is a public blockchain platform, which allows anyone to participate. A child blockchain can be either public or private, depending on the use case. Child blockchains may be designed according to the prerequisites, such as for implementing trade fundamentals for a particular industry or use case. The newly created child blockchain uses a unique exchange that is documented in the parent blockchain. Waltonchain uses a half-and-half agreement calculation known as Waltonchain Proof of Contribution (WPoC). WPoC is a mixture of three distinct agreement calculations: Proof of Work (PoW), which is used in Bitcoin; Proof of Stake (PoS), a stake-based agreement calculation; and Proof of Labor (PoL), a formula for cross-chain information transmission.

5.3.2 Origin Trail

Origin Trail is a decentralized, open blockchain information sharing platform for multiorganizational conditions. The platform integrates the blockchain technology with supply chains to allow or permit supply chain immutability and integrity. The aim is to provide a typical blockchain-based arrangement with a boosted convention to guarantee item norms and the security of clients or purchasers.

Origin Trail uses an Electronic Product Code Information Service (EPCIS) system to encourage a layered, extensible, and measured plan over the whole structure. The Origin Trail environment can be viewed as a four-layered system: At the top of the blockchain layer, there are two system layers—the system and information layers, which actualize an off-chain decentralized shared system known as the Origin Trail Decentralized Network (ODN). On the head of the system layer, there is a decentralized application layer, which interfaces between the clients and the structure to provide information input. The current version of Origin Trail executes a PoW that runs on the head of the Ethereum blockchain. Future versions are anticipated to include distinctive blockchains with various agreement calculations.

5.3.3 IBM Watson

IBM Watson is a coordinated innovation consolidating Watson IoT stage and blockchain. Watson use a Hyperledger Fabric system to provide blockchain administrations. It can capture information continuously by IoT gadgets and provides data analyses to the client.

The Hyperledger venture is a two-way application between IBM and Linux Foundation to create a venture-grade, open-source disseminated record system and code base. The aim of this undertaking is to provide an open-standard blockchain platform with the goal that any venture can assemble its own solution. There are a few dynamic ongoing tasks in the Hyperledger venture, such as Burrow, Fabric, Sawtooth, Iroha, and Indy. Texture is the most pertinent stage in this group; it is a permissioned blockchain framework with measured design that arranges various kinds of algorithms. It also assists with the execution of smart contracts (called "chain codes" in Fabric) and enrollment administrations given by a Certificate Authority, overseeing X.509 endorsements that are used to validate roles and jobs.

There are three types of hubs in the Hyperledger Fabric:

1. **Orderer node:** This node provides a correspondence channel to customers and peers, over which messages can be communicated. Being a permissioned blockchain, it can uphold an impressive number of transactions. However, the number is dependent on the utilization case and how they are sent.
2. **Peer node:** This node maintains the records and obtains requested updates for submitting new exchanges to the record. "Endorsers" support an exchange by confirming whether it satisfies requirements.
3. **Client node**: Customer hubs follow up in the interest of end-clients and facilitate exchanges.

5.3.4 Slock.it

Slock.it is an IoT platform on the head of the Ethereum blockchain. Its objective is to build up a genuinely decentralized sharing economy that will empower immediate communication between a maker or proprietor and a customer of IoT objects. A sharing economy allows individuals to share their unused physical or virtual assets, such as housing, vehicles, power, or even time, for financially motivated reasons. The customary methodology requires a great deal of human mediation, with major issues regarding trust and transparency. In the current utilizations of a sharing economy, for example, Uber and Airbnb are not decentralized; rather, they depend on their monopolistic incorporated suppliers, which charge a significant expense. Security, trust, and transparency issues are predominant in such applications.

Slock.it intends to address these issues by providing a platform composed of IoT objects. The Slock programming platform and smart contracts based on the Ethereum blockchain have introduced completely computerized machine-to-machine, machine-to-human, and human-to-human connections. The IoT items interact with each other through a smart contract or agreements conveyed in the Ethereum blockchain. An individual can communicate with each IoT object using a preferred gadget, such as their cell phone. Of note, Slock.it does not have its own blockchain. Rather, it uses Ethereum as its underlying blockchain stage. Subsequently, it depends on Ethereum's current consensus mechanism, compensating measures, and different properties.

5.3.5 *NetObjex Platform*

NetObjex is a decentralized advanced resource platform that utilizes IoT and blockchain to offer assistance to four significant market segments: supply chain and coordination, manufacturing, smart cities, and the automotive industry. The platform uses IoT for information exchange and supports a wide range of correspondence conventions, such as cellular, mid-run conventions (LoRA, Sigfox, NB-IOT), wi-fi, Ethernet, BLE (Bluetooth Low Energy), and other specific conventions (DSRC). Moreover, it empowers ventures to share data safely through blockchain and to authorize business rules through smart agreements. To guarantee authenticated access to delicate data, NetObjex stores them in cryptographically secure records using blockchain technology. The NetObjex platform provides an adaptable framework for clients to create and convey their own smart items. It coordinates various enormous databases, appropriates record advancements, and manages devices with its center blockchain middleware component to encourage interoperability and cross-correspondences among various segments. The NetObjex stage executes a normalized instrument for smart gadgets to communicate internationally with one another. To interface between the advanced resources used by various associations within a solitary environment, the platform uses an innovation layer through its IoToken system. It additionally supports IoToken local cryptographic money for interdevice exchanges.

5.4 Smart Contracts

Smart contracts are reshaping traditional industry and business practices. When integrated with blockchain, smart contracts allow the authoritative terms of an agreement to be upheld without the intervention of a third party. Ethereum is an extensively used platform for the execution of smart contracts. Thus, these smart agreements or contracts can reduce administration time and cost, improve productivity, and decrease risk. Smart contracts can be stored and managed in the appropriate blockchains in a peer-to-peer marketplace.

5.5 Tourism Industry

Digital innovation has caused rapid development in the travel industry. The capabilities of blockchain can reduce costs and increase the efficiency of processes, while also alleviating the risk of information double-dealing and increasing the degree of trust among the colleagues. In this manner, it is imperative to concentrate on all parts of blockchain innovation and its associations within and among businesses in order to foresee future changes in the travel industry [26, 27]. This innovation could affect several areas of the transportation business, including action plans, money transfer frameworks, security, execution, and trust [20].

Travellers need to display proof of identity in various stages during their trip, from booking through boarding and lodging registration. With the assistance of blockchain, this strategy can be disentangled so that tourists only need to show identification one time [21, 25]. SITA is a technology undertaking that gives IT backing and broadcast communications to the airline business. This organization has proposed a strategy that can simplify the traveler's journey and streamline their identifications. It utilizes blockchain technology to actualize a solitary and secure biometric personality framework that permits travelers to demonstrate their identity using a wearable or portable device during their trip. Therefore, identification cards, travel papers, and driver's licenses would not be required. For example, a customer may have a token that contains their own biometrics and various confirmations or verifications attached to it. Hypothetically, the traveler could then be identified using a biometric check combined with a token verification. None of the traveler's data would be shared or visible to organizations, as the entirety of the checks would happen in the organization's machinery. Blockchain innovation could be especially helpful in monitoring baggage, particularly during international travel when a customer's baggage may change hands more than once during their visit. The utilization of a decentralized database would make it easy to share records among organizations [23, 24].

5.6 Future Works

Military Use
Digital technologies have changed modern warfare. Nowadays, soldiers use connected devices for air strikes, and drones on the battlefield are controlled from remote locations. In the past, hackers could take control of the operator's terminal and could see, in real time, whatever the operators saw on their screens. Hackers could then compromise the system and send a pop-up alert on the user's screen. A London-based non-governmental organization warned in a January 2018 report that nuclear weapons systems are becoming increasingly vulnerable to cyberattacks. The Nuclear Threat Initiative, a US non-profit organization, published a report on cyberthreat to nuclear weapons. They concluded that there is a high Block Chain

Platforms and Smart Contracts possibility that US nuclear weapons systems could be compromised [13]. Considering these cyberattacks, a new paradigm is needed to address the vulnerabilities of defence systems. Blockchain can be a key player in rectifying these weaknesses. The potential benefits of blockchain to protect defence systems against cyberattacks can be presented as distinct use cases:

- Defending basic weapons frameworks
- Managing robotized, swarm frameworks
- Defending critical weapons systems

The system operator receives data from many sensors. These data notify command authorities about an incoming threat. Command authorities then direct the weapon to respond to threats. In a centralized system, there is one point of vulnerability that can be breached by external bad actors. Therefore, command authorities of the weapon system may receive deceptive information, which could lead to either illegal use of weapons or even failure to respond to a legitimate threat. Alternatively, when using blockchains, data transmissions from sensors to the operator are validated using a consensus system. Because a transaction is approved by most of the nodes within the blockchain network, any hacker would have to hack all nodes in the chain simultaneously. The computing power needed to hack such a system is magnificent [5].

Managing Automated Swarm Systems
Swarm robotics is a way to coordinate many robots as a system. It can implement a desired combined behavior from the connection between many robots, as well as the interaction of robots with their surroundings. The dependence of robots on communication and interaction opens a loophole for hackers. Blockchain proposes a mechanism to protect intraswarm coordination. In this system, each robot of the swarm acts as a node in their blockchain. In such an implementation, the swarm can exchange information and protect itself from cyberattacks [6].

Intelligent Transportation Systems
The intelligent transportation system (ITS) has potential in various fields, such as communication frameworks for vehicles and decentralized transportation frameworks. It is fundamental for modern smart vehicles to have continuous Internet access, allowing them to speak with one another with respect to their environmental factors, along with other transportation design improvements. In present-day ITS, smart vehicles are ready to speak with one another through different system interfaces, such as Bluetooth, wi-fi, and so on. Therefore, the decentralized and circulated nature of blockchain can make this framework more proficient. Moreover, the coordination of blockchain with ITS also highlights their security risks due to its start-to-finish encryption. The combination also improves security and trust, as well as protection against hazards in the transportation arrangements. For example, the vehicles can be associated with each another in a vehicular system of ITS; each vehicle will trade distinctive sensor data with one another. This correspondence can be made safer by utilizing the key encryption innovation of blockchain, so that no one from outside the system can view the transmission message. Moreover, clients

inside an open blockchain can obtain data with respect to different clients. In this way, the reconciliation of a safeguarding technique on the head of blockchain-based ITS design is required, and differential security can be the most appropriate decision for it due to its ever-changing nature.

Real Estate

Real state transactions should be straightforward and transparent. However, agents are commonly used to facilitate these arrangements. The transaction may include many intermediaries, such as specialists, assessors, and legal officials, which can be difficult and costly. To improve these circumstances, a blockchain-based land arrangement may be used to eliminate the need for brokers. A decentralized open blockchain would allow vendors to promote their properties using the broadcast facility in the system; in addition, purchasers can select their ideal properties, make exchanges, contact dealers, register properties with their names, and broadcast the offer to the system using a blockchain-based site. Along these lines, blockchain would eliminate the utilization of mediators, thus reducing the overall cost.

This framework would work like Bitcoin, which has been effectively operating for a decade. So far, a few projects integrating blockchain for real estate have been done by analysts, such as MultiChain and South African Blockchain models. These blockchain-based land frameworks are quite secure and productive. For example, after the successful acquisition of any property or while promoting a particular property, the identity of the purchaser and vendor would not be exposed [22].

Blockchain technology enables a trade or exchange between the purchaser and vendor, without the involvement of any third party. Making false name (pseudonym)-based identities will provide a sense of insecurity for the people who are often trading and earning good returns. This being the case, then the protection using open key cryptography is not sufficient as there are experiments which show that the identity can be tracked using hash and public keys. To ensure this cycle, and so as to make it safer and private differential security based blockchain land framework will be a feasible arrangement.

Health Care Systems

Because of population growth, conventional healthcare frameworks are incorporating more mechanically advanced approaches, with numerous health-related gadgets for well-being support and continuous vital sign monitoring. These smart healthcare services can contain patient data that can help specialists and caregivers to monitor and investigate a particular ailment, even from remote locations. Because medical records are private and critical, where a minor change in any attribute may constitute a high health risk to a patient, it is important to protect these systems. Therefore, to improve security and trust, blockchain-based smart healthcare frameworks are expanding quickly. The secure nature of blockchain can support patients and emergency clinics in controlling the use and sharing of their information just to certain specialists. A few approaches coordinating blockchain-based security approaches have been proposed in the literature.

References

1. G. Wang, Z. Shi, M. Nixon, S. Han, ChainSplitter: Towards blockchain-based industrial IoT architecture for supporting hierarchical storage, in *Proceedings of the 2nd IEEE International Conference on Blockchain (Blockchain), Blockchain 2019*, (2019), pp. 166–175
2. J. Shen, T. Zhou, D. He, Y. Zhang, X. Sun, Y. Xiang, Block design-based key agreement for group data sharing in cloud computing. IEEE Trans. Dependable Secur. Comput. **16**(6), 996–1010 (2019)
3. K. Wang, H.S. Kim, FastChain: Scaling blockchain system with informed neighbor selection, in *Proceedings of the 2019 2nd IEEE International Conference on Blockchain (Blockchain), Blockchain 2019*, (IEEE, Atlanta, 2019), pp. 376–383
4. Y. Xu, Q. Li, X. Min, L. Cui, Z. Xiao, Collaborate computing: Networking, applications and worksharing, in *12th International Conference, CollaborateCom 2016*, Beijing, China, November 10–11, 2016, Proceedings, vol. 201, no. July 2016, pp. 490–496, 2017
5. T.M. Choi, S. Guo, N. Liu, X. Shi, Optimal pricing in on-demand-service-platform-operations with hired agents and risk-sensitive customers in the blockchain era. Eur. J. Oper. Res. **284**(3), 1031–1042 (2020)
6. V. Hassija, V. Chamola, S. Zeadally, BitFund: A blockchain-based crowd funding platform for future smart and connected nation. Sustain. Cities Soc. **60**, 102145 (2020)
7. A. Farouk, A. Alahmadi, S. Ghose, A. Mashatan, Blockchain platform for industrial healthcare: Vision and future opportunities. Comput. Commun. **154**, 223–235 (2020)
8. H. Feng, X. Wang, Y. Duan, J. Zhang, X. Zhang, Applying blockchain technology to improve agri-food traceability: A review of development methods, benefits and challenges. J. Clean. Prod. **260**, 121031 (2020)
9. L. Guo, H. Xie, Y. Li, Data encryption based blockchain and privacy preserving mechanisms towards big data. J. Vis. Commun. Image Represent. **70**, 102741 (2020)
10. O. Pal, B. Alam, V. Thakur, S. Singh, Key management for blockchain technology. ICT Express, 4 (2019). https://doi.org/10.1016/j.icte.2019.08.002
11. B. Teufel, A. Sentic, M. Barmet, Blockchain energy: Blockchain in future energy systems. J. Electron. Sci. Technol. **17**(4), 100011 (2019)
12. B. Guidi, When blockchain meets online social networks. Pervasive Mob. Comput. **62**, 101131 (2020)
13. S.K. Lo, X. Xu, M. Staples, L. Yao, Reliability analysis for blockchain oracles. Comput. Electr. Eng. **83**, 106582 (2020)
14. D.C. Nguyen, P.N. Pathirana, M. Ding, A. Seneviratne, Blockchain for 5G and beyond networks: A state of the art survey. J. Netw. Comput. Appl. **166**, 102693 (2020)
15. H. Wang, H. Qin, M. Zhao, X. Wei, H. Shen, W. Susilo, Blockchain-based fair payment smart contract for public cloud storage auditing. Inf. Sci. (NY) **519**, 348–362 (2020)
16. S. Xuan et al., An incentive mechanism for data sharing based on blockchain with smart contracts. Comput. Electr. Eng. **83**, 106587 (2020)
17. X. Liu, K. Muhammad, J. Lloret, Y.W. Chen, S.M. Yuan, Elastic and cost-effective data carrier architecture for smart contract in blockchain. Futur. Gener. Comput. Syst. **100**, 590–599 (2019)
18. J.L. Ferrer-Gomila, M. Francisca Hinarejos, A.P. Isern-Deyà, A fair contract signing protocol with blockchain support. Electron. Commer. Res. Appl. **36**, 100869 (2019)
19. D. Unal, M. Hammoudeh, M.S. Kiraz, Policy specification and verification for blockchain and smart contracts in 5G networks. ICT Express **6**(1), 43–47 (2020)
20. A. Sah, S.J. Bhadula, A. Dumka, S. Rawat, A software engineering perspective for development of enterprise applications, in *Handbook of Research on Contemporary Perspectives on Web-Based Systems*, (IGI Global, Hershey, 2018), pp. 1–23
21. A. Sah, A. Dumka, S. Rawat, Web technology systems integration using SOA and web services, in *Handbook of Research on Contemporary Perspectives on Web-Based Systems*, (IGI Global, Hershey, 2018), pp. 24–45

22. A. Sah, S. Rawat, S. Pundir, Design, implementation and integration of heterogeneous applications. Int. J. Comput. Appl. **54**(5), 11–16 (2012)
23. V. Fore, A. Khanna, R. Tomar, A. Mishra, Intelligent supply chain management system, in *Proceedings—2016 3rd International Conference on Advances in Computing, Communication and Engineering, ICACCE 2016*, (2017). https://doi.org/10.1109/ICACCE.2016.8073764
24. A. Khanna, A. Sah, T. Choudhury, Intelligent mobile edge computing: A deep learning based approach, in *Communications in Computer and Information Science: Vol. 1244 CCIS*, (Springer, Berlin, 2020). https://doi.org/10.1007/978-981-15-6634-9_11
25. D.S.R. Krishnan, S.C. Gupta, T. Choudhury, An IoT based patient health monitoring system, in *Proceedings on 2018 International Conference on Advances in Computing and Communication Engineering, ICACCE 2018*, (2018). https://doi.org/10.1109/ICACCE.2018.8441708
26. T. Wasson, T. Choudhury, S. Sharma, P. Kumar, Integration of RFID and sensor in agriculture using IOT, in *Proceedings of the 2017 International Conference on Smart Technology for Smart Nation, SmartTechCon 2017*, (2018). https://doi.org/10.1109/SmartTechCon.2017.8358372
27. M. Khurana, T. Choudhury, P. Malik, A review on network security challenges and the internet of things (IoT), in *Proceedings of the 4th International Conference on Contemporary Computing and Informatics, IC3I 2019*, (2019). https://doi.org/10.1109/IC3I46837.2019.9055675

Chapter 6
Blockchain Technology: Concept, Applications, Challenges, and Security Threats

Charu Gandhi, Nitin Shukla, Gagandeep Kaur, and Kusum Yadav

6.1 Introduction to Blockchain

Blockchain technology is a combination of various digital mechanisms like cryptography, data management, and networking which supports capturing, validation, and execution of transactions between the communicating users. Blockchains are basically tamper evident and tamper resistant distributed digital ledgers implemented with a decentralized repository and has no central authority. It is a distributed and public database of all transactions executed and shared among participating parties. Basically, they enable a group of users to record transactions in a shared ledger and do not allow transactions to be changed once published [11, 15]. The blockchain contains a verifiable record of every transaction ever made which is verified by consensus of the participants in the system. Blockchain can be technically defined as (Fig. 6.1),

Blockchain is a peer-to-peer, distributed ledger that is cryptographically secure, append only, immutable (extremely hard to change), and updateable only via consensus or agreement among peers [3].

In blockchains the cryptographically signed transactions are grouped into blocks. Each block is cryptographically linked to the previous block making the chain tamper evident. New blocks are added after validation using the consensus mechanism and replicated across the ledger of the network. Blockchain can create a shared reality across non-trusting entities where the participating nodes in the network do

C. Gandhi (✉) · N. Shukla · G. Kaur
Department of Computer Science and Engineering, Jaypee Institute of Information Technology, Noida, India
e-mail: charu.gandhi@jiit.ac.in; nitin.shukla@jiit.ac.in; gagandeep.kaur@jiit.ac.in

K. Yadav
College of Computer Science and Engineering, University of Hail, Hail, Saudi Arabia

T. Choudhury et al. (eds.), *Blockchain Applications in IoT Ecosystem*,
EAI/Springer Innovations in Communication and Computing,
https://doi.org/10.1007/978-3-030-65691-1_6

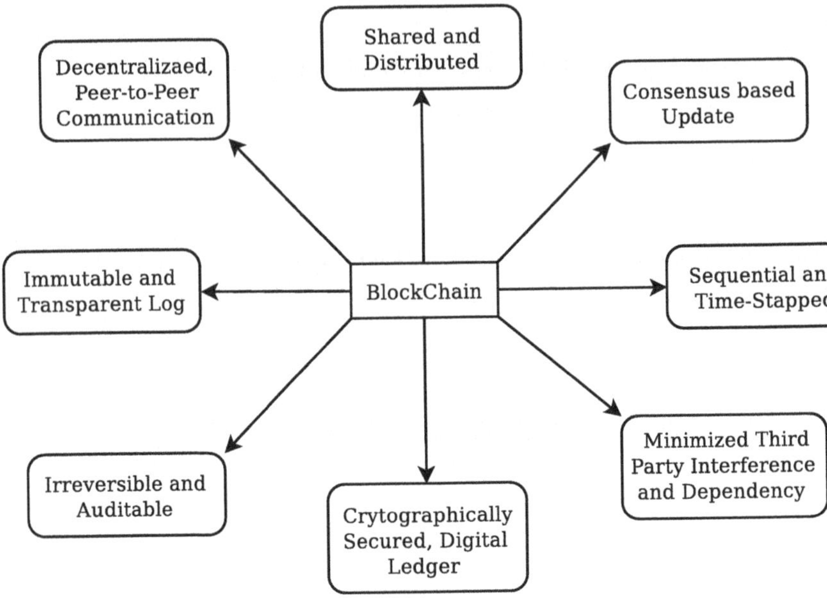

Fig. 6.1 Characteristic features of a blockchain

not need to know or trust each other. Each participant has the ability to monitor and validate chain for themselves. Prior to the use of Blockchain technology, this trust was typically delivered through intermediaries trusted by both parties. The key features of Blockchain can be summarized as:

(a) **Increased Capacity:** Due to the decentralized working principle, the blockchain technology increases the capacity of the whole network. Because of the multiple shared computing systems working together which in total offers much larger computation capacity as compared to individual systems in a centralized environment.
(b) **Better Security:** Ledger is cryptographically secure, which makes it immune to tampering and misuse. Cryptographic services ensure non-repudiation, data integrity, and data origin authentication. Blockchain in its functioning allows access to the past transactions and at the same time protects the identity of the individuals associated with the transaction. The identity theft of a user during the execution of a transaction is therefore infeasible.
(c) **Immutability:** Immutability is ability of a blockchain ledger to remain unaltered and indelible. Data in the blockchain cannot be altered. Every block contains a unique hash or digital signature for itself as well as for the previous one. Thus, the blocks are tightly coupled together and any intrusion into the systems and data modification is very difficult. Immutability along with the consensus approach makes the data auditing process more integral, efficient, cost-effective, and trust worthy.

(d) **Faster Settlement:** Network-based infrastructure of blockchain allows for the instantaneous recording and retrieval of information. It is expected to increase the efficiency of operations by improving the speed of execution and reducing resource requirement and cost. This is the major reason that the modern financial systems are currently experimenting with the use of the blockchain aiming to facilitate intermediation between banks, clearing houses, and central banks.

(e) **Distributed System:** Distributed ledger, the basis of blockchain, can be shared among a group of users connected through the internet. To ensure that all users have a latest version of the ledger, a message is relayed on creation of every new block. This feature eliminates the need of a central authority to record and validate the information. Since the ledger is stored in multiple different locations, any form of data loss due to system failure is protected. Further, other users can still access and add information on the blockchain, until there is at least one online device having the latest version.

(f) **Minting:** Crypto Minting is an integrated platform for staking cryptocurrency. Staking is the process where the Proof Of Stake (PoS) cryptocurrency wallet stakes a sum of coins when broadcasting to the network that the transaction they are processing is true and non-malicious. If the network agrees on the verdict, the verifying wallet will receive a reward in the form of coins on the associated blockchain network. With enormous growth in number of cryptocurrencies, several blockchain maintenance methods are investigated for better support.

6.2 Blockchain Origin

The blockchain generally comprises several fields like software engineering, distributed computing, cryptography, and game theory. Real-world blockchain applications are usually defined discussed under the umbrella of what is known as cryptoeconomics. Cryptoeconomics [26] is defined as, "a discipline concerned with the production, consumption and transfer of wealth using computer networks, cryptography, and game theory to enhance prosperity of groups in current and future digital market economies."

Blockchain is a basis of Bitcoin. Satoshi Nakamoto is considered as the inventor of blockchain. He published a research article to a cryptography forum outlining technique to counter the double-spend scenario. Early cryptocurrencies suffered from the problem of double spending. He described his technique by using a series of hashed timestamps, without explicitly defining blockchain. "Each timestamp includes the previous timestamp in its hash, forming a chain, with each additional timestamp reinforcing the ones behind it" [23]. The original method was modified to adapt the needs of Bitcoin, and was defined as—*cryptographically linked chain of blocks, where each block uses hash digest of previous block for security*. Figure 6.2 depicts the simple structure of a blockchain.

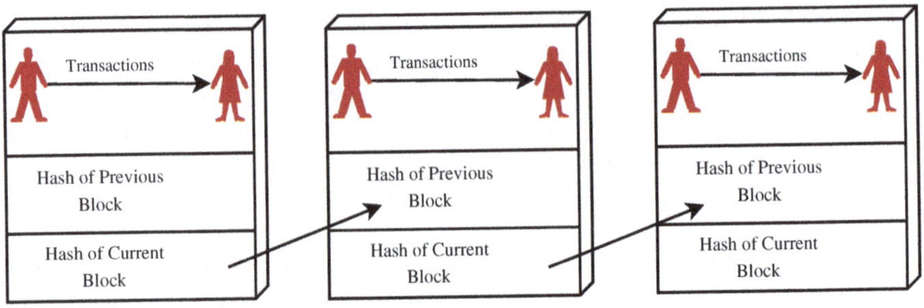

Fig. 6.2 Simple blockchain structure

To protect itself from double spending through blockchain, Bitcoin needs a methodology for its network to reach a consensus for business valuation. Therefore, a proof of work model was introduced, in which agents repeatedly hash a block with a random number until it reaches a value below the specified value. Only then will the block be added to the existing chain. Blockchain technology provided an ever-expanding collection of blocks to simplify the need for Bitcoin to record transaction order, verify the order, and secure access. Each block has a pool of transactions, and it gets linked to a high-level block after computing the cryptographic Hash Digest. The Bitcoin Network relies on a decentralized distribution and consensus model for proof of work to organize the addition of new blocks and update existing copies. Blockchain in Bitcoin acts as a database to store transactions. Bitcoin balance is not centrally managed, and no coins are printed and serialized. Users can calculate the available credit by going through the blockchain. Any attempt to change the blockchain will fail because the hash needs to be recalculated. Bitcoin was first used in January 3, 2009, by Nakamoto. He created the first block of blockchain known as genesis block. He then issued the first 50 bitcoins to himself [34].

6.2.1 *Tiers of Blockchain Technology*

The rapid developments in blockchain technology have led to the evolution of many applications. Based on these evolution areas and usage of blockchain, the various tiers have been categorized [34]. These functionalities and their applications are supported by majority of blockchain platforms usually with minor exceptions.

1. **Blockchain 1.0:** Introduced in 2009 along with the Bitcoin, this generation lasted till 2010. It has found its primary usage in cryptocurrencies like and has been used in applications involving payments.
2. **Blockchain 2.0:** This second generation of blockchain emerged in 2010 and finds its applications in financial sectors and smart contracts. However, its applications have expanded beyond currencies, stock markets, and finances. Different financial assets which come under Blockchain 2.0 application domain

are the derivatives, swaps, and bonds. Blockchain platforms like Ethereum and Hyperledger are considered to be the part of Blockchain 2.0.

3. **Blockchain 3.0:** Blockchain 3.0 emerged in 2012 and has found its applications in various sectors like government, healthcare, media, and justice. Since this blockchain technology tier has ability to code smart contracts, it has led to the evolution of many new blockchains apart from Ethereum and Hyperledger.
4. **Blockchain X.0:** It represents a future blockchain concept, where, public blockchain services will be available and that can be used by anybody like the Google search engine. It is expected to be a public distributed ledger with generic agents executing on the blockchain. These agents shall have the decision-making capability and can interact with intelligent and independent agents acting on user's behalf. It is also expected to be regulated by law and contracts implementable in code instead of paper.

6.3 Blockchain Categorization

Various categories of blockchain networks have been identified on the basis of their applications and access permissions. Depending on their applications, there are public, private, and hybrid variants of blockchain:

1. **Public blockchain:** Public blockchains are visible by anyone and do not have any single owner. They are fully decentralized and their consensus mechanism is open for all the users to participate in validation process. For example, Bitcoin.
2. **Private blockchain:** These blockchains use access control to manage read and write privileges for the blockchain network. A single entity has control over the ownership and block creation process. Hence, private blockchains do not usually require consensus algorithms and mining.
3. **Hybrid blockchain:** Hybrid blockchains or consortium blockchains are public only for a particular group and a few privileged servers control the consensus process. A set of agreed upon rules are used in the validation process using consensus. Additionally, only the entitled participants receive the updated copies of the blockchain. Thus, the network is only partially decentralized.

Based upon who can maintain and publish blocks in a blockchain, the two categories defined are permissionless and permissioned:

1. **Permissionless blockchains:** These networks use decentralized ledgers which allow anyone to publish the blocks, with no need of any permission from the authority. Thus, any user can read from and write transactions to the blockchain. As, these are implemented using open source software, they are highly vulnerable to the attackers, attempting to modify the transactions in the blocks. Thus, a multiparty agreement known as consensus system is utilized to prevent unauthorized access by rewarding the publishers of valid blocks with cryptocurrency.

Table 6.1 Characteristic analysis of basic blockchain

	Public blockchain	Private blockchain	Hybrid/consortium blockchain
Accessibility	Anyone	Owner organization only	Few selected and multiple organizations
Participant type	Anonymous	Known entity	Known entity
Access type	Permissionless	Permissioned	Permissioned
Security mechanism	Consensus mechanisms, Proof of Work, and Proof of Stake	Multiparty consensus and voting with pre-approved participants	Multiparty consensus and voting with pre-approved participants
Transaction speed	Slow	Light weight and fast	Light weight and fast

2. **Permissioned blockchains:** These are the networks where a centralized or decentralized authority allows the users publishing the block. Thus, it is feasible to restrict the access permissions to read and write transactions. The software used to implement these platforms can be either open or closed source. However, both permissioned and permissionless blockchain networks have the same capability to trace the digital assets passing through the blockchain. Since, they require the authentication of every user to participate as a member of the network, the consensus models used in publishing blocks do not require the expense of cryptocurrency and are faster and computationally cheaper.

Table 6.1 presents the characteristic analysis of these basic types of blockchain on various key parameters like, accessibility, type of participants and access, security levels, and transaction speed.

Based on their overall system and applications, a novel categorization of blockchains into three types has been identified [8]:

1. **Only cryptocurrency blockchain (C2C):** This type deals with only cryptocurrency chain and is totally reserved for payments or money decentralization decentralization. Bitcoin [14] introduced in 2009 is the most widely used (C2C) based cryptocurrency.
2. **Cryptocurrency to business blockchain (C2B):** This type of blockchains makes use of smart contracts, the logic tier which provides a multi-purpose programmable infrastructure. The public ledger stores financial transactions in C2B, and also facilities to deploy and execute programs on the blockchain. The tamper-proof nature of smart contracts, reduces the verification and execution costs and prevents any malicious activity. Ethereum [7] is the most widely used example of this type of blockchain.
3. **Business to business blockchains (B2B):** These blockchains do not support any currency, however, to support business logic, software execution is required. To cater to variable and distinct needs of each industry or business personalized

blockchains are required. Thus, organizations have started deploying blockchains designed to cater to their specific needs allowing them to overcome the challenges like privacy, scalability, and lack of governance, faced with other types of blockchain, These are usually implemented using Ethereum and Hyperledger.

6.4 Blockchain Working

Blockchain implements a centralized and distributed system, the distributed ledger, for storage and verification of transactions. Distributed ledger is a shared database as shown in Fig. 6.3. It is decentrally synchronized and replicated among various network users. Hence, Distributed Ledger Technology (DLT) [27], distributed across multiple computing nodes forms the backbone of blockchain. Before a transaction can be stored, it must be consented upon by majority of the users in the blockchain network. Each user in the blockchain has the most recent copy of the ledger which is updated and synchronized with any changes that occur.

A blockchain consists of several blocks with each block having many transactions. Every newly added block elongates the blockchain representing a full ledger of the transaction records. Every block in a blockchain comprises a timestamp, a random number known as nonce for verification and hash value of the previous block. These security parameters of a blockchain ensure the integrity of the overall blockchain down to the genesis block which is first block in the chain. Because of the uniqueness of the hash values, any modification made to any block in the chain will immediately get identified. Hence, any manipulation attempt in the chain is reflected and this feature makes blockchain less vulnerable to malicious attacks. All the transactions as well as the blocks are validated by most of the user nodes before adding them to the chain. This validation and consensus process is performed by special peer nodes called as miners.

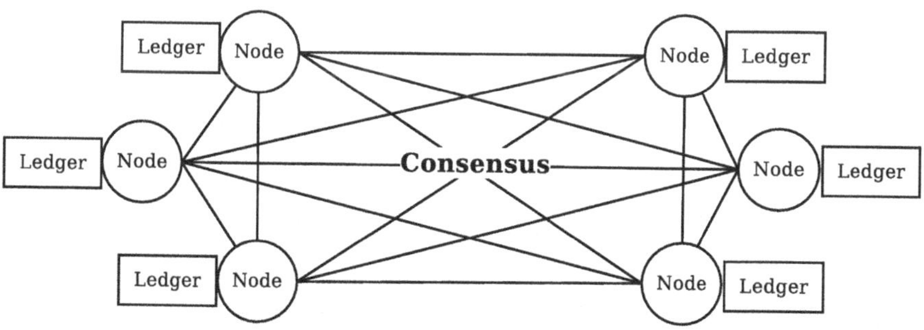

Fig. 6.3 Distributed Ledger in blockchain

6.5 Blockchain Concepts

Blockchain technology uses various digital concepts such as signatures, hashing, public-key cryptography, and transaction recording like attach-only ledgers. These key underlying functional concepts used in blockchain technology are detailed as follows.

6.5.1 *Cryptographic Hash Functions*

Blockchain technology employs cryptographic hashing as its underlying operational technique. Hashing is performed by implementing a cryptographic hash function on the input, to calculate a distinct message digest. It generates the identical result for the given input for each application of the hash function. Altering a single bit in data outputs an entirely dissimilar message digest. Hence, the correctness of input can be easily proven using hashing.

Cryptographic hash functions are preimage resistant, i.e., they are one-way functions. Also, these are designed to be both second preimage resistant and collision, that is, it is mathematically unattainable for dissimilar inputs to generate same output. These perform various tasks in a blockchain operation like: address derivation and creation of unique identifiers. These also secure the block data by calculating the hash value of the block data, and inserting the digest in its header. Since, a block header's digest is passed on to the succeeding block's header, its data is inherently protected when the block's header digest is passed on to the succeeding block [13, 18, 29].

A Secure Hash Algorithm's (SHA) [35] variant, having an output of 256 bits, (SHA-256), is used for blockchain implementations. It produces a 32 bytes long output which is presented as a string of 64 hexadecimal characters. SHA-256 is widely supported by many machines in hardware, which makes it fast to compute.

6.5.2 *Cryptographic Nonce*

A nonce is a single use number generated randomly. It is created for a specific security purpose and often modifies the result of a cryptographic function in a secret communication. Nonces are usually numbers that change over time. This prevents some values from being reused. Nonce is commonly used to send varied input to a hash function, authentication, identification, digital signatures, etc. [36]. The basic blockchain block hashing process and the associated nonce are shown in Fig. 6.4.

In the blockchain, miners create a block, verify it, and are rewarded for using their processing power. The block that receives more than 50% of the consensus is appended to the blockchain. While verification, miners provide PoW covering all

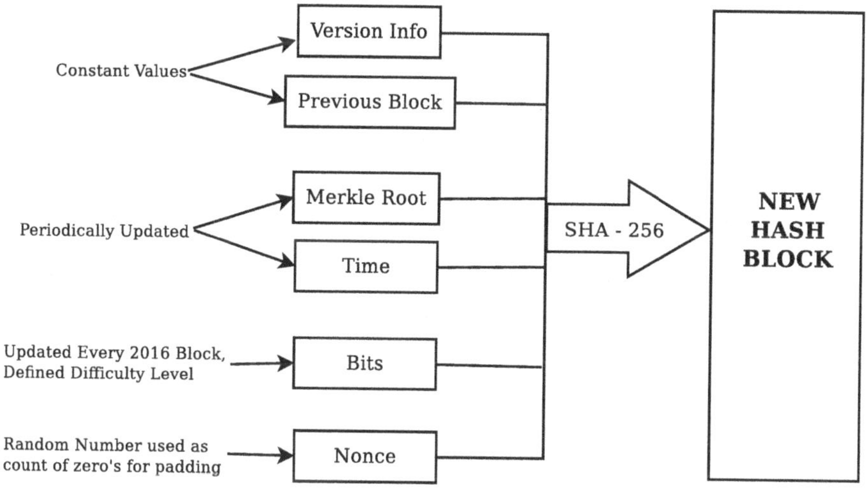

Fig. 6.4 Blockchain block hashing and the nonce

the available transactions in the block. It also checks the hash of the present block to be less than the predetermined value. To create a block that can be accepted by as many participants as possible, minors compete among themselves to provide PoW as early as possible.

The nonce is the most important for PoW. Nonce is a 4-byte random number, continuously adjusted by the miners, till it becomes valid for finding the hash value of a block. Once the perfect nonce is found, it is entered in the hashed block header. The hash value of that block is rehashed along with this number and creates a difficult algorithm. This eliminates the chances of duplicating, or exploiting the same crypto currency twice [37].

6.5.3 *Transactions in Blockchain*

Transactions represent the interaction between participants in a blockchain. For example, a transaction can represent the cryptocurrency transfer between the users. Every block in a blockchain has several transactions which are used by the users to send information to the blockchain network. The information sent typically includes the identifier of the sender, its public key, digital signature, and transaction inputs and outputs. A typical blockchain transaction is shown in Fig. 6.5.

The input to a transaction is the list of the digital assets to be sent. Digital assets are referenced by their sources either through the previous transaction available with the sender, or the originating event if it is a new asset. Sending the input to a transaction as a previous event reference restricts the addition or deletion of any value from the existing assets. However an asset can be bifurcated into multiple

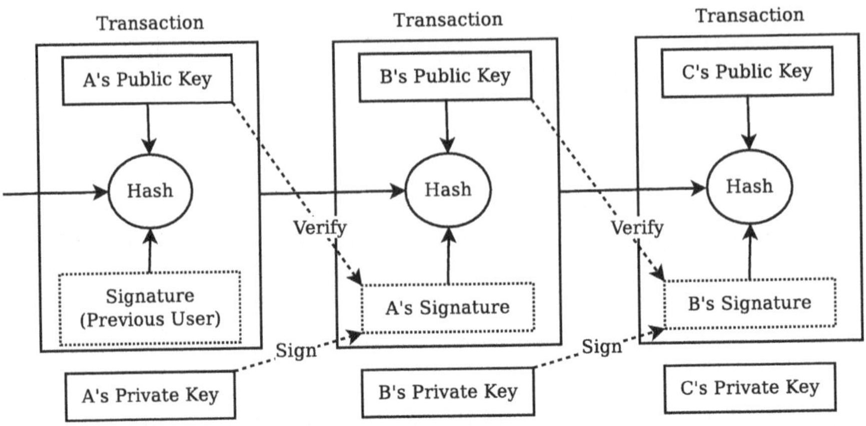

Fig. 6.5 A typical blockchain transaction

lesser value assets and, vice versa. The splitting or aggregation of assets is reflected in the transaction output correspondingly.

The outputs of a transaction are the receiving accounts' identifier, and the quantity of the asset transferred. It also specifies the various conditions to be satisfied by the new reliever/owners to be eligible to spend the received value. However, the funds received in excess of the requirement must be returned to the sender.

Furthermore, it is very crucial to validate and authenticate a transaction. Validating a transaction is a surety that the transaction adheres to the underlying protocol and data format. The authenticity of a transaction ensures that the sender is authorized to access the sent digital assets. Each sender uses its private key to sign the transactions. These digital signatures can be verified by using the public key of the sender.

6.5.4 Asymmetric Key Cryptography

Asymmetric key or public key cryptography [30], the core of the blockchain technology, uses two types of keys: public and private. The private key is available to the sender only whereas the public key is accessible to all the parties [38]. Both keys are mathematically related, however, either of them cannot be deciphered if the other one is known. User can encrypt the data with one key and needs the other one to decrypt the same.

Asymmetric key cryptography by default establishes trust between two unknown users to authenticate transactions while keeping them public. The transactions are signed using private key which can be decrypted with only the public key. As public key is accessible to all, encryption using private key confirms that the sender is authorized to use the private key. In blockchain transactions, private keys just sign them whereas, the public keys extract the addresses and prove the signature generated by a private key.

6.5.5 Addresses

Addresses are used by the blockchain implementations to act as the sending and receiving endpoints in a transaction. An address is an alphanumeric string obtained by hashing the user's public key and other values like version number, checksum, etc. As permissionless blockchains do not require user identification for account creation, it allows a user to generate multiple addresses using multiple pairs of keys.

Blockchain users are not lone source of addresses within the network. For example in Ethereum networks, smart contracts can be accessed through a special address, the contract account address [39]. This account address facilitates the access to the smart contract after deployment. It is calculated using the smart contract owner's address and permits the execution whenever a transaction is received by it, and in turn, creates additional smart contracts.

6.5.6 Wallets

Some permissionless blockchains use a software, referred as a wallet, to secretly store their private keys. The wallet is used to securely record private and public keys, and their associated addresses. It is also used to record the number of digital assets a user is having. As recreation of private key is computationally impossible, if a user's private key is lost, all digital assets associated with it are also lost. This may result in an attacker gaining full access to all the assets that use the lost private key.

6.5.7 Digital Ledgers

A digital ledger is accumulation of transactions, oftenly stored using large databases. Owner of the ledger is a trusted third party, which manages and operates it centrally, on user's behalf. These ledgers can be centralized using single server or distributed using coordinating group of servers. However, distributed ownership of the ledger is becoming popular these days. Blockchain technology makes use of distributed ownership along with the distributed architecture for implementing the digital ledger. Distributed architecture of these networks need higher number of computers as compared to the traditional centrally managed distributed architecture.

A blockchain network is distributed by design i.e. it creates multiple copies which are all updated and synchronized to the same ledger data between the peers.

New nodes joining the blockchain network, request the latest copy of the ledger from the other users only, thereby, preventing any loss or destruction of the ledger. Blockchain network is heterogeneous, having varied types of software, hardware, and infrastructure. This heterogeneity of the nodes, guarantee for an attack on one node to have same effect on the other nodes. To provide tamper evident and resistant ledgers a blockchain network uses digital signatures and hash functions.

Distributed Ledger Technology (DLT) [16] platforms are classified mainly as permissionless and permissioned or public and private ledgers. In permissionless DLT platforms, the ledger is maintained by joint action of the nodes in the public network accessible to everyone. Here, any user can enter the network and perform block confirmation to create consensus. On the other hand, a permissioned DLT platform controls the users participating in the consensus process of the system state. Thus, the ledger is controlled by only authorized nodes and is accessible to specified users only. Permissioned DLTs allow rapid validation of transactions and provide improved privacy. Hyper Ledger Fabric [9] and R3 Cooda [5] are the popular permissioned DLTs.

6.5.8 Blocks in Blockchain

In the blockchain users put forward their transactions through various software, like desktop, web and mobile applications, e-wallets, etc. These software forward them to one or more nodes in the network. The entered transactions are further dispatched to other nodes in the network, keeping the transaction pending. A pending transaction waits in a queue before getting added to the blockchain by a publisher. A block consists of a header and data as depicted in Fig. 6.6.

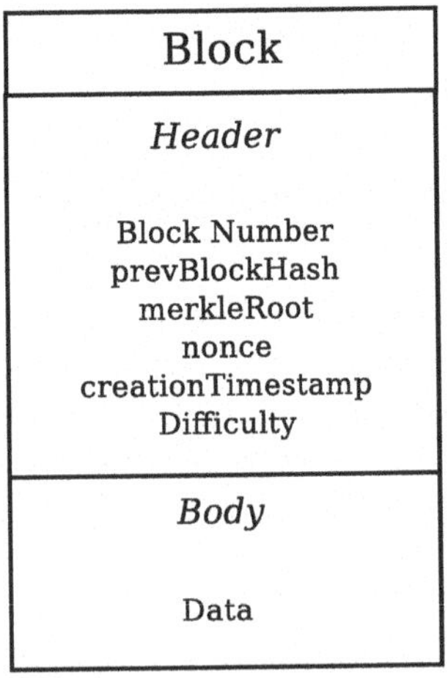

Fig. 6.6 Basic block in the blockchain

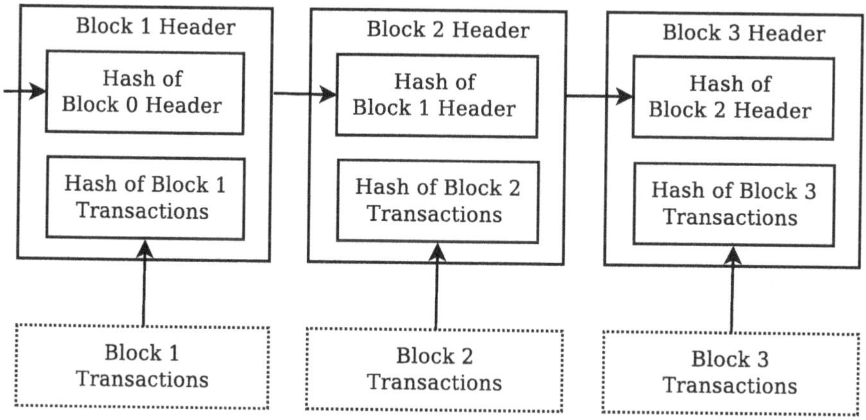

Fig. 6.7 Chaining of blocks in a blockchain

The header in the block represents the metadata for a block consists of the block number or height, hash value of preceding block header, hashed value of block data, timestamp, block size and the nonce value. Chaining of different successive blocks in the blockchain is shown in Fig. 6.7.

Block data is a list of confirmed transactions that have been published to the blockchain. The authenticity of the transaction is verified by checking the proper format and ensuring that the digital asset providers have signed the transaction. It also proves the authenticity of providers of assets to access the private key for a transaction which is used to sign them. Full nodes check the validity of all published transactions for a block and reject blocks with invalid transactions.

6.6 Consensus in Blockchain

Consensus models [32] are used in blockchains to decide upon the user who publishes the next block. In a permissionless blockchain many nodes compete simultaneously to publish the next block, in order to win the cryptocurrency. The users generally know each other only through their public addresses and have a mutual distrust among them. In such a situation, to sort out the conflicts between multiple nodes publishing a block concurrently and to allow the working of the distrusting users together, consensus models are used.

When a new user enters a blockchain, it must retrieve the system's initial state from the genesis block, which is first and the only pre-composed block. Every block is attached to the blockchain only after the genesis block is created and must agree upon the consensus model. Users can autonomously consent upon the state of the blockchain by using the initial state and verifying each block being added since then. Thus, trusted third party is not required for providing the system state. However, in permissioned blockchain networks, there exists trust level among publishing nodes. Hence, a consensus model is not needed to decide which participant shall append the next block to the chain.

6.6.1 Poof of Work (PoW) Consensus Model

According to the PoW principle, a miner that publishes a block after solving a computationally complex problem is considered as the first miner. The result of the puzzle is considered as the proof that the user has accomplished the work. Solving this puzzle is difficult but validating the correctness of the output is straightforward. This permits all full nodes to easily authenticate any submitted block. A commonly used puzzle requires checking if the block header's hash value is lesser than a pre-specified threshold. The key aspect of PoW is that the puzzles are independent. This means that the work input to a puzzle has no impact on the probability of solving the puzzles. However, there are some major issues with the Proof of Work consensus mechanism:

- **The 51% or majority risk:** If a miner gets hold of 51% or more than 51% of the blocks, it can corrupt the blockchain by gaining the control of the majority of the network.
- **Time intensive:** Miners have to check several nonce values till they find the accurate solution to the puzzle that is required to mine the block. Since, the complexity of the puzzle is computationally very high, this process becomes very time-consuming. Also confirming a transaction is not instantaneous as it takes some time (50–60 min approx.) to mine the transaction and adding it to the blockchain.
- **Resource intensive:** Solving the puzzle to mine the block requires high computing power. Thus, miners need the machinery having high computing power. Such a machinery also consumes lot of energy.

6.6.2 Proof of Stake (PoS) Consensus Model

According to the PoS principle, the larger shares a user invests in the blockchain, the higher likelihood that it wants the system to succeed. The stake is the amount of cryptocurrency invested by a blockchain node in the system. If staked, the cryptocurrency is not available for consumption. PoS model uses this share of the user's stake as a determinant to publish the blocks. The percentage of a user's stake in the total amount staked in the network defines the probability of a user's success in publishing the block.

In PoS, since all the nodes are not competing against each other to attach a new block to the blockchain, and no computation is to be done, this save a lot on resources and energy. Also, PoS is completely decentralized. That is, in Proof of Stake, rewards are proportional to the amount of stake. Hence, it provides absolutely no extra rewards to join a mining pool. It also ensures a secure network as a person attempting to attack a network must own 51% of the stakes which is highly expensive.

6.6.3 Round Robin Consensus Model

This model is implemented by a Permissioned Blockchain network in which nodes alternately create blocks. These systems set a timer that allows the current node to publish the block so that unavailable nodes do not delay the block's publishing. Also, it ensures that a single node does not create the bulk of the block. The Round Rodin model has the advantage of a simple, low power approach. However, they do not use crypto puzzles, and Round Robin does not work efficiently on the blockchain without permission.

6.6.4 Proof of Authority/Identity Consensus Model

The proof of authority or proof of identity consensus model is implemented in permissioned blockchains as they require very high levels of trust. In this model the publishing nodes establish trust through their real-world identities like identification documents that are usually verified and publically notarized. Publishing nodes must have their identities provable in the blockchain network. Thus, the publishing node has placed its reputation on publishing new blocks. Blockchain users can directly affect a publishing node's reputation, where it can lose or regain its reputation by acting in a mode to which other users agree or disagree. The lower the reputation of a publishing node, the less likely it is to publish a block.

6.6.5 Proof of Elapsed Time (PoET) Consensus Model

PoET is the fairest and widely used consensus model in a permissioned blockchain. In this model, every validating node on the network adds proof of their wait in the block. Based on this random waiting time only, a node gets the fair chance to create their own block. Newly created blocks are then sent to other nodes for consideration. Validator having the lowest timer value in the proof wins the consensus and its block is appended to the blockchain. A dishonest publishing node can stay back for minimum time and thus, control the system. This model uses a random wait time for the publishing node and also that it must wait for actual time and did not start early.

6.6.6 *Proof of Burn (PoB) Consensus Model*

In Proof of Burn (PoB) [19], valuators burn coins by transferring them to an address where they cannot be claimed again. Thus, they obtain the autonomy to mine on the basis of a random selection process. "Burning coins" provides authenticators a long term dedication instead of short term loss. The more coins the miner burns, the more likely he is to be selected for the mining of subsequent blocks. However, the protocol involves an unnecessary waste of resources since the mining power is directly supplied to those who burn more money.

6.7 Smart Contracts

A smart contract consists of code and data implemented by digitally signed transactions and executed by blockchain network nodes. All nodes executing a smart contract should receive the same results, which are then written to the blockchain, as shown in Fig. 6.8. Smart contracts provide publicly available functions that are performed with data provided by the user to perform a service. Since the code runs from the blockchain, it is transparent and tamper-proof and is used in various applications as a trusted third party.

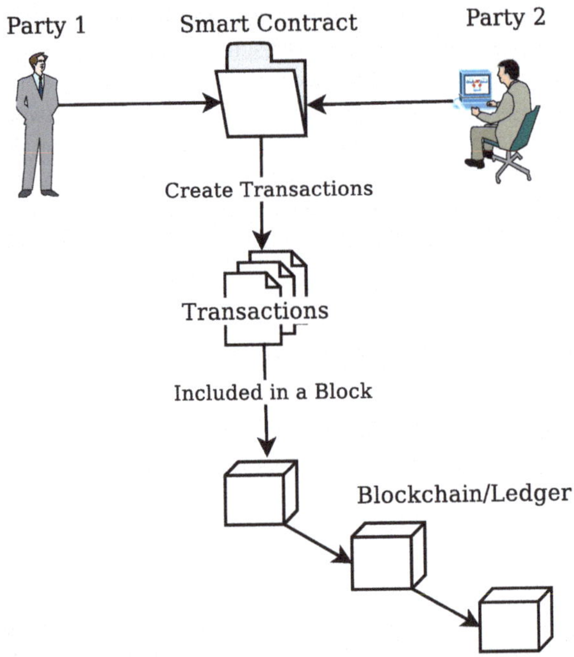

Fig. 6.8 The blockchain smart contract

A smart contract can perform many business enterprise functions including calculations, storing information, exposing properties, and spontaneous fund transfers. It also represents business process transactions involving many parties. The output of smart contracts must be predetermined for a given input. In addition to this, new state after the execution of smart contract must be accepted by all the nodes. A smart contract that operates on the data outside its system context uses an 'Oracle' [6].

In many blockchain implementations, publishing nodes execute the smart contracts concurrently while publishing the new blocks. In other implementations, the results of contract execution by other nodes are validated by only publishing nodes. In permissionless blockchain networks, the user sending the transaction to a smart contract pays the code execution cost. Further, depending upon the complexity of the code, the execution time duration dissipated by a call smart contract is restricted. The execution halts as soon as this limit is exceeded and the transaction is then discarded. This method not only remunerates the publishers for executing the smart contract, it also prevents malicious users from installing and then accessing smart contracts. Since, a malicious user can attack the publishing node and consume all the resources by deploying a denial of service.

However, in permissioned blockchain networks implementing smart contracts, trusted network avoids the users to pay towards code execution for smart contracts. The trusted participants and other mechanisms like revocation of access rights to prevent malicious behavior.

6.8 Challenges in Blockchain

In spite of having huge potential, blockchain faces multiple challenges which restrict its usage widely. Some of the major challenges faced by the blockchain network are as follows.

1. **Scalability:** As the volume of transactions is increasing, each node on a blockchain has to store and validate all the transactions to the blockchain. Further, because of the limited block size and the time interval in generating a new block, a limited set of transactions can be processed at an instance. For example, a typical Bitcoin blockchain processes approximately 7 transactions per second. Thus, it does not fit for the real time applications which require processing billions of transactions. Also, since the capacity of blocks is very less and the miners favor the transactions having higher transaction fee, several small transactions are therefore delayed.
2. **Privacy Leakage:** Blockchains can protect their privacy through asymmetric key cryptography. Users transact using private and public keys without revealing their real identity. However, blockchain does not undertake transactional privacy [33], because for each public key, the values of the transactions and balances are openly visible. Also a user's alias can be associated with its IP address even when it is behind Network Address Translation (NAT) and the firewall. In certain

cases, a client can simply be identified by a group of nodes it is connected to and which can be learned and be utilized to track the origin of a transaction.

3. **Selfish Mining:** A selfish mining or block withholding attack [21] is a malevolent attempt to disrupt the integrity of the blockchain network. In selfish mining, miner in a mining pool withholds a validated block to be broadcasted to the rest of the pool. It then continues with mining the next block, resulting in the selfish miner to present more proof of work as compared to others in the mining pool. Thus, the selfish miner affirms to the block and all financial rewards while the rest of the network take on their block solutions. A selfish miner maintains its own chain, and releases it publicly speculatively, in order to get higher rewards that would otherwise be granted based on their real contributions to the mining pool.
4. **Security:** Security aims to maintain the integrity, availability, and confidentiality of a system. Confidentiality and integrity are major concerns in distributed networks like blockchain. In such systems, availability is not an issue as replication due to replication. Since, all the transactions happening on the Blockchain have to be validated by the major chunk of miners present in the network, it makes it highly vulnerable to 51% or the majority attack as, in this situation, one miner can get the control of the chain completely.
5. **Energy consumption:** Blockchain networks, use Proof of Work mechanisms that validate transactions and add them to the network. These require highly complex mathematical calculations and hence, PoW requires large amount of energy to provide power to computers doing this task. To better understand this issue, blockchain proponents have developed many efficient consensus algorithms, like Proof of Stake (PoS), that are less taxing on energy.
6. **Integration with legacy systems:** The major challenge for corporate these days is to merge the blockchain with already existing systems. In most cases, the use of blockchain requires either complete restructuring of the existing system, or designing the ways to successfully integrate the two technologies. To replace existing system with blockchain based system will involve high cost and time. Furthermore, the blockchain technology suffers the insufficiency of specifically trained and qualified people for developing and managing peer-to-peer networks. As a result, organizations are unable to approach the desired group of blockchain professionals to take part in transition process.
7. **Regulation and standardization problems:** A major challenge faced in the implementing blockchain is the regulatory of various countries because, the decentralized structure is expected to loosen the control of centralized banks in terms of economic policies and transaction amounts. With so many different types of networks being implemented in various industrial domains, the blockchain space suffers a major issue of disarray due to absence of common standards to allow the communication between networks. This disparity across blockchain protocols also restricts technology in basic processes like security and mass adoption. Thus, creation of industry standards for different blockchain protocols can help businesses to collaborate on various fronts like application development, validating proof of concept, and sharing blockchain solutions and its integration with existing systems.

6.9 Applications of Blockchain

Blockchains have recently been adopted worldwide in different applications as decentralized mechanisms to fraud defiant computing and without a requirement of a trusted central entity. Various sectors where blockchain is contributing in a larger degree are described below [1, 4, 10, 22, 24, 31]:

1. **Financial Services:** Traditional financial systems tend to be extremely slow, cumbersome, and prone to errors. They require an Intermediary in the process mediation and conflict resolution. This results in high levels of stress and expenditure, both in terms of time and money. In contrast, users and businesses these days find the use of block chain technology as cheaper, more transparent, and cost-effective. The prominent applications of blockchain in financial sectors can be detailed as:

   (a) **Assets Management:** The asset management, where parties trade and manage resources, needs remarkable processing of confidential and expensive transactions involving more than two parties, usually in a cross boundary scenario. The parties maintain copy of their transactions. Leading to the unwanted usage of space and processing resources. Additionally, it makes the system prone to human errors, inconsistencies in the records. The distributed ledger system in the blockchain reduces such risks of errors by encrypting the records. The use of ledger further reduces the procedural complexity by omitting the need for intermediaries.

   (b) **Insurance:** Claims processing is often a difficult task for the insurance officials because of falsified cases, mounted incidents, tough clients, and many more. These significantly increase the chances of errors and misclassifications. In such scenarios, blockchain provides a perfect framework for safer processing of the cases. Use of cryptographic tools enables guarantors to make sure that processing is safe and reliable.

2. **Health Care:** Several characteristics like transparency, auditability, collaborations, no intermediation, and designing of customized models make the blockchain suitable for healthcare organizations. The scattered medical records of a patient due to visits to different medical institutions is a major issue that hinders the effective IT based healthcare. The blockchain comes to the rescue in such situations by providing platform for distributed and reliable health records maintenance and monitoring. It enables the scattered healthcare records to be integrated for tracking personal health records. Also, medical applications require stringent integrity tracing potential, privacy assurance. And the inherent intricacy of medical records makes historical diagnosis dearer. Blockchain-based solutions can provide continuous tracing of the sequence of treatment and hence, expenditures can be worked out at reduced levels.
3. **Manufacturing, Supply Chain, and retail services:** Various manufacturing, retail, and logistics industries have taken on blockchain to perform different tasks in a supply chain network. Blockchains have found their usage in transparent

movement of a product from producer to the customer in supply chains. These are also being used for product tracking and tracing during transportation and intermediary payments are implemented using various crypto currencies such as Bitcoin, etc. During the manufacturing process, blockchain-based techniques can be used to make sure that standards are adhered to and the environmental affect of the product is within limits. A decentralized blockchain-based market platforms are underway to facilitate the trading process without any mediator. In e-retail systems, blockchain-based payment gateways, loyalty programs, and gift coupons are also being designed.

4. **Intellectual Property Rights (IPR) and copyright protection:** The Internet based technologies have led to copyright issues in last two decades. From file sharing applications to photographs on the Web, copyright rules have not always been followed. As a file is replicated across the network, regular updation and reconciling of all the copies need to be done to maintain the consistency of all records. In blockchain network, with the absence of a central storage location and central authority, manipulating or corrupting the files becomes a tedious task. Since, the modifications made in the ledger can be traced down to the starting record, any illegal use of a copyrighted file can be easily detected.
5. **Governance and Identity management:** Governments at national, state, and local levels are responsible for maintaining citizen records like birth and death certificates, Universal ID's, property transfers, etc. The use of blockchain shall aid in reducing paper based record management. The efforts and time of citizens expended to visit government offices for any renewal, updation, and issuance of any such document can be saved. Further, since, the records can be traced back, applying for any new document will require least amount of information. This can further aid in filing of taxes and other financial transactions very convenient and free of unauthorized access. Blockchain can play a very crucial role in voting systems too, as the government can now ensure a tamperproof casting and counting of the votes.
6. **Smart city development:** Smart cities are the upcoming frameworks to undertake the modern day challenges of urbanization by integrating new technology planning, energy conservation, and transportation management. However, due to lack of standardization and varied requirements the technology infrastructure design can cause some challenges. Blockchain systems can be used to connect these technologies together to achieve the automation of smart cities. The key potential smart city development applications of blockchain are the smart transportation management (ride sharing and vehicle leasing), energy management(trading of energy, billing, promoting renewable energy, tracking emission footprints, etc.), waste management (financial rewards in form of cryptocurrencies for depositing non-recyclable products), governance (voting, tracking find transfers, tax management, etc.), and identity management (preventing illegal migration, identity record keeping, etc.).
7. **Blockchain for Internet of Things:** Blockchain technology has provided solutions to some of the key challenges of IoT like scalability, data management, privacy, and reliability. It has provided an efficient means to track and monitor

millions of connected devices, thereby enabling the sharing and processing of transactions between them. Being a decentralized system, it eliminates the single point of failure, generating a more flexible ecosystem for devices. With help of an efficient P2P application, large volume of transactions between interconnected devices can be processed reducing the installation and maintenance costs of large centralized data centers.

6.10 Blockchain Security Threats and Attacks

Blockchain technology is inherently secure since distributing data using a ledger across several computers, blockchains prevent any single point of failure. Moreover, use of cryptographic constructs like hashing, private keys, etc. and game theory based consensus mechanisms make a blockchain difficult to access and tamper. However, these basic safety features do not ensure the non-vulnerability of blockchain to security threats. Blockchains have exposure to their own specific set of security issues as defined below [20, 25].

6.10.1 Blockchain Structure Attacks

The potential vulnerabilities of the blockchain structures can be exploited to generate structural attacks and these can compromise the entire blockchain-based application.

1. **Blockchain Forks:** Forking is a condition arising in the network where nodes have different viewpoint about a particular state of the network which usually persists. Protocol malfunctioning and client software upgradation related incompatibilities result in creation of forks unintentionally. These can also be created by malicious actions like inserting nodes, which keep to the varying validation rules, known as Sybil nodes or by performing selfish mining. Fork is a representation of an inconsistent state which is used by the attackers to create ambiguous system state, illicit transactions, and mistrust in the network.
2. **Stale and Orphaned Blocks:** Consensus process can result in two major types of inconsistencies leaving valid blocks out of blockchain, stale blocks and orphaned blocks. A stale block was well mined; however, it was not adopted in the blockchain. These occur often in the public blockchains because of the race condition where, miners compete to look for the next valid block with two or more miners being able to find a valid solution. In such situation, the network goes for one of the winner blocks and scraps the remaining. Hence, remaining blocks that are valid but unaccepted and added to the current blockchain become stale blocks. On the other hand, a block where hash of the parent block pointing to an invalid block and is rejected from the blockchain, becomes an orphan block. Orphaned blocks can either be inserted by an attacker or is the resultant of the race conditions among the miners.

6.10.2 Blockchain's Peer-to-Peer System Attacks

Blockchain network can ensure security and accessibility due to its peer-to-peer architecture. On the contrary, this architecture makes a blockchain network vulnerable to different attacks like selfish mining, 51% attack, Distributed Denial of Service (DDoS), and DNS attacks, eclipse attacks, and consensus delay.

1. **Selfish Mining:** Here malicious miners try to increase rewards by privately holding their blocks. In this attack, the dishonest miners hide the data by keeping back a mined block and harm truthful miners by a twofold process: (a) Acquiring an unfair higher compensation than deserved, and (b) confusing others by eventually forcing them to deplete their resources. Selfish mining affects all applications since the miners are the only ones to add the blocks into the blockchain. Thus, addition of genuine blocks becomes difficult if miners are involved in withholding valid blocks and adding invalid blocks.
2. **The Majority or 51% Attack:** It is the most common security threat to the blockchain system. This is realized when a set of Sybil nodes, a mining pool or a single attacker, achieves most of the network's hash rate to control the blockchain. By gaining the maximum hash rate, these nodes can harm the network by:

    (a) Making blocks invalid by preventing transactions being verified,
    (b) Allowing double spending by reversing the transactions while they are under their control,
    (c) Splitting the network by forking the main blockchain, and
    (d) Preventing rest of the miners from finding any blocks for a brief time.

3. **DNS Attacks:** A new node joining the blockchain network discovers its peers identified by the IP addresses using the Domain Name System (DNS). The reply to a DNS query returns type A-records containing the addresses of available peers ready to connect to the blockchain. As the upcoming node connects with the peers, it sends its IP address and port number to establish connections with other peers. The DNS resolution process is prone to man in the middle attacks, cache poisoning, and stale records at the resolver side.
4. **BGP hijacks and spatial partitioning:** Most blockchain applications have two types of nodes: full nodes and lightweight nodes. Full nodes perform the relaying of blocks and transactions and maintenance of an updated copy of the blockchain. Whereas, the services of the full nodes are utilized by the lightweight nodes, in accessing the network and creating their picture of the blockchain. Therefore, exploiting a full node compromises all its associate lightweight nodes. Spatially concentrated nodes within an Autonomous system (AS) or an Internet Service Provider (ISP) are highly vulnerable to the routing attacks, like BGP hijacking. Here, an adversary AS hijacks the traffic to a destination AS hosting the blockchain application nodes. Thus, the information being sent to other nodes in the target AS is disrupted. If targeted nodes are among the miners, the

malicious node can reduce the rate of the blockchain hashing, hence, hitting the smooth network functioning.

5. **Eclipse Attacks:** In eclipse attack [28], a set of attacker nodes isolate its neighbor nodes by using their IP addresses and compromising the traffic. When, such malicious nodes surround the honest nodes, they become susceptible to the eclipse attack. Malicious nodes feed these nodes with fraudulent transactions and blocks resulting in them developing the wrong picture of blockchain state and becomes the part of the cluster having malicious nodes. When a trusted node connects to this cluster, it also propagates fake transactions and blocks, without knowledge.
6. **Distributed Denial of Service Attacks (DDoS):** In blockchain system, DDoS attacks [28] are realized in many ways as per the structure of the network, applications and participating peers. Out of total attacks, 51% attacks are denial of service attacks. Some miners gain access to immense hashing power and restrict other sets of miners from publishing new blocks. Also, the intentional forks can convert into hard forks, causing denial of service. Another possible attack, known as stress testing, limits the number of transactions per block processed by a blockchain application in a given time. Mempool flooding is also a DDoS attack that is performed at the cryptocurrency memory with an intention to shoot up the mining fee.
7. **Block Withholding Attacks:** In this attacker, a malicious node creates a contradictory view of the blockchain. With this, the attacker can forge, mask, or hold back the data that is required to be communicated. This type of attacks are also known as Finney Attacks. The Finney attack is a form of the double-spending attack where, a miner induces additional delays to double his share of cryptocurrency benefit for a transaction [40]. Fork after withholding (FAW) [19] is more rewarding than block withholding attacks, since, a malicious miner attaches itself to two mining pools and computes a valid Proof of Work in one mining pool. However, he holds back its solution and publishes this block after the second mining pool adds the block. One of the available block gets selected by the network and the dishonest miner is rewarded in any case.
8. **Consensus Delay:** Here, an attacker places the incorrect blocks in the blockchain and introduces the latency intercepting the peers from reaching consensus. The delays introduced in blockchain include authentication time delay, transmission delay and propagation delay in messages and block transmission. While the former category is dependent on the size of block, the latter one depends upon link bandwidth between the nodes. Thus, attackers can introduce intentional delays in the network by sending stale blocks and double spent transactions.
9. **Timejacking:** An attacker can alter a node's network timer by encompassing majority of peers and broadcasting the incorrect timestamps in the network. This speeds up the peers and cuts off the target out of the network without intervention of the legitimate nodes.

6.10.3 Application Based Attacks

In this section, we discuss the possible attacks on blockchain applications. Depending upon the type of application, blockchain networks suffer from various threats and a significant number of attacks exist which can be detailed as:

1. **Blockchain Ingestion and Anonymity:** Public blockchains are usually less anonymous, and provide easier data access for the public. Thus, analyzing the public blockchain can let out sensitive information to an opponent, which is commonly called as blockchain ingestion. Ingestion of blockchain is usually not useful to underlying application and the users. Anonymity in cryptocurrencies also gives worthwhile chances to attackers to carry out fraudulent activities. The tamper-proof, append only and decentralized nature of blockchain where a committed transaction is reversible, results in several irreparable scam activities online. Here, the users are misled to send money through wallets and the lack of central authority further makes it difficult to notify the fraud and look forward to payment reversal.
2. **Double Spending:** In this attack, the same digital currency is spent more than once. That is, it is an instance in which two transactions use the same input and one of them has already been broadcast on the network. A malicious user can realize double spending by sending two contradicting transactions in prompt succession. Hence, same bitcoins are spent redundantly in multiple transactions. These two transactions are carried out in a very close time interval. For example, an unfair customer can transact at time t1 using some bitcoins with a recipient address of a producer. The client publishes this transaction information at time t2 considering that t1 and t2 are close enough. It makes another redundant transaction at t2 with same set of coins and recipient address as his own address or the address of the wallet under his control. Hence, if the unfair customer is able to dupe the merchant with this transaction, the purchased products are delivered to him, with merchant never receiving the payment.
3. **Cryptojacking:** In cryptojacking or covert mining, cloud and web-based services are exploited to illegitimately perform the Proof of Work (PoW) consent [41]. The in-browser cryptojacking, websites are turned into the mining pools which further enhance their hashing capacity by collaborating with other miners and buying high-performing hardware with ample resources. Consequently the mining process becomes more expensive and competitive, preventing small scale miners from mining the blocks singly.

   In cloud-based cryptojacking, malicious miners attack the system by restricting the working of trusted nodes and virtual machines for mining and thus, exhausting cloud resources. Web-based cryptojacking is performed by attackers by injecting harmful JavaScript code in websites. In browser-based cryptojacking, browser on the client machine executes a JavaScript code which creates a Web Socket with a remote system. The server then asks client to compute hashes for the PoW and transmitting them back. This computation task requires a lot of resource, resulting in excessive usage of CPU and battery exhaustion.

4. **Wallet Theft:** Digital wallets store credentials of the nodes present in the blockchain network. In many blockchains, the wallet is stored non-encoded, thereby, permitting an opponent to access the credentials linked with it and also type of transactions issued. Although a wallet is secured, instigating an attack on the host allows the opponent to perform wallet theft. Moreover, there are various third party services that enable storage of wallets, compromising them can leak the wallets to an adversary.
5. **Replay attacks:** It involves implementing same transaction on two separate blockchains [41]. For example, when forking occurs in a blockchain, it separates into two different chains and the users hold equal resources on both the ledgers. It can make a transaction on either of the available chains. Here, the attacker gains access to the transaction information available in one ledger and repeats the same transactions on other, thereby making a user lose its assets on both the chains.

6.11 Blockchain Prospects for Internet of Things (IoT)

The Internet of Things (IoT) interconnects people, products, and places and provides freedom for value creation. Several micro level sensing elements, chips, and actuators are inserted into physical items, each sending data to the IoT network, which is analyzed to convert insights into action. However, there are several technical and security issues in IoT systems that need to be addressed. Security concerns in IoT systems have shadowed its mass deployment. IoT devices usually suffer from security threats which makes them an easy victim of Distributed Denial of Service (DDoS) attacks. Scalability is yet another key issue with IoT networks. With the exponentially growing number of devices communicating via an IoT network, centralized systems used currently to connect, authenticate, and authorize different nodes will turn into a processing bottleneck.

Blockchain with its underlying Distributed Ledger Technology (DLT) is capable to address these challenges in the following ways [12]:

- Tamper-proof nature of the distributed ledger in a blockchain system alleviates the need to establish trust among the users. Moreover, no standalone entity shall control the enormous volume of data created by IoT devices.
- Storing IoT data over blockchain shall add an extra layer of security for the adversaries to bypass before accessing it. Since, blockchain provides a robust level of encryption deleting or modifying existing data records is nearly impossible.
- Blockchain is transparent as it allows only authorized users to gain access to the network and enable them to trace the transactions that occurred in the past. It also aids in identification of the source of any data leakages.
- Blockchain can facilitate the speedy processing of transactions and synchronizing billions of connected devices. With the increase in number of such devices, the DLT technology provides a feasible solution for processing large number of transactions efficiently.

- Enabling the trust among users, blockchain can help IoT based enterprises to reduce their costs by reducing the processing and communication overheads associated with IoT gateways.

Thus, the future blockchain enterprise areas where IoT can be combined with blockchain can be outlined as in [2, 17, 42], supply chain management and logistics and transportation, automated industry, smart homes, cities and agriculture, sharing economy and pharmacy industry.

6.12 Conclusion

With the progressing computer technologies, several things that seemed impossible have been realize, the prominent ones being the contactless operations, cashless payments, e-commerce, and cryptocurrencies. Secured online payments with no added fee have influenced all the economic industries and have been made possible due to blockchain solutions. Blockchain is expected to prove its worth in modifying traditional industry owing to its main characteristics like decentralization, persistence, anonymousness, and verifiability. Thus, the blockchain technology needs to be introduced with an objective to bring operational efficiencies and the proper implementation of the blockchain technology is expected to have wider and more progressive implications.

References

1. A.B. Ayed, M.A. Belhajji, The blockchain technology: applications and threats, in *Securing the Internet of Things: Concepts, Methodologies, Tools, and Applications* (IGI Global, New York, 2020), pp. 1770–1781
2. A. Banerjee, Blockchain with IoT: applications and use cases for a new paradigm of supply chain driving efficiency and cost, in *Advances in Computers*, vol. 115 (Elsevier, Amsterdam, 2019), pp. 259–292
3. I. Bashir, *Mastering Blockchain: Distributed Ledger Technology, Decentralization, and Smart Contracts Explained* (Packt Publishing Ltd, Birmingham, 2018)
4. U. Bodkhe, S. Tanwar, K. Parekh, P. Khanpara, S. Tyagi, N. Kumar, M. Alazab, Blockchain for industry 4.0: a comprehensive review. IEEE Access **8**, 79764–79800 (2020)
5. R.G. Brown, The Corda platform: an introduction. Retrieved **27**, 2018 (2018)
6. J. Buck, Blockchain oracles, explained. Cointelegraph, October, vol. 18 (2017)
7. V. Buterin et al. *A Next-Generation Smart Contract and Decentralized Application Platform.* White Paper **3**(37) (2014)
8. V. Buterin, A. Todd, G. Nguyen, A. Rosic, P. Westerhof, J. Beranger, A. Guerra, C. Mulder, M. Urling, S. Andonov et al. *What are Smart Contracts? A Beginner's Guide to Smart Contracts* (2016)
9. C. Cachin et al. Architecture of the hyperledger blockchain fabric, in *Workshop on Distributed Cryptocurrencies and Consensus Ledgers*, vol. 310 (2016)

10. W. Chen, Z. Xu, S. Shi, Y. Zhao, J. Zhao, A survey of blockchain applications in different domains, in *Proceedings of the 2018 International Conference on Blockchain Technology and Application* (2018), pp. 17–21
11. A.S. Chhabra, T. Choudhury, A.V. Srivastava, A. Aggarwal, Prediction for big data and IoT in 2017, in *Proceedings of the 2017 International Conference on Infocom Technologies and Unmanned Systems (Trends and Future Directions)(ICTUS)* (IEEE, New York, 2017), pp. 181–187
12. H.N. Dai, Z. Zheng, Y. Zhang, Blockchain for internet of things: a survey. IEEE Internet Things J. **6**(5), 8076–8094 (2019)
13. V. Fore, A. Khanna, R. Tomar, A. Mishra, Intelligent supply chain management system, in *Proceedings of the 2016 International Conference on Advances in Computing and Communication Engineering (ICACCE)* (IEEE, New York, 2016), pp. 296–302
14. W. Gao, W.G. Hatcher, W. Yu, A survey of blockchain: techniques, applications, and challenges, in *Proceedings of the 2018 27th International Conference on Computer Communication and Networks (ICCCN)* (IEEE, New York, 2018), pp. 1–11
15. G. Garg, S. Sharma, T. Choudhury, P. Kumar, Crop productivity based on IoT, in *Proceedings of the 2017 International Conference On Smart Technologies For Smart Nation (SmartTechCon)* (IEEE, New York, 2017), pp. 223–226
16. S. Kadam, Review of distributed ledgers: the technological advances behind cryptocurrency, in *Proceedings of the International Conference Advances in Computer Technology and Management (ICACTM)* (2018)
17. M.A. Khan, K. Salah, IoT security: review, blockchain solutions, and open challenges. Future Gener. Comput. Syst. **82**, 395–411 (2018)
18. A. Khanna, A. Sah, T. Choudhury, Intelligent mobile edge computing: a deep learning based approach, in *Proceedings of the International Conference on Advances in Computing and Data Sciences* (Springer, Berlin, 2020), pp. 107–116
19. Y. Kwon, D. Kim, Y. Son, E. Vasserman, Y. Kim, Be selfish and avoid dilemmas: fork after withholding (FAW) attacks on bitcoin, in *Proceedings of the 2017 ACM SIGSAC Conference on Computer and Communications Security* (2017), pp. 195–209
20. X. Li, P. Jiang, T. Chen, X. Luo, Q. Wen, A survey on the security of blockchain systems. Future Gener. Comput. Syst. **107**, 841–853 (2020)
21. M.H. Miraz, M. Ali, Applications of blockchain technology beyond cryptocurrency (2018). arXiv preprint arXiv:1801.03528
22. A.A. Monrat, O. Schelen, K. Andersson, A survey of blockchain from the perspectives of applications, challenges, and opportunities. IEEE Access **7**, 117134–117151 (2019)
23. S. Nakamoto, Bitcoin: A Peer-to-peer Electronic Cash System. Technical report, Manubot (2019)
24. M. Pilkington, Blockchain technology: principles and applications, in *Research Handbook on Digital Transformations* (Edward Elgar Publishing, Cheltenham, 2016)
25. M. Saad, J. Spaulding, L. Njilla, C. Kamhoua, S. Shetty, D. Nyang, A. Mohaisen, Exploring the attack surface of blockchain: a systematic overview (2019). arXiv preprint arXiv:1904.03487
26. K. Sultan, U. Ruhi, R. Lakhani, Conceptualizing blockchains: characteristics and applications (2018). arXiv preprint arXiv:1806.03693
27. F. Tschorsch, B. Scheuermann, Bitcoin and beyond: a technical survey on decentralized digital currencies. IEEE Commun. Surv. Tutorials **18**(3), 2084–2123 (2016)
28. M. Vasek, M. Thornton, T. Moore, Empirical analysis of denial-of-service attacks in the bitcoin ecosystem, in *International Conference on Financial Cryptography and Data Security* (Springer, Berlin, 2014), pp. 57–71
29. A. Verma, A. Khanna, A. Agrawal, A. Darwish, A.E. Hassanien, Security and privacy in smart city applications and services: opportunities and challenges, in *Cybersecurity and Secure Information Systems* (Springer, Berlin, 2019), pp. 1–15
30. L. Wang, X. Shen, J. Li, J. Shao, Y. Yang, Cryptographic primitives in blockchains. J. Network Comput. Appl. **127**, 43–58 (2019)

31. J. Xie, H. Tang, T. Huang, F.R. Yu, R. Xie, J. Liu, Y. Liu, A survey of blockchain technology applied to smart cities: research issues and challenges. IEEE Commun. Surv. Tutorials **21**(3), 2794–2830 (2019)
32. D. Yaga, P. Mell, N. Roby, K. Scarfone, Blockchain technology overview (2019). arXiv preprint arXiv:1906.11078
33. Z. Zheng, S. Xie, H.N. Dai, X. Chen, H. Wang, Blockchain challenges and opportunities: a survey. Int. J. Web Grid Serv. **14**(4), 352–375 (2018)
34. A. Zohar, Bitcoin: under the hood. Commun. ACM **58**(9), 104–113 (2015)
35. https://en.bitcoinwiki.org/wiki/SHA-256
36. https://paybis.com/blog/what-is-a-blockchain-nonce/
37. https://www.tutorialspoint.com/what-is-a-nonce-in-block-chain
38. https://www.tutorialspoint.com/difference-between-private-key-and-public-key
39. https://ethereum.stackexchange.com/questions/760/how-is-the-address-of-an-ethereum-contract-computed
40. https://bitcoin.stackexchange.com/questions/4942/what-is-a-finney-attack
41. https://hackerbits.com/programming/what-is-cryptojacking/
42. https://www2.deloitte.com/content/dam/Deloitte/de/Documents/Innovation/IoT-powered-by-Blockchain-Deloitte.pdf

Chapter 7
IoT-Integrated Blockchain in the Drug Supply Chain

Rehab A. Rayan and Muhammad Asim Masoom Zubair

7.1 Introduction

Counterfeiting pharmaceuticals in the drug supply chain is a critical public health issue worldwide. Pharmaceutical products are vital elements in healthcare to prevent and treat diseases, hence saving lives. Lately, counterfeiting pharmaceuticals has grown evidently leading to poor healthcare delivery and more deaths yearly [2]. The World Health Organization (WHO) describes counterfeit pharmaceuticals as falsified or substandard drug products. Falsified drug products have illegal or/and intentionally misrepresented identity, contents, or origin, while substandard drug products are allowed products, which could not fit the quality specification standards [23].

The Organization for Economic Cooperation and Development (OECD) considers trading counterfeit pharmaceuticals among the highly profitable markets for illicitly traded goods, valued yearly by billions of dollars. Professionals forecast that the rate of trading counterfeit pharmaceuticals and hence its revenues would double the legal ones. Meanwhile, drug companies are afflicting the results of falling income worth hundreds of billions of dollars with the added prices of adopting the costly safeguarding drug supply chain technologies, and hence, lower resources are there for research and development of novel drugs [1, 3].

Universally, counterfeit pharmaceuticals get in the market via the drug supply chain, which grows more challenging for the globalization of manufacturing

R. A. Rayan (✉)
Department of Epidemiology, High Institute of Public Health, Alexandria University, Alexandria, Egypt
e-mail: rayanr@alexu.edu.eg

M. A. M. Zubair
Department of Pharmacy, The Islamia University, Bahawalpur, Pakistan

T. Choudhury et al. (eds.), *Blockchain Applications in IoT Ecosystem*,
EAI/Springer Innovations in Communication and Computing,
https://doi.org/10.1007/978-3-030-65691-1_7

processes where the pharmaceutical products are usually distributed in other countries than those who produce active substituents. Missing transparency in the drug supply chain cycle lets illicit players interfere with corruption or replace pharmaceuticals across the drug supply chain, compromising medications prior to reaching the patients. In the United States, most medications are produced with active substituents from either China or India. China, followed by India, is the major country for manufacturing counterfeit pharmaceuticals where millions of counterfeit antibiotic tablets were confiscated in 2009 [10]. A famous incidence is the made-in-China impure heparin that was distributed in 11 countries and caused the death of 81 individuals in the United States in 2008 [15].

Securing the drug supply chain should involve monitoring pharmaceuticals from the laboratory to the patient. The Falsified Medicines Directive (FMD), a track and trace project adopted by the European Union, encodes all pharmaceuticals via an automated drug supply chain in the pharmaceutical companies. Internet of Things (IoT) techniques like the blockchain could trace and remember items in the drug supply chain minimizing both interruptions like thieving and counterfeiting pharmaceuticals. Adopting the IoT-integrated blockchain to trace pharmaceuticals across the drug supply chain from the active constituents' vendors to the producer to the patient would enhance managing pharmaceuticals. This chapter examines the drug supply chain and its limitations. It then explores the IoT-integrated blockchain framework in the pharmaceutical industry [15].

7.2 The Drug Supply Chain and Counterfeiting Pharmaceuticals

A drug supply chain is linking between an agency and the different vendors for properly manufacturing and distributing products to end users. In the pharmaceutical industry, managing the drug supply chain could enable the agencies to well use properties and investments for both making profits and meeting the end users' needs. Smoothly managing the drug supply chain could genuinely affect and prosper the entire approach for a business involving less complicated processes such as selecting vendors, storage, and distribution and precise statistics. In the drug supply chain, counterfeit pharmaceuticals are a significant problem endangering patients and costing the pharmaceutical industry vast sums of money. Counterfeit pharmaceuticals are not marketed under the original trade name and violate the patents of pharmaceutical companies [22, 23].

There are several gates where counterfeit pharmaceuticals can be introduced into the drug supply chain. Gate-1 is where the vendors might provide impure, expired, poor-quality, or different raw materials for manufacturing pharmaceuticals, hence compromising the quality of the manufactured products and the safety of the consumers. Gate-2 is where most of the counterfeit pharmaceuticals or active constituents are introduced into the drug supply chain via the manufacturer. Counterfeiting a pharmaceutical product could involve using improper constituents, dose, or labeling and even marketing a placebo rather than the authentic drug. A counterfeit pharmaceutical products' manufacturers develop products that resemble the authentic ones but with poor quality.

Gate-3 is where the manufacturers of the counterfeit pharmaceuticals target marketing their products, hence selling them directly to the distributor via fake networks. Usually, this process could range from simply shipping fake pharmaceuticals to applying very sophisticated trailing techniques for breaching current legal networks with poor security [20]. Gate-4 is where the pharmacies are the core of the whole issue. It might happen to receive a counterfeit pharmaceutical while buying a certain brand medication from a local or even a big pharmacy. Lately, more individuals are interested in online purchasing of medications for less money. Typically, such medications come with diverse labels from various places and a promise of the exact drug while it might not. In places with poor laws, online pharmacies could market counterfeit or fake pharmaceuticals to gain more for lower costs.

7.3 Blockchain Technology

In this digital era, applying technology, such as blockchain, to the drug industry could enhance all procedures enabling better services. Blockchain technology is a sort of distributed ledger (DLT), which keeps a changeless record of all data on transactions and devoid manipulations. Unlike conventional storage techniques, DLT applies nodes and separate computers to record, share, and synchronize the data of a transaction. Afterward, blockchain cluster data in blocks linked together in a chain. This framework could be used in the distributed setting of the drug supply chain where vendors, healthcare facilities, pharmacies, and patients produce separate but yet to be coordinated data.

Exposing the data on transactions in the markets might lead to leakage and manipulations; hence data should be housed securely in repositories. Blockchain ensures a confidential environment and protects anonymous public data via decentralization applying encoding asymmetrically and algorithms for hashing [18]. The algorithms of hashing map differently lengthened input data into similarly constant length output hashes uniquely representing their blocks, thus preventing any backward mapping targeting retrieving the original message's data and allowing approvals from all entities.

Modifications in the block's source data yield an output of diverse hash, hence interrupting the chain. Following the hashing procedure, every block would have a unique digital signature, and the chain is established based on the prior block signature hence an anonymous signature for the upcoming block guaranteeing all

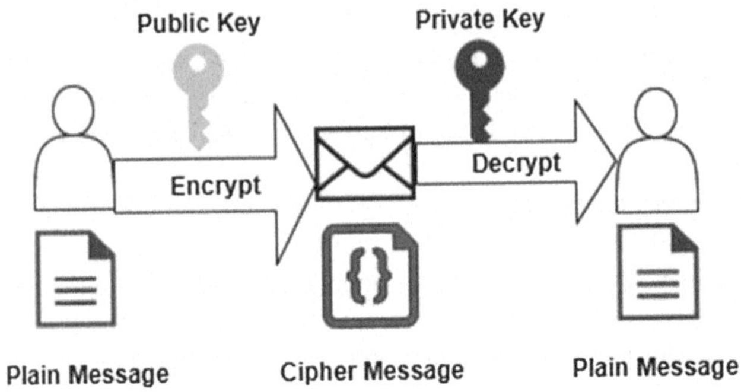

Fig. 7.1 Asymmetric encryption

transactions executed via the proper entity. Such a hashing algorithm would prevent manipulating the blockchain, smoothly locating a hashed record, and creating random threads eliminating housed data redundancy in the database.

Added to the hashing encryption, another secure technique for authentication and confidentiality is called asymmetric encryption or private key cryptography where every blockchain party has a public and private key. The transactions' encoding keys differ from the decoding ones where the public key is accessed by all parties, while the private one is only accessed by the party generating it for encoding contents to be then decoded by the other party through its private key. Figure 7.1 shows that one entity uses the public key to send a specific content, which gets encrypted and then decrypted by the other entity's private key to read it.

In the blockchain, decentralization, a transparent medium, enables exchanging and recording the data; therefore, entities looking up records in such a credible distributed system could find solid and transparent data on transactions [11]. Hence, securing explicitly open and trustworthy repositories that are needed for the drug supply chain in the pharmaceutical business worldwide where the required data is smoothly accessed and tracked by all involved entities is shown in Fig. 7.2 [17].

The transparency offers readily available data on the source of elements and aspects of the procedure, with different levels of accessibility, to all entities [7]. Hence, the social accountability of all entities for their activities affecting the behavior of consumers and allowing competing parties to keep their credibility. Yearly, the drug supply chain is challenged progressively by counterfeit pharmaceuticals, deficiency, or cancellation of drug products that adversely affect patients and healthcare systems; therefore regulating entities are moving ahead to protect the drug supply chain [5]. Adopting blockchain has evidently optimized transparency for achieving operational objectives regarding the source of raw materials and reframed end services or products hence, the more supply chain transparency, producing changeless, persistent, and distributed records to enable monitoring [7].

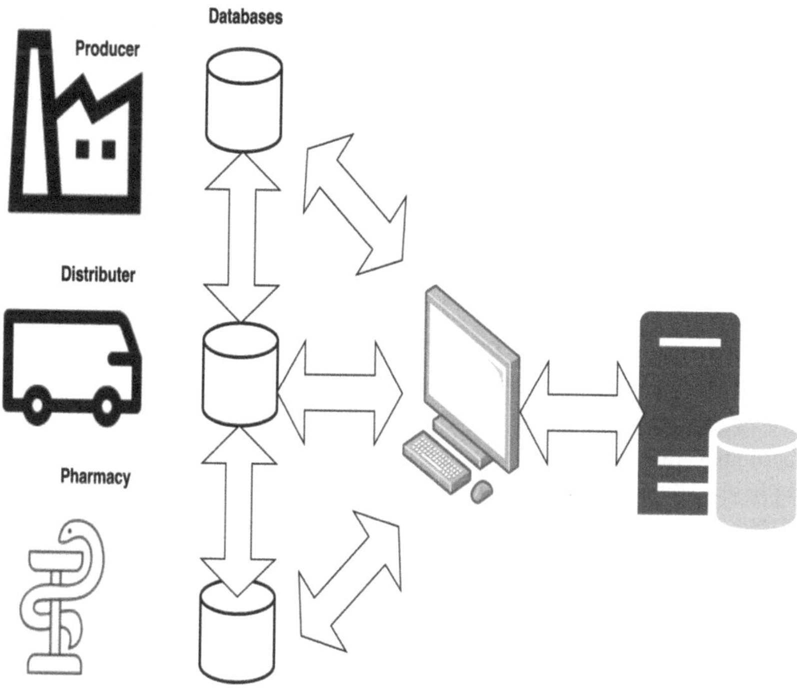

Fig. 7.2 Centralized system

7.4 The Applications of Integrating IoT with the Blockchain

The IoT has an influencing role in the logistics department of the pharmaceutical industry and presented many opportunities. IoT integration in the blockchain has improved the logistic operations in a big way [12]. The IoT brings ease for supply chain operations and reduces the risks that can cause a long-lasting disaster. Some applications of the IoT in the drug supply chain are the following:

7.4.1 Overcoming the Shortage of Medicines

The employment of the IoT from the digital technologies can ensure a synchronized supply of medicine resulting in patients getting the medicines on time. The IoT can help manage the best inventory based on company-defined business rules so that it can make planned decisions about manufacturing and assist in sending related products to the market [12].

7.4.2 Information Transmission Network in the Drug Supply Chain

Due to improper drug handling and counterfeiting issues, manufacturers and governments are facing the challenge of being consistent with regulations and ensuring the safety of customers' lives. IoT devices and solutions can be used to record drug information throughout the supply chain [12].

7.4.3 Enhance Supply Chain Security

Safety of the supply chain is improved with the help of the IoT in a process of facilitating two-way communication between information seekers and devices, tracking inventory about the location and current status of packaging. Supply chain security can be safeguarded by RFID format labels, two-dimensional barcodes, and smart labels for packaging. The IoT can easily track the movement of drug inventory at each checkpoint [25]. It saves stakeholders in the supply chain from losing millions of dollars and at the same time ensures that while delivering real products to customers, it can also stop counterfeiters.

7.4.4 Freight Tracking

According to the report of Freight Watch International Supply Chain Intelligence Center, pharmaceutical companies suffered heavy load losses due to theft of goods. On the other hand, due to the use of intelligent freight tracking and data collection methods, the average loss value of the US pharmaceutical industry has dropped by about 55% from 2010 to 2012. This clearly shows the importance of tracking the goods just in case [8].

7.4.5 Temperature Control

In addition to managing temperature peaks during transportation, the quality of medicines should also be ensured. The temperature of the drugs being transported can be tracked to ensure that they remain within the specified stability range. If the drug storage conditions deviate from the specified temperature range, an environmental sensor will be embedded, and a programmed sensor will be used to sound an alarm. In addition to reducing drug waste and ensuring effectiveness, these are the key solutions that guarantee acquiescence with international and regional laws and regulations [19].

7.5 The Proposed IoT-Integrated Blockchain Framework

Today, the growing counterfeit pharmaceuticals have turned problematic since black marketplaces are supplying likewise medications with no monitoring for quality. For monitoring all undesirable acts in the drug supply chain until arriving the consumer, an IoT-integrated blockchain framework should be adopted for tracking purity, delivery, fraud, and making data about contaminated and counterfeit pharmaceuticals available for verifying the source and originality of the provided medications [5].

Following exporting a medication, all the chemical and technical data are warehoused on the producer system; meanwhile, the IoT links between the received medications all along until reaching the consumer. Yet, the unavoidable and undesirable acts of fraud happen but also detectable. Figure 7.3 shows a radio-frequency identification (RFID)-based drug supply chain that monitors the system for securing the chain and ensuring the delivery of quality healthcare services. The blockchain-based IoT framework would secure the drug supply chain against the yearly ever encountered network breaches worldwide that threaten data about both companies and patients [14] where it benefits from the vital blockchain feature in securing transactions that, upon verification, are not liable for falsification.

Figure 7.4 shows connected parties via a decentralized network where all are contacting the verified and approved medications' transactions. When a party executes an activity, all the other parties get alerted and could assess verification via an authorized algorithm. Once a new block is verified, it would be added to the chain, hence minimizing sharing files where all transactions and verifications are substitute evidence of conventional paperwork.

Counterfeit pharmaceuticals could imply wasting money on products that might contain allergenic impurities, hence threatening health and causing an economic load. Discovering counterfeit products depends on validating data by all bodies

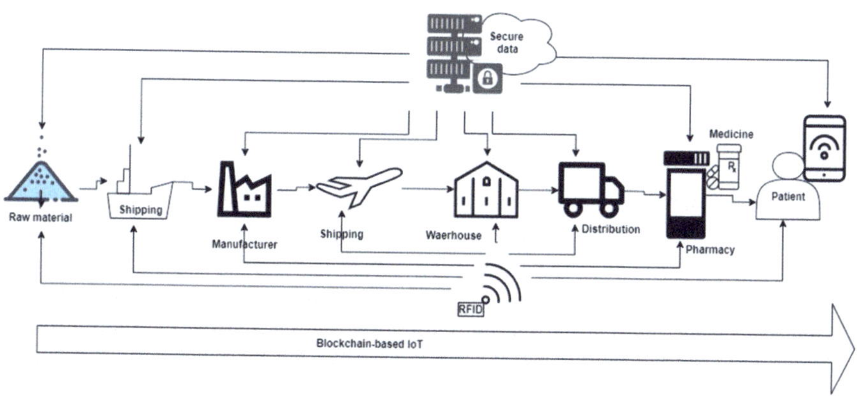

Fig. 7.3 Blockchain-based IoT drug supply chain

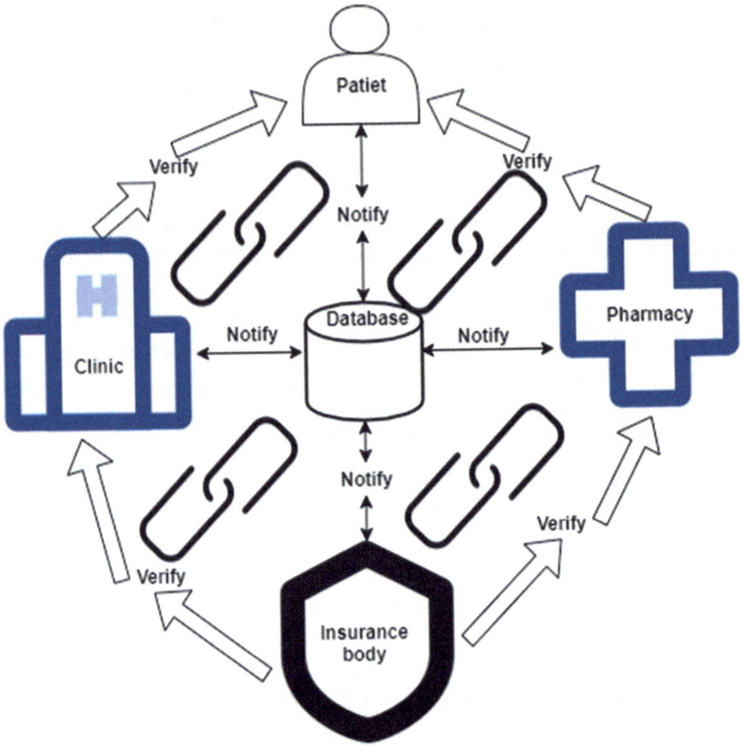

Fig. 7.4 Decentralized blockchain framework

at the drug supply chain. Adopting blockchain coupled with integration with IoT devices would facilitate management via tracking, reporting ownership, and intelligent contract rewarding, which are promising outputs eliminating fake data and counterfeit products.

7.6 The Case of Modum.io

The IoT technology is being used in many sectors including the transmission of real-time information and the monitoring of assets throughout the chain [24]. Also, other technologies integrated with the IoT (such as blockchain) can advance the security and reliability of these resources. There are many start-ups in which the IoT is integrated into blockchain technology for supply chain networks. Among several start-ups, Modum.io is a start-up that uses IoT sensor device blockchain technology to accentuate data invariability and make "temperature records" accessible for the public. Modum results in a decrease in operational costs in the pharmaceutical industry. This section attempts to explain how these technology combinations work better. We will also discuss the case of Modum, which mainly works in this field.

Modum is a use case that illustrates the blockchain and the IoT in the same application, which was established in cooperation with the University of Zurich (UZH). These companies seek to establish a drug sharing network integrated with blockchain technology. The company's goal is to integrate the concept of the IoT to monitor changes in the status of drugs [9]. The technology checks whether certain transport standards are met to ensure that the quality is maintained until the goods reach their destination. Modum.io AG works closely with the UZH to jointly develop sensor equipment and its related software program. The planning of this project began in 2015, leading to the development phase in 2016, and was established in July 2016.

Modum.io monitors transportation of medicinal products and gathers all the necessary data by integrating IoT sensors with blockchain technology [13]. The integrity of data is thus guaranteed by the use of this technology, thus making it impossible to change the records. After delivery of the supply, a smart contract is initiated that guarantees compliance with the temperature. The data can be verified by any party once it enters the blockchain system and thus remains unchanged. The data/results are reported to the recipient and distributor and can be accessed publicly. It is foreseeable that customers can also check the temperature in the future. However, the serial number of each pharmaceutical package must be executed first.

The architecture of Modum.io AG is designed as front-end, back-end, and IoT-integrated sensor devices. The temperature records that are registered/stored in the front-end are verified by the Ethereum network of blockchain. Smart contracts are regulated by the use of Ethereum Virtual Machine, and data can be verified through smart contracts [16]. To meet the compliance with temperature data, for every new shipment, a smart contract is issued that maintains this responsibility. To store the raw temperature data and credentials of the user, an interactive database is used. Linkage of blockchain networks and the front-end users is established via a server. This linkage can modify or create a smart contract and the database stores the data. The new shipments can be registered through mobile devices, and the temperature data is sent to the server in a trackable manner. A thermal device (sensor), compatible with Bluetooth technology, is configured to send data to the mobile device at a fixed polling interval. A Bluetooth-integrated thermal device (sensor) sends the data at defined intervals to the mobile device.

SensorTag4's IoT sensor provides the temperature data and can be placed at a critical location for consignment. The sensor has the functions of identification and can communicate a particular temperature at specific points. In addition, it requires a low energy Bluetooth connection, which is nowadays available in most of the mobile devices. Modum.io AG has established a prototype, and the first pilot project is accomplished with a pharmaceutical distributor. From July to August, they conducted a pilot project to deliver medical products from a trader to a medical retailer every week. 55 batches in total were sent. 29 batches were selected by the retailer to 1 site, 21 batches to another setting, and 05 batches of goods were sent within [4].

7.7 Discussions

The presence of counterfeit drugs in the global supply chain is causing a public health crisis. Pharmaceutical drugs are important components of the healthcare setup, bear crucial importance in diagnosing and curing diseases, and are a valuable asset in saving lives. However, counterfeit drugs are increasing over the years, resulting in deaths. According to estimates, there are millions of deaths because of counterfeit drugs ending the efficacy of healthcare. Complex and insufficient drug supply chain leads to the worst scenario of counterfeit drugs According to experts, it is estimated that the counterfeit pharmaceutical drug trade is growing double as compared to legitimate pharmaceuticals. However, it is still difficult to estimate the economic loss due to counterfeit drugs around the world.

Many traditional technologies have been suggested for the tracking and tracing of medicines, such as the barcode scanning system, RFID, mobile technology, and others. Still, counterfeit drug debacles happen on a global scale. The best way of preventing counterfeit drugs in the supply chain is the use of blockchain technology. It safeguards an absolute chain of transaction ledger, tracking each step of the drug supply chain.

Many serious issues are encountered by the pharmaceutical supply chain because of drug shortage and counterfeit drugs. Law enforcement and related agencies are concerned with the increased number of cases every year resulting in a burden on the public health system. To cover all the issues in the drug supply chain, an IoT-integrated blockchain system should be implemented to trace and keep track of transportation, contamination, and theft. Besides, the IoT-integrated system will provide information regarding falsified and contaminated drugs to authenticate the source and validity of the supplied medicines.

Nowadays, there are high regulations in the distribution of medicinal products for human use. The transfer of these products from the manufacturer to those who will use requires the utmost importance and involves various mediators. Now the supply chains are responsible to report any deviations such as temperature to the distributor and also to the recipient of the medicinal product, and the temperature of every parcel should also be monitored at all times. Thus, pharmaceutical companies are enforced to order special services from the logistics department. Blockchain technology provides a solution to the decentralized process in which data of the medicinal products can be stored and accessed during the logistic process by both parties ensured through a smart contract.

7.8 Insights for the Future

With the blockchain traceability and transparency features, the hash of the transferred pharmaceutical product, the physical asset, could validate the drug supply chain at all phases [21]. In the temperature-monitored drug supply chain, the so-called cold chain, several pharmaceuticals are highly affected by temperature and need extreme care during transport. The IoT-integrated blockchain could be applied in managing the cold chain where a sensor for heat could be included in each pharmaceutical batch for ongoing monitoring of the temperature. If the temperature highly differs from the limit, the pharmaceutical batch should be terminated in the blockchain, where it could no longer be marketed to anybody in the drug supply chain and should be wasted [6].

7.9 Conclusions

This chapter explored the IoT-integrated blockchain technique for managing the drug supply chain where it emphasized implementing the blockchain to minimize counterfeit pharmaceuticals in the drug supply chain. First, it described key entities in the drug supply chain, then highlighted the issues in the pharmaceutical industry, and identified the likely breaches in the drug supply chain facilitating leakage of counterfeit pharmaceuticals. However, regarding the richness in research covering the problem of the counterfeit pharmaceuticals in the drug supply chain, the technological aspects were yet inadequately covered. Next, it proposed the technical solution involving blockchain, RFID, and the IoT to enhance monitoring the pharmaceutical products traveling across the drug supply chain. Consequently, there are several advantages of adopting the IoT-integrated blockchain, which could address the issue of counterfeit pharmaceuticals in the drug supply chain along with other domains in the pharmaceutical industry including integrity and controlling misusing and fraud.

References

1. K. Acri, *They Cost Us Billions and They Can Kill: Counterfeit Drugs Are Invading Canada | Financial Post* (2 March 2018). https://business.financialpost.com/opinion/they-cost-us-billions-and-they-can-kill-counterfeit-drugs-are-invading-canada
2. V. Ahmadi, S. Benjelloun, M. El Kik, T. Sharma, H. Chi, W. Zhou, Drug governance: IoT-based blockchain implementation in the pharmaceutical supply chain, in *2020 Sixth International Conference on Mobile and Secure Services (MobiSecServ)*, (IEEE, Piscataway, 2020), pp. 1–8. https://doi.org/10.1109/MobiSecServ48690.2020.9042950
3. E.A. Blackstone, J.P. Fuhr, S. Pociask, The health and economic effects of counterfeit drugs. Am. Health Drug Benefits **7**(4), 216–224 (2014) https://www.ncbi.nlm.nih.gov/pmc/articles/PMC4105729/
4. T. Bocek, B.B. Rodrigues, T. Strasser, B. Stiller, Blockchains everywhere – A use-case of blockchains in the pharma supply-chain, in *2017 IFIP/IEEE Symposium on Integrated Network and Service Management (IM)*, vol. 2017, (IEEE, Piscataway), pp. 772–777. https://doi.org/10.23919/INM.2017.7987376
5. J.B. Choi, J. Rogers, E.C. Jones, The impact of a shared pharmaceutical supply chain model on counterfeit drugs, diverted drugs, and drug shortages, in *2015 Portland International Conference on Management of Engineering and Technology (PICMET)*, (IEEE, Piscataway, 2015), pp. 1879–1889. https://doi.org/10.1109/PICMET.2015.7273165

6. F. Tian, A supply chain traceability system for food safety based on HACCP, blockchain internet of things, in *2017 International Conference on Service Systems Service Management*, (IEEE, Piscataway, 2017), pp. 1–6. https://doi.org/10.1109/ICSSSM.2017.7996119
7. K. Francisco, D. Swanson, The supply chain has no clothes: Technology adoption of blockchain for supply chain transparency. Logistics **2**(1), 2 (2018). https://doi.org/10.3390/logistics2010002
8. A. Jablokow, How Will IoT Revolutionize Pharmaceutical Manufacturing? *IoT For All* (17 April 2020). https://www.iotforall.com/drugs-continuous-manufacturing-and-iot/
9. N. Kshetri, 1 Blockchain's roles in meeting key supply chain management objectives. Int. J. Inf. Manag. **39**, 80–89 (2018). https://doi.org/10.1016/j.ijinfomgt.2017.12.005
10. H. Lock, B. Team, Fight the Fakes: How to Beat the $200bn Medicine Counterfeiters. *The Mail & Guardian* (10 June 2019). https://mg.co.za/article/2019-06-10-00-fake-medicine-makers-blockchain-artificial-intelligence/
11. Y. Lu, The blockchain: State-of-the-art and research challenges. J. Ind. Inf. Integr. **15**, 80–90 (2019). https://doi.org/10.1016/j.jii.2019.04.002
12. S. Mishra, A. Dash, B.K. Mishra, Chapter 9—An insight of Internet of Things applications in pharmaceutical domain, in *Emergence of Pharmaceutical Industry Growth with Industrial IoT Approach*, ed. by V. E. Balas, V. K. Solanki, R. Kumar, (Academic, London, 2020), pp. 245–273. https://doi.org/10.1016/B978-0-12-819593-2.00009-1
13. S. Norton, CIO Explainer: What Is Blockchain? *WSJ* (2 February 2016). https://blogs.wsj.com/cio/2016/02/02/cio-explainer-what-is-blockchain/
14. K.V.O. Rabah, Challenges & opportunities for blockchain powered healthcare systems: A review. Mara. Res. J. Med. Health Sci. **1**(1), 45–52 (2017)
15. V. Rees, The impact of counterfeit drugs in south and south-east Asia. Eur. Pharm. Rev. (2019). https://www.europeanpharmaceuticalreview.com/article/92194/the-impact-of-counterfeit-drugs-in-south-and-south-east-asia/
16. M. Sahu, 10 Best Tools for Ethereum Development Every Blockchain Developer Should Know About. *UpGrad Blog* (24 March 2020). https://www.upgrad.com/blog/tools-for-ethereum-development/
17. C.G. Schmidt, S.M. Wagner, Blockchain and supply chain relations: A transaction cost theory perspective. J. Purch. Supply Manag. **25**(4), 100552 (2019). https://doi.org/10.1016/j.pursup.2019.100552
18. T. Scott, A. Post, J. Quick, S. Rafiqi, Evaluating feasibility of blockchain application for DSCSA compliance. SMU Data Sci. Rev. **1**(2), 4 (2018) https://scholar.smu.edu/datasciencereview/vol1/iss2/4
19. U. Sharma, Transforming Pharma Logistics with IOT. *Express Pharma* (1 November 2018). https://www.expresspharma.in/pharma-logistics/transforming-pharma-logistics-with-iot/
20. M. Snyder, *Keeping Counterfeit Medicines Out of the Supply Chain* (Pharmaceutical Processing World, 28 January 2016). https://www.pharmaceuticalprocessingworld.com/keeping-counterfeit-medicines-out-of-the-supply-chain/
21. S. Tendulkar, A. Rodrigues, K. Patel, H. Dalvi, System to fight counterfeit drugs, in *Advanced Computing Technologies and Applications*, ed. by H. Vasudevan, A. Michalas, N. Shekokar, M. Narvekar, (Springer, Singapore, 2020), pp. 465–470. https://doi.org/10.1007/978-981-15-3242-9_43
22. M. Tremblay, Medicines counterfeiting is a complex problem: A review of key challenges across the supply chain. Curr. Drug Saf. **8**(1), 43–55 (2013). https://doi.org/10.2174/1574886311308010007
23. WHO, *WHO Global Surveillance and Monitoring System for Substandard and Falsified Medical Products* (World Health Organization, 2017). http://www.who.int/medicines/regulation/ssffc/publications/gsms-report-sf/en/

24. B. Yan, G. Huang, Supply chain information transmission based on RFID and internet of things. 2009 ISECS Int. Colloq. Comput. Commun. Control Manage. **4**, 166–169 (2009). https://doi.org/10.1109/CCCM.2009.5267755
25. W. Zhou, E.J. Yoon, S. Piramuthu, Simultaneous multi-level RFID tag ownership & transfer in health care environments. Decis. Support. Syst. **54**(1), 98–108 (2012). https://doi.org/10.1016/j.dss.2012.04.006

Chapter 8
The Desiderata of Blockchain and IoT in Medical and Pharmaceutical Enterprises

M. Manikandan, R. Subramanian, S. Nagajothi, S. Karthik, and Anand Paul

8.1 Introduction

Enterprises are typically known to be interpreting substantial data on an enormous scale which are made use of for heterogeneous processing. In the event of being extremely hard-pressed for time, conditioning voluminous records of patients locally and manually in a single node within an organization paves the way for less privacy and frequent chaos owing to the considerable hike in the number of patients in a hospital [1]. The demands for storage can be met with ease when the appropriate cloud computing service models are chosen as the solution for managing the workloads and computing instances of the respective enterprises. Cloud storage is regarded as the giant leap from the traditional way of handling IT resources. The expenditure on upgrading hardware and software and equipping data centers can be acutely minimized [2]. Cloud storage promotes remote access, data backup, and disaster recovery of indispensable information of the patient's data with enhanced

M. Manikandan (✉) · S. Nagajothi
Assistant Professor, Department of CSE, Sri Krishna College of Engineering and Technology, Coimbatore, Tamil Nadu, India
e-mail: manikandanm@skcet.ac.in

R. Subramanian
Department of EEE, SNS College of Technology, Coimbatore, Tamil Nadu, India
e-mail: deaneee@snsct.org

S. Karthik
Department of CSE, SNS College of Technology, Coimbatore, Tamil Nadu, India
e-mail: deancse@snsct.org

A. Paul
Department of CSE, Kyungpook National University, Daegu, South Korea
e-mail: anand@knu.ac.kr

T. Choudhury et al. (eds.), *Blockchain Applications in IoT Ecosystem*, EAI/Springer Innovations in Communication and Computing,
https://doi.org/10.1007/978-3-030-65691-1_8

reliability than the formal methods of data storage as this data can be mirrored at multiple sites on the cloud provider's network. This article provokes how the amalgamation of cloud storage, blockchain, and IoT can render an unparalleled service to the pharmaceutical enterprise community to have a modicum of command over the sensitive data and transactions [3].

The blockchain technology promises to be bridging the gap between the establishments of a mutual conventional manner without any dispensary to rate the IoT transactions. It is predicted that many IoT services will extend its global reach across billions of devices. The current IoT ecosystems rely on centralized, brokered communication models [4]. The devices are authenticated and authorized via extremely robust servers that are capable of handling huge computations. Even after the confrontation of the prevailing pitfalls, cloud servers tend to turn out to be erroneous. Moreover, the alternating spectrum of ownership among devices and their respective cloud servers makes it too cumbersome to interact with their counterparts. Not all services offered by different organizations are neither completely interpretable nor portable to bring about the devices and their controls under a stand-alone roof [5]. But blockchain could effectively lend a straightforward infrastructure for devices to directly transfer paramount entities like money or data clustered with a secured and an enhanced time stamp. Utilizing the latter would also eliminate the need for centralized authority constraints to enable the autonomous functioning of smart devices resulting in a completely distributed worthwhile digital infrastructure [6].

Effectuating blockchain for IoT data yields new dimensions to automate healthcare processes among the medical and pharmaceutical enterprises without setting up a complex and expensive IT infrastructure [7]. The data protection in blockchain fosters a stronger working relationship with the partners and greater efficiency as enterprises take up the advantage of the information provided. Incorporating blockchain would be an eye-opener to settle vulnerability and privacy concerns in IoT. Single points of failure could be knocked out, and a more adaptable environment for the devices to run can be accomplished. Blockchain can keep track of unvarying reports of connected devices in the IoT network [8]. Blockchain holds an impeccable role in transforming the unadorned aspects of IoT into reality. This was made possible owing to its centralized authority and as the nodes in the blockchain do not extend the scope of validating and verifying the transactions to any other functional component.

Cloud storage has evolved to be one of the emanating technologies which primarily aid in storing data and eliminating storage constraints with an appreciable extent of security and privacy [9]. The ample responsibility of counteracting multiple storage requests and service discrepancies completely relies on the cloud service provider. The outcomes of the requests from the cloud server are typically shielded with a robust encryption schedule thereby turning out to be less erroneous than the existing data storage methodologies. Equipping a medical enterprise with a private cloud enables access to hosted services to a pre-assigned number of authorized people (doctors and other medical representatives) [10]. Cloud-based solutions also bring about opportunities for more portability and higher productivity and efficiency for the physicians as everyone is assured access to the same updated information within a few mouse clicks.

The security attributes provided by the cloud service organizations naturally make the data ultimately devoid of cautionary threats. The transactions and security of the sensitive patient's data can be enhanced by implementing blockchain technologies [11]. Blockchain is an ingenious invention originally framed for crypto-currency, but it has now found many other applications in the world. Since the data held in a blockchain exists as a common and continually streamlined database, the medical reports added by a wide range of computers turn out to be verifiable with ease as the entire entity is backed up in a secure domain of space. The peculiar factor of a blockchain is that none could attempt to alter the data that was previously appended to the distributed ledger. The transactions stored in the nodes of the blockchain network cannot be falsified as attempting so would require the intruder's machine to overpower the entire network which is practically impossible.

8.2 IOT in Medical and Pharmaceutical Enterprises

8.2.1 Digitizing Patient's Records Using Blockchain with Electronic Health Records (EHR)

Electronic health records are regarded as a digital form of a patient's database. They are real-time monitored, patient-centered records that deliver information instantly and securely to physicians. They are usually composed of patients' medical history, medical transcriptions, diagnoses, radiology images, laboratory, and test results. The records stored in the EHRs can be effectively shared among other healthcare organizations which can avail access to evidence-based tools to conclude decisions about a patient's healthcare [12].

8.2.2 Precedence of Electronic Health Records (EHR) over Electronic Medical Records (EMR)

Electronic medical records are digitized structures of the patient's record in the physician's enterprise. An EMR comprises the medical history of the patients in due course of time. They keep track of data over time and identify the patients who are to undertake medical checkups. The quality of care within the practice can be delved into accordingly. In case of the requirement of a hard copy of a patient's record, it might be delivered to veteran physicians and other medical professionals of the care team. In that regard, EMRs cannot be handled with much ease than the traditional paper records, making it considerably less consistent [13].

Electronic health records, on the other hand, do satisfy all the parameters which the EMR failed. When EHR is chosen to undertake the patient's records, they could reach out beyond the healthcare enterprises that initially collect and record information. Electronic health records can be created, maintained, and shared by the concerned personnel across more than one medical enterprise.

The data is coordinated with the patient to the concerned representatives, the hospital, or even across the continents. Electronic health records are extremely handy when sharing medical and treatment records among associates and comprehending the levels of care undertaken by the individual.

The patient's records shared with encryption turns out to be powerful. The culminating value from the enterprises is an outcome of productive interaction of information from one recipient to another and the capability of interchanging the information to indulge in an interactive communication of the curative particulars which can only be accomplished by cloud storage and blockchain [19].

8.2.3 Advantages Associated with Electronic Health Records (EHR)

1. Individual patients can log on to his/her record and keep track of the laboratory results over their past arrivals, which can present a clear and concise picture of the medications that he/she takes in and the routines he/she maintains to keep up the scale [14].
2. At times of emergency, electronic health records (EHR) can reveal the patient's medical threats, so that medications and healthcare processes are adapted accordingly even if the patient is insensible.
3. Precise conclusions can be drawn from the previous test results by the specialists despite running duplicate tests to ensure the current status of the patient.
4. The physicians' descriptions from the patient's period of hospital stay can reveal the requisite details and pave the way for the portability of patients [14].

8.2.4 Incorporating Privacy and Security into EHR and IoT

Health information breaches can lead to a plethora of disastrous consequences. The data breaches might result in harming patients, downfall in finance, and degradation of an enterprise's reputation. It is rather indispensable for the healthcare records to be furnished with a considerable extent of privacy and security implications than making this sensitive information immune to destructive cyber threats. Encrypting patient privacy and seclusion of electronic health information is of paramount importance and is an arbitrary responsibility. Clinging to constructive security attributes greatly enhances trust among patients. The patient must trust the enterprise that their sensitive medical records will remain confidential, accurate, and secure in safe hands [15].

This trust can sum up the patient's overall health, in a nutshell, paving the way to better-structured decisions. The advantages associated with the security of EHRs over voluminous paper records include encoding medical information, backup, and automated control. The privacy guidelines enable the patients to gain a modicum of control over their medical records and let them determine their ways of utilizations. This also provides safeguard measures that make the medical professionals and their fraternity to abide by these rules that are set by the patients. Blockchain's capability to keep an indestructible and decentralized log of all patient data makes it a technological ruse for potentially scalable security solutions. The delicacy of the medical records can be greatly preserved as the blockchain promotes the anonymity of the individual. This disseminated nature of the latter enables rapid exchange of healthcare records among the authorized medical professionals as well.

8.2.5 Impediments Involved in Implementing EHR and Blockchain

In the course of technological development, several incentives have been provided by the respective democracies to deploy electronic health records to improve and coordinate quality care across the healthcare environment. Cost prevails to be the subjective concern for the organizations that are yet to incorporate blockchain. To establish an EHR system regardless of whether in an enterprise or a medical environment is a high-priced and comprehensive process that demands an enormous amount of manual workload and monetary requirements. On the other hand, security vulnerabilities for digital and classified information remain to be the predominant cause for why the enterprises hesitate to adapt to blockchain technology [16].

8.2.6 Medical Transcription and EHR

Organizations equipped with EHR have always looked forward to getting deprived of the transcriptions and opting for electronic records from the lucrative generated from the transcriptions. This stereotype existed until the advent of EHR. It was then realized that traversing through the EHR was time-consuming when clustered with the concerns about the standards of the documentation, and minimization of errors in documentation allowed the veteran medical associates to acknowledge that there is still much more scope for such services to evolve. Medical transcriptions help in the accumulation of the medical narrative, which prevails to be the medical memorandum in the physician's aspect. This narrative is often regarded as the first-person perspective of the patient's medical history. This narrative being circumstantial is impossible to replicate with the finest EHR testimonial features. These transcriptions furnish hassle-free and significant documentation which is the much-expected factor to the physicians in an EHR. However, highly compliant documentation tools are provided by transcription service companies for the physicians and the authorities [17].

8.3 Existing Model

Electronic medical records (EMRs) also known as electronic health records (EHRs) are the foundational components of all healthcare systems. Despite the growing literature on boons of various EHR interoperabilities, there are proven potential banes accompanying this emanating technology. These include monetary loss, workflow chaos, and interim forfeiture of efficiency, security and privacy concerns, and various unprecedented aftermaths. Since the medical information is electronically exchanged extensively, the likelihood of violating a patient's privacy is at stake. A poorly designed EHR leads to elevated medical fallacies, disrupted misconceptions, and overdependence on technology [18]. Though geared with arbitrary security measures, the authorities must be trained in basic digital security to ensure they do not leave their organizations vulnerable to unauthorized access. Any discrepancy in the medical records eventually leads to shrunken efficiency thereby making tasks of ease to be tiresome. Yet another cause of shrunken efficiency relates to inadequate tutoring on the EHR infrastructure which emerges when a system is renovated or modernized. The EHR of each patient must be frequently updated to enhance the precision of their data, while non-frequent reconditioning can lead to misconceptions in diagnoses of the patients.

8.4 Proposed Model

The existing healthcare facilities are known to be still solely relying on traditional systems for maintaining medical and healthcare records and processing payments thereby making it extremely tedious for the healthcare professionals to diagnose as the records are locally stored. Yet another painstaking threat prevailing in the medical industry is the exchange of the patient health information. The records are more susceptible to the chances of impersonation and fraudulent monetary crimes as the patients don't own any control over their records/data. The availability of acutely handy devices had still not let the patients collect, analyze, secure, and exchange healthcare information seamlessly. Adopting a blockchain and IoT leveraged solution can help accomplish a smooth, transparent, economically efficient, and easily operable system combined with secure payment options. As far as healthcare is concerned, none of the enterprises were found to be possessing a universally recognized patient identifier. An intriguing factor is that a unique patient identifier is capable of solving the disputes of mismatched EHRs (electronic health records) which have in the past led to numerous errors in patient care and increase the likelihood of patient harm. The medical and pharmaceutical industries are subjected to indispensably handle and store a colossal amount of data leading to potential security breach prospects. Patient records which are recorded and stored in a blockchain network are persistent and sturdy. Either to revamp or to refurbish the records in a block, it solely requires the prerogative of the patients, hence making them the owners of their respective medical information. The effectuation of blockchain also facilitates authentic and genuine payments through cryptocurrencies. The implementation of blockchain technology could mitigate these errors in the following ways:

1. A unique identifier is imparted to each patient thereby establishing a smart contract between patients and medical wellness enterprises. Doing so would not only ensure that the data shared is genuine and precise but also extend coherence with rapid diagnoses.
2. The courtesy of the inherent transparency of the blockchain technology guarantees hospitals that all patients' health records are "tamper-proof" while still ensuring the data's privacy. This would potentially eliminate the security hurdles that come along with IoT.
3. Enacting cryptocurrencies in place of cash or fiat money would ease bill processing automation and henceforth obliterate the third parties from the chain and increase the lucrative revenue of the organizations. Deploying such technologies would enable the tracking of each penny paid to the healthcare enterprise and verify that no errors are encountered.

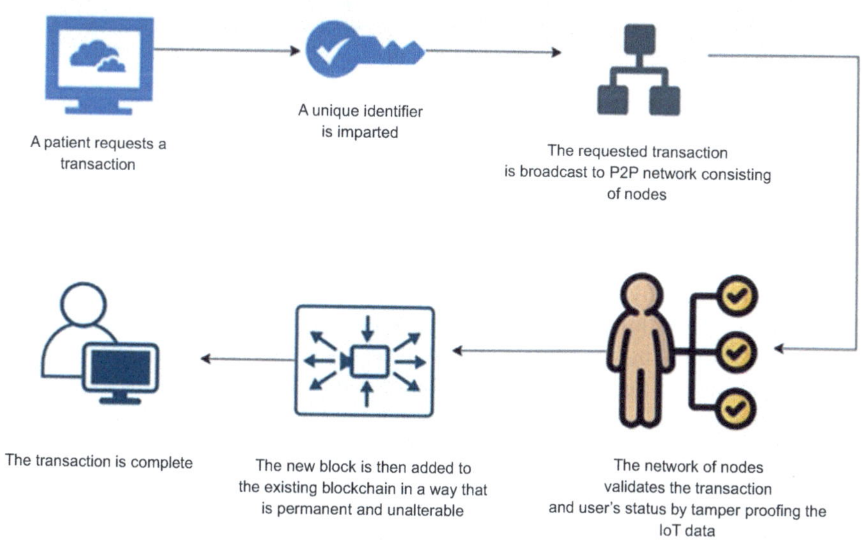

Since healthcare is a convoluted system with miscellaneous structures, it necessitates a patient to share their healthcare records across the medical environment.

The abrupt increase in the number of patients had led to the need for intensified data management by medical enterprises. Thus, maintaining patient information within hospital premises would be a big deal.

Financial aspects of medical care are of paramount importance. This domain is usually clogged with inefficiencies which can be optimized by the utilization of blockchain. In the view of the establishment of smart contracts between medical care professionals and patients, the latter can be made use of in premium negotiating phases. Data which deals with current health status, types of medications intake, etc. entangled to the blockchain to evolving premiums, through smart contracts. The intrusion of too many mediators or intermediaries turns out the claim handling process to be enduring and burdensome for the end customer. Taking these counterproductive effects into account, few propositions had been employed into blockchain for billing claims management and broader financial aspects of care delivery. The Ethereum or Hyperledger framework used in streamlining claim management in medical care is known for its versatility in delivering the patients, providers, and the insurers together into one ecosystem for real-time insights into the patients' health journey and ease of health claims management.

The cluster of blockchain and IoT can hopefully replace cash or government-backed currencies. Innumerable nodes in a blockchain network utilize cryptographic strategy to establish an irreversible and collective record of all transactions that had ever taken place. Processing transactions with a decentralized currency provides added security and minimizes the erroneous instances of fraud. The transparent nature of the blockchain leads to a robust repository of databases and secure payments subsequently instituting falsification or corruption of records impossible. Healthcare organizations who had succeeded in administering this emanating technology would witness consequent ease of access to their billing and accounting thereby ignoring the fear of fraud being committed by employees of the enterprise or from the patients themselves.

8.5 Conclusion

Blockchain has the inherent knack to solve several disputes storming the medical care industry today. It poses a boundless dimension among various healthcare stakeholders, namely, patients and providers. A blockchain which is a decentralized network may minimize stakeholder lock-in problems in healthcare. Despite the hype and the deal of interest circling over blockchain and IoT, its implication and effect on healthcare had been quite low and are still in the early days. Medical organizations and enterprises which have deployed blockchain applications are working with a limited user base. Yet a mammoth growth is anticipated with a significant positive impact of blockchain in healthcare in the future. Handling patients' records and the dome of secured payments is the key that blockchain holds which the existing entities are devoid of. The processes will not only be trustworthy and persistent, but also the quality of healthcare will be extended cost-effectively. The prioritization of data is allocated solely to the patients. The idea of an encrypted public ledger via cryptocurrency among medical agencies, patients, and networks brings about an intriguing future that could change the way data is used to treat patients. The transactions of both the enterprise and patients turn out to be more explicit after the successful effectuation of blockchain and cryptocurrency than maintaining incommodious physical records and ingenuine payments. By incorporating IoT with tampered-proof protocols, its security threats are eliminated, thereby paving the way for a smooth process curve, and the prioritization of data is allocated solely to the patients.

References

1. A. Firdaus, N.B. Anuar, M. Faizal, I.A.T. Hashem, S. Bachok, A. Kumar, Root exploit detection and features optimization: Mobile device and blockchain-based medical data management. J. Med. Syst. **42**(6), 1 (2018 June). https://doi.org/10.1007/s10916-018-0966-x
2. T. Bocek, B.B. Rodrigues, T. Strasser, B. Stiller, *Blockchains Everywhere – A Use-Case of Blockchains in the Pharma Supply-Chain* (2017 May). https://doi.org/10.23919/INM.2017.7987376
3. E. Chukwu, L. Garga, Systematic review of blockchain in healthcare: Frameworks, prototypes, and implementations. IEEE Access **8**, 21196 (2020 January 7)
4. WHO, *World Health Statistics 2018: Monitoring Health for the SDGs* (WHO, Geneva, 2018)
5. AAMC, Careers in medicine (2019). Accessed: Jan. 29, 2019. [Online]. Available: https://www.aamc.org/cim/specialty/exploreoptions/list/
6. P. Zhang, D.C. Schmidt, J. White, G. Lenz, Blockchain technology use cases in healthcare. Adv. Comput. **111**, 1–41 (2018 January)
7. X. Zheng, R.R. Mukkamala, R. Vatrapu, J. Ordieres-Mere, Blockchain-based personal health data sharing system using cloud storage, in *Proceedings of the IEEE 20th International Conference on e-Health Networking, Applications Services (Healthcom), September 2018*, (2018), pp. 1–6
8. X. Liang, J. Zhao, S. Shetty, J. Liu, D. Li, Integrating blockchain for data sharing and collaboration in mobile healthcare applications, in *Proceedings of the IEEE 28th Annual International Symposium on Personal, Indoor, Mobile Radio Communication (PIMRC), October 2017*, (2017), pp. 1–5
9. Y. Du, J. Liu, Z. Guan, H. Feng, A medical information service platform based on distributed cloud and blockchain, in *Proceedings of the IEEE International Conference on Smart Cloud (SmartCloud), September 2018*, (2018), pp. 34–39

10. C. Ananth, M. Karthikeyan, N. Mohananthini, A secured healthcare system using private blockchain technology. J. Eng. Technol. **6**(2), 42–54 (2018)
11. [49] G. Srivastava, A.D. Dwivedi, R. Singh, Automated remote patient monitoring: Data sharing and privacy using blockchain. arXiv:1811.03417 (2018). https://arxiv.org/abs/1811.03417
12. S. Rahmadika, K.-H. Rhee, Blockchain technology for providing an architecture model of decentralized personal health information, *Int. J. Eng. Bus. Manag.***10**, Art. no. 184797901879058 (2018 January)
13. A.K. Talukder, M. Chaitanya, D. Arnold, K. Sakurai, Proof of disease: A blockchain consensus protocol for accurate medical decisions and reducing the disease burden, in *Proceedings of the IEEE SmartWorld, Ubiquitous Intelligence and Computing, Advanced and Trusted Computing, Scalable Computing and Communications, Cloud Big Data Computing, Internet People Smart City Innovation (SmartWorld/SCALCOM/UIC/ATC/CBDCom/IOP/SCI), Oct. 2018, pp. 257–262*
14. Z. Shae, J.J. Tsai, On the design of a blockchain platform for clinical trial and precision medicine, in *Proceedings of the IEEE 37th International Conference on Distributed Computing System (ICDCS), June 2017*, (2017), pp. 1972–1980
15. A. Al Omar, M.S. Rahman, A. Basu, S. Kiyomoto, MediBchain: A blockchain-based privacy-preserving platform for healthcare data, in *Security, Privacy, and Anonymity in Computation, Communication, and Storage (Lecture Notes in Computer Science)*, (Springer, Cham, 2017)
16. T.-T. Kuo, L. Ohno-Machado, ModelChain: Decentralized privacy-preserving healthcare predictive modeling framework on private blockchain networks, arXiv:1802.01746 (2018). [Online]. Available: https://arxiv.org/abs/1802.01746
17. T. Heston, A case study in blockchain health care innovation. Int. J. Curr. Res. **9**(11), 60587–60588 (2017)
18. A. Azaria, A. Ekblaw, T. Vieira, A. Lippman, MedRec: Using blockchain for medical data access and permission management, in *Proceedings of the 2nd International Conference on Open Big Data (OBD), August 2016*, (2016), pp. 25–30
19. D. Ichikawa, M. Kashiyama, T. Ueno, Tamper-resistant mobile health using blockchain technology. JMIR Mhealth Uhealth **5**(7), e111 (2017 July)

Chapter 9
Microchain: A Light Hierarchical Consensus Protocol for IoT Systems

Ronghua Xu, Yu Chen, and Erik Blasch

9.1 Introduction

With the proliferation of Internet of Things (IoT), a large volume of smart devices is connected to the Internet at an unprecedented scale. The prevalence of IoT devices has changed human lives by ubiquitously providing applications and services that have revolutionized transportation, healthcare, industrial automation, emergency response, and so on. For instance, thanks to the rapid advances in IoT and edge-fog-cloud computing technologies, which are among the hot research topics in Smart Cities, the Smart Public Safety (SPS) system has become feasible by integrating heterogeneous computing devices and different types of networks to collaboratively provide seamless public safety services for communities and the society [18, 27].

With an ever-increasing presence of IoT-based smart applications and their ubiquitous visibility from the Internet, the highly connected smart IoT devices with a huge volume of generated transaction data incur more concerns on security and privacy [4]. IoT systems are deployed in a distributed network environment that consists of a large number of devices with high heterogeneity and dynamics. The heterogeneity and resource constraint at the edge networks necessitate a scalable, flexible, and lightweight system architecture [17] that supports fast development and easy deployment with multiple application vendors using non-standard development technologies. Furthermore, those smart devices are geographically scattered across the on-site/near-site edge networks and managed by fragmented service providers that enforce different security policies. Thus, traditional security policies on a

R. Xu · Y. Chen (✉)
Binghamton University, SUNY, Binghamton, NY, USA
e-mail: rxu22@binghamton.edu; ychen@binghamton.edu

E. Blasch
The U.S. Air Force Research Laboratory, Rome, NY, USA

T. Choudhury et al. (eds.), *Blockchain Applications in IoT Ecosystem*, EAI/Springer Innovations in Communication and Computing,
https://doi.org/10.1007/978-3-030-65691-1_9

centralized authority basis, which suffer from the performance bottlenecks or single point of failures, are not efficient and suitable to address the performance and security challenges in IoT systems.

Recently, designing new decentralized security mechanisms for distributed network applications becomes one of the most intensively studied topics both in academia and industry. Blockchain, which acts as the fundamental protocol of Bitcoin [16], has demonstrated great potential to revolutionize the fundamentals of information technologies (IT) due to many attractive properties, such as decentralization and transparency [19]. Essentially, the blockchain is a public ledger based on consensus rules to provide a verifiable, append-only chained data structure of transactions. Blockchain uses a decentralized architecture that does not rely on a centralized authority, such that the data can be stored and updated distributively under a peer-to-peer network. It improves system availability and mitigates the single point failure problem compared to a centralized architecture.

In a blockchain network, a consensus mechanism is enforced on a large amount of distributed nodes called *miners* to maintain the sanctity of the data recorded in the blocks. The transactions are approved by miners and recorded in the time-stamped blocks, where each block is identified by a cryptographic hash and chained to preceding blocks in a chronological order. Therefore, multiple participants can access the shared public ledger stored worldwide on distributed nodes maintained by "miner-accountants," as opposed to establishing and maintaining trust with a transaction counter-party or a third-party intermediary. Thus, blockchain is an ideal decentralized architecture to ensure distributed transactions among all participants in a trustless environment, like edge-fog-edge computing based IoT applications under heterogeneous network environment [38].

Recently, there are many reported efforts that address security issues in IoT systems leveraging the blockchain and smart contract enabled mechanisms. For example, public safety system [33], smart surveillance system [15, 17], social credit system [11, 12, 32], decentralized data market [34], space and avionics systems [1, 36, 38], biometric imaging data processing [35], identification authentication and access control [30, 31]. All these reported researches have verified that blockchain and smart contract together are promising to provide a decentralized security mechanism to IoT systems. They have also shown that, however, directly integrating existing cryptocurrency-oriented blockchain technologies into IoT systems is hindered by several challenges in terms of scalability, computing intensity, storage capacity, data security, and privacy preservation.

The efforts in blockchain-enabled services for IoT system face critical challenges in designing a blockchain network in terms of high quality of service, data confidentiality, and privacy-awareness. Particularly, the performance of blockchain networks significantly relies on the efficiency of the consensus mechanisms, e.g., in terms of data consistency, speed of consensus finality, robustness to arbitrarily behaving nodes, and network scalability [26]. Unfortunately, existing blockchain protocols are mainly designed for cryptocurrency, and they are not suitable to be directly embedded into IoT scenarios.

To evaluate the challenges in designing blockchain protocols for IoT systems, this chapter provides a comprehensive overview on present blockchain networks regarding cryptographic technologies and incentive mechanisms, followed by a case study.

The rest of the chapter is organized as follows. Section 9.2 provides an overview of blockchain fabric, and the basics of classic fault tolerant consensus in distributed systems are explained in Sect. 9.3. Given the analysis on existing issues of consensus protocols, general challenges on integrating blockchain with IoT are identified in Sect. 9.4. In Sect. 9.5, Microchain, a hybrid blockchain architecture, is introduced as a case study on designing scalable, lightweight blockchain protocols for IoT systems. Section 9.6 concludes this chapter and summarizes the future research opportunities for integrating blockchain technology within the context of IoT systems.

9.2 An Overview of Blockchain Fabric

Compared to the traditional distributed computing paradigm with a clear client–server model, a blockchain network allows every participant to be both a client (to issue transactions) and a server (to validate and finalize transactions) [29]. Each participant could maintain a local view of the distributed ledger, which contains valid transactions and data. The ledger should be consistent with other nodes across the network given an underlying consensus protocol in the blockchain. The core task of a blockchain network is to ensure that the trustless nodes in the network reach agreements upon a single tamper-proof record of transactions [26]. The decentralized network architecture and the fault tolerance enabled consensus protocol allow the blockchain to be a prospective infrastructure for distributed services and applications.

Figure 9.1 provides an overview of the blockchain infrastructure from the perspective of system level design and implementation. The application layer, which is on top of the global state machine replication (SMR) layer, exerts smart contracts to build a wide range of applications. The global SMR layer serves as a basic service layer of the blockchain to support distributed computing functions for upper applications. The consensus layer acts as the core in the whole system by executing the consensus protocol to ensure tamper-proof of the distributed ledger and the SMR. The main function of the data organization and network layer is to identify an optimized data representation and an efficient cryptography to improve the performance of the blockchain given a certain network environment.

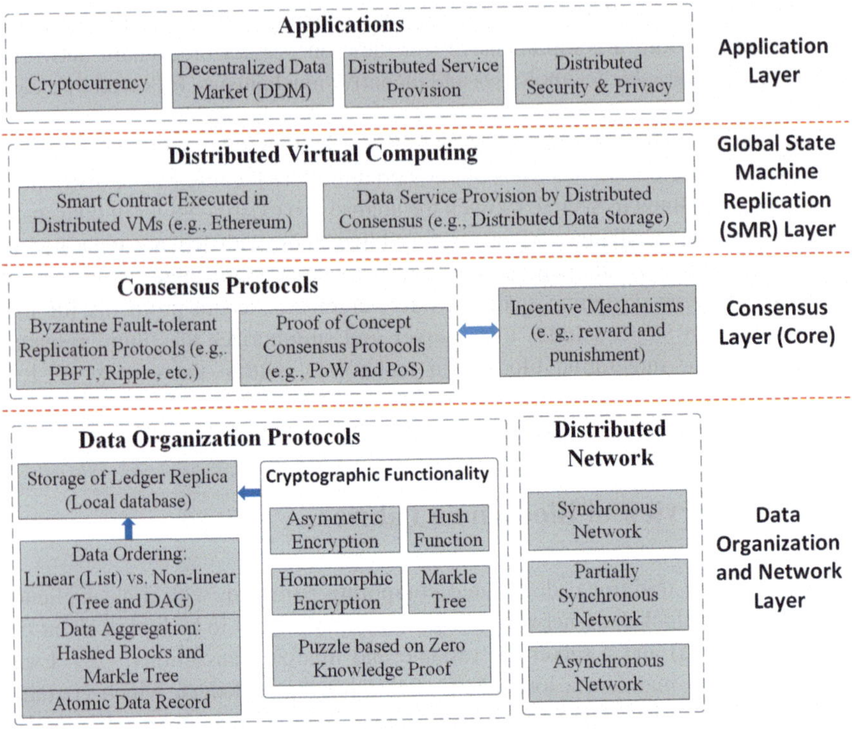

Fig. 9.1 A layered overview of blockchain fabric implementations

9.3 Distributed Consensus Protocols and Algorithms

In a distributed system, all participants cooperate with each other to achieve a common goal in spite of being deployed at geographically separated locations. Since each node could be prone to system faults and communication channels suffer from adversarial attacks, a consensus mechanism allows that the participants still can reach agreement on a global state in the presence of component failures, either *crash failure* or *Byzantine failure*. The crash failure happens when the host system of participant abruptly stops functioning and cannot resume by itself. A Byzantine failure is often caused by system malfunctions or malicious behaviors, such as sending contradictory messages to partners or withholding messages. Therefore, the consensus protocol is aimed to solve fault tolerant problems in distributed system scenarios.

A consensus protocol defines a set of rules for message passing and processing for all networked components to reach agreement on a common subject [28]. A messaging passing rule specifies the way of messages broadcasting and relaying among system components. A processing rule defines how a component changes its internal state as receiving these valid messages. The goal of consensus is reached as long as all non-faulty participants make agreement on a target subject. In general, the tolerant number of faulty nodes in a network is used to measure the strength of a consensus protocol from security's perspective. Given two failure types, the fault tolerant problems are divided into Crash-Fault Tolerant (CFT) and Byzantine Fault Tolerant (BFT). A consensus protocol that tolerates at least one crash failure is called CFT, while BFT requires that a consensus protocol can tolerate at least one Byzantine failure. In terms of failures, crash failure is considered as a benign case while Byzantine failure is considered the worst case. Therefore, a BFT consensus is naturally CFT [28]. More precisely speaking, BFT consensus protocol must satisfy the following properties:

- *Validity (Correctness)*: If a honest node receives a trusted common replicate proposed by other nodes, this common replicate should be accepted into the blockchain.
- *Agreement (Consistency)*: All the honest nodes should update their local replicates of the blockchain with the block header of confirmed global blockchain.
- *Termination (Liveness)*: Every honest node should either discard or accept new transactions into the blockchain, and all transactions originated from the honest nodes will be eventually confirmed.
- *Integrity (Total Order)*: All honest nodes should accept the same chronological order of transactions which are correctly appended to the hash-chained blockchain.

In a consensus algorithm, validity, integrity, and agreement define the consensus safety properties, while termination defines its liveness property. Given variant consensus protocols, the blockchain networks could be categorized into permissionless blockchain (e.g., Nakamoto Consensus Protocols) or permissioned blockchain (e.g., Practical Byzantine Fault Tolerant (PBFT) Consensus). The remainder of this section will focus on underlying consensus protocols for blockchain in terms of the consensus goal and network model.

9.3.1 Byzantine Fault Tolerant Consensus

The classic consensus in a distributed system can be expressed abstractly as a Byzantine General Problem [10], which copes with a single-value agreement among different parts of a system given failure of communication or conflicting information. Formally, the Byzantine General Problem can be explained in a message-passing system with N participants $p_i \in P$, where $i \in (1, 2, \ldots, N)$. They are geographically distributed and inter-connected by communication links. They communicate with each other only through broadcasting messages across the network to make agreement on a common plan of action: a (attack) or r (retreat). To achieve an agreement on a single value, each participant broadcasts his/her vote for a or r and makes a decision locally based on the received votes.

Due to the Byzantine failure, some dishonest participants f attempt to prevent consensus by sending contradicting votes to different nodes. Therefore, in a network including N nodes, the ultimate goal is that all loyal participants are still able to agree on the consistent action in spite of the false information. It requires that the super-majority of the participants must be honest, which means $N - f > 2f$.

Although the classical BFT provides a solution to single-value consensus in a synchronous network, the correctness of a typical distributed system not only requires every single data message is processed correctly, but it also means the processed results should satisfy the total ordering requirement. Moreover, the real-world distributed computing systems rely on a partial synchronous network, or even an asynchronous network. The sequential operations could be defined as state machines, which consists of state variables, encodes the state and commands, and transforms its state [22]. Therefore, SMR is widely used as an active replication to ensure the data ordering consensus. Here, two SMR based fault tolerant consensus protocols for distributed network are discussed: Viewstamped Replication (VR) and Piratical Byzantine Fault Tolerance (PBFT).

Viewstamped Replication (VR) The original Viewstamped Replication (VR) protocol was firstly developed in 1980s [20], and an updated version was presented in 2012 [13]. The VR protocol is aimed to use state machine replication to address fault tolerance issues in distributed systems. VR works in asynchronous networks like the Internet and handles failures in which nodes fail by crashing [20]. In a VR system that includes N replicas, one replica works as the *primary* and other $N - 1$ replicas are *backups*. The primary is responsible for ordering clients' requests while the backups simply accept orders collected by the primary. Each replica operates a local state machine with pre-defined state variables. VR ensures reliability and availability when no more than a threshold of f replicas are faulty [13]. The total replicas N in VR system should be no less than $2f + 1$, which is the minimum number of replicas to ensure CFT in an asynchronous network.

VR uses three sub-protocols to provide correctness:*normal*, *view change*, and *recovery*. Figure 9.2 describes the normal operation workflow in a VR system with three replicas. A client sends an operation request to the primary at the beginning of a protocol run. On receiving a request message, the primary starts the *Prepare* stage by firstly updating its local state, then it forwards the request to all backups using *Prepare* message. Upon receiving the *Prepare* message, each backup locally executes the operation request from the primary and then updates its state with the given executing results. Finally, every backup sends a *PrepareOK* message to the primary to notify that it has finished the operation and updated the state. After receiving f *PrepareOK* messages, the primary starts the *PrepareOK* stage by executing the operation and updating its state accordingly. Then, the primary sends a *Reply* message back to the client to wrap up the replication session.

Since VR is only a CFT based replication protocol which requires that total number of replicas should satisfy $N \geqslant 2f + 1$ to prevent crash failures, it does not handle Byzantine failures as basic BFT consensus does. Given analysis on communication overhead on normal operation, the message complexity for VR is $O(N)$.

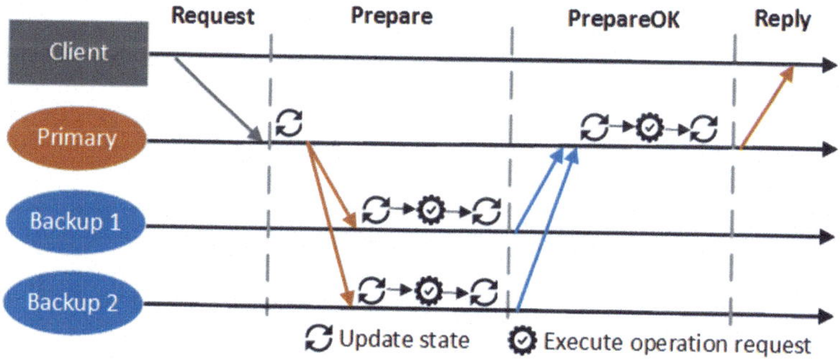

Fig. 9.2 The normal operation sub-protocol of VR in a 3-replicas system

Piratical Byzantine Fault Tolerance (PBFT) Since malicious attacks and software failures are increasingly common in distributed networks, both the primary and backups in the replication system are vulnerable to Byzantine failures. However, the traditional consensus protocols relied on a synchronous network assumption that requires high communication complexity, like BFT-OM algorithm [10]. Meanwhile, the VR based consensus algorithms [13, 20] cannot tolerant Byzantine failures in distributed networking environments. To provide a practical and efficient BFT protocol under asynchronous environments, a Practical Byzantine Fault Tolerance (PBFT) solution is proposed that advances the VR protocol to tolerant Byzantine failures in asynchronous networks [2, 3].

The PBFT protocol can be used to implement any deterministic replicated services with a state and some operations [3]. Given a replication system which has N replicas and f of them are Byzantine faulty nodes, the PBFT algorithm guarantees *safety* under condition $N \geqslant 3f + 1$, which means at most f replicas are Byzantine faulty. To provide *liveness*, PBFT assumes a synchronous network. Therefore, clients could evenly receive replies to their operation requests only if at most f replicas are Byzantine faulty and the process delay does not grow beyond the upper bounded time. Like the VR, PBFT utilizes sub-protocols to implement a BFT-based distributed file system: normal case operation, view changes, and checkpoint protocol.

The normal case operation uses a three phase protocol to automatically broadcast request among replicas, and it executes as session cycles with an increasing view number. Figure 9.3 describes the normal case operation workflow in a PBFT system with four replicas. Replica 0 acts as a primary while other replicas are backups. A client sends operation request to the primary at the beginning of a view session to launch the three phase consensus protocol.

(1) Pre-prepare: After receiving a request message from the Client C, the Replica 0 starts the phase one *Pre-prepare* by firstly updating its local state, then it multicasts the request to other replicas using a *Pre-prepare* message. Upon receiving the *Pre-prepare* message, each backup checks whether the message is with valid

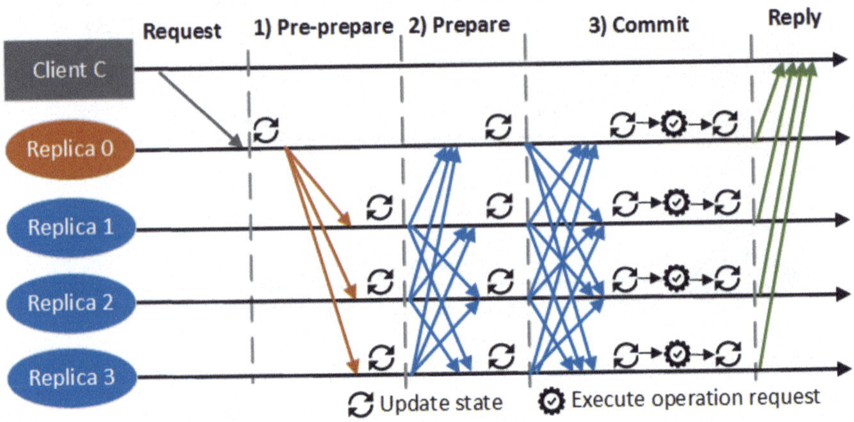

Fig. 9.3 The normal operation protocol of PBFT in a 4-replicas system

signature and state information. If yes, the replica locally updates its local status and moves to the Prepare phase by sending a *Prepare* message to other replicas.

(2) Prepare: If a backup agrees with the operation request, it votes for agreement by multicasting a *Prepare* message to all replicas including the primary. After receiving at least $2f + 1$ *Pre-prepare* messages with the same view number and state information, a replica updates its local status accordingly and proceeds to following Commit phase by sending a *Commit* message to other replicas. This phase ensures that all replicas achieve a common state before executing assigned operation requests.

(3) Commit: Similar to phase two *Prepare*, each replica firstly checks the received *Commit* message until there are $2f + 1$ valid messages. In phase three *Commit*, each replica firstly updates its status to "commit," then locally executes the assigned operation request from the client and then updates its state given executing results. This phase ensures that executed operation requests could be totally in order across cycle views.

Reply: When a replica completes the commit process, it sends a *Reply* message to the client. The *Reply* message includes the total ordered operation request executions results and state information. The client accepts the execution results after receiving at least $2f + 1$ *Reply* messages.

The PBFT assumes that total replicas $N \geqslant 3f+1$ when f replicas are Byzantine failures. Therefore, messages from $2f + 1$ non-faulty replicas are sufficient to achieve a super-majority consensus in the voting process, like prepare and commit in normal operation. Regarding the communication overhead, since only the primary multicasts messages to all replicas in the pre-prepare phase, the complexity is $O(N)$. In Prepare and Commit stages, every replica broadcasts a message to other replicas, such that the complexity is $O(N^2)$.

9.3.2 *Nakamoto Consensus Protocol*

To jointly address several critical issues such as pseudonymity, scalability, and poor synchronization in an open-access network environment, the Nakamoto consensus protocol [16] is implemented as the consensus foundation of Bitcoin. The Nakamoto consensus is based on a cryptographic hash value discovery racing game called Proof-of-Work (PoW), which is widely adopted by many cryptocurrency-based blockchain networks.

Consensus Protocol The goal of Nakamoto consensus is to ensure all participants agree on a common network transaction log as a serialized blockchain. In a distributed network, each node maintains a local replica and executes Nakamoto consensus protocol independently. Security of the consensus protocol requires that the majority of nodes are honest and they can correctly execute the consensus protocol. The Nakamoto protocol can be summarized as following rules:

- *Message Gossiping Rule*: All newly received and locally generated transactions and blocks should be multicasted to peers in a timely manner. This ensures that all nodes receive transactions and blocks in spite of the asynchronous network environment.
- *Agreement Rule*: After receiving a block, all honest nodes should either accept or discard it based on the block's validity. In other words, all honest nodes should agree on the same blockchain if each node accepted the same number of blocks in its local replica.
- *Validation Rule*: All received transactions and blocks need to be validated before being appended to the blockchain or broadcasted to peers. Only valid transactions could be saved into new blocks and valid blocks encapsulating valid transactions could be accepted by the blockchain network.
- *Proof-of-Work (PoW)*: Every node has to solve a computing-intensive, time-consuming hash puzzle as a Proof-of-Work for block generation. In brief, PoW solution requires exhaustively querying a cryptographic hash function for a partial preimage generated from a candidate block [26]. The hash code of a candidate block is expected to satisfy a pre-defined difficulty condition parameter h, like fixed length of bits are 0 s. The PoW puzzle can be formally defined as the following equation:

$$hash\_block = \mathcal{H}(block\_data|nonce) \leqslant D(h), \tag{9.1}$$

  where for some fixed length of bits L and difficulty condition $D(h) = 2^{L-h}$. $\mathcal{H}(\cdot)$ is a pre-defined collision-resistant cryptographic hash function that outputs hash string $L \in \{0, 1\}^{\lambda}$, and λ is the length of hash string.
- *Longest Chain Rule*: All honest nodes always accept the longest chain as the main chain and append their collected valid blocks on the main chain. The longest chain rule ensures the consensus in an asynchronous network, such that all honest miners are working on a common main chain. This rule ensures probabilistic finality given a certain length of sequential blocks.

The above rules provide safety for achieving Nakamoto consensus in a distributed network. However, incentive mechanism is also indispensable to a public blockchain network, especially for those who act as financial infrastructure, like cryptocurrency and digital payment systems. Nakamoto consensus uses block rewards and transaction fees as an incentive mechanism to encourage participants to invest computation power to join the network and make contributions.

Chain Finality and Complexity Analysis The PoW process defined by Eq. (9.1) is essentially a verifiable process of a weighted random coin-tossing, where the probability of winning is no longer uniformly associated with the nodes' identities but in proportion to the resources, e.g., hash rate casted by the nodes [26]. In the PoW-like leader election process, the probability of a node for winning the block generation follows:

$$p_{win\_block}(i) = \frac{w_i}{\sum_{i \in N} w_i}, \tag{9.2}$$

where w_i is the shared verifiable resource node i can have, such as computational power, memory and storage, etc.

According to the longest chain rule, if blocks are appended to a chain branch that is not suffix of the longest chain, those blocks shall be discarded or "orphaned." As defined by Eq. (9.2), if attackers have more than 50% of the whole network's gross hash computing power, they will have higher hash generating rate so that producing blocks faster than rest of the participants in the network. The probability of an attacker to win the longest chain by continuously generating m blocks is:

$$P_{win\_chain} = \left(\frac{p_{win\_block}}{1 - p_{win\_block}} \right)^m. \tag{9.3}$$

The P_{win_chain} drops exponentially as m increases if $p_{win\_block} < 0.5$. Therefore, if more than half of the miners are honest, it is computationally impossible for attackers to revoke a block from the blockchain. Bitcoin network specifies $m = 6$ as the longest chain confirmation. Since the Nakamoto consensus protocol uses a gospel style message delivery without using all-to-all message phase like BFT, it also produces smaller communication complexity $O(N)$.

Constraints and Vulnerabilities The Nakamoto consensus protocol demonstrates good scalability in a trustless, open-access network environment. However, PoW also incurs several performance issues, such as limited throughput, high demand of computation and storage resources as well as unsustainable energy consumption. Furthermore, verifiable random block generation and probabilistic finality make Nakamoto consensus protocol vulnerable to several security problems, like majority attack and selfish mining.

9.4 Challenges on Integration Blockchain with IoTs

Thanks to the distributed ledger, blockchain has the potential to enrich the IoTs by providing a trusted sharing service, where information is reliable and traceable. Blockchain and smart contract technologies are more and more accepted as the key to solve scalability, privacy, and reliability problems related to the IoT paradigm. However, directly incorporating blockchain into the IoT is infeasible. Since the traditional blockchain networks, like bitcoin, were designed for Internet-based scenarios, where rich-resource devices like servers, desktops, or laptop computers are participants, and the network environment is stable. Consequently, it cannot meet the requirements of IoT reality, such as constraints in computation and storage resources, dynamic network topology, communication efficiency and energy consumption, etc. Although several recent efforts that focus on either improving the performance of PoW blockchain like Bitcoin-NG [6] or scaling classical BFT protocols through parallelization like sharding [14], those cryptocurrency-based blockchain solutions bring up other issues when being introduced into IoT systems. The identified challenges are presented as follows:

(1) *The trade-off between scalability and efficiency*: The IoT applications, such as smart surveillance systems, involve a large volume of generated transaction data among users and service providers, the efficient throughput and lower latency become key metrics of designing blockchain protocols for IoTs. Utilizing BFT and state machine replication protocols can potentially improve the efficiency with the trade-off of poor scalability, which causes system security issues like being vulnerable to Sybil attack. Furthermore, IoT systems generally rely on identity registration and authentication process to enroll known participants due to legal and compliance reasons, implementing permissioned blockchain for IoTs ensures a certain level of security with the supplementary node identity management.

(2) *The cost for transaction confirmation and storage*: Since IoT devices are resource-constrained with limited computation and storage capacity, the high complexity consensus based on computing-intensive cryptographic algorithms like PoW is not affordable to IoT applications. In addition, storing the entire blockchain history to validate the current state is not only overwhelming for storage constrained IoT devices, but also introduces longer bootstrap time when new nodes join the network. Furthermore, blockchain runs on a peer-to-peer network and consensus protocol requires frequent data transmissions and exchanges to ensure consistent records in distributed ledger. It will bring significant communication overhead on light IoT networks and extra energy consumption by data transmission. Thus, lightweight considerations, such as efficient transaction processing, optimized chain data organization, and energy efficiency, etc., are critical to design new blockchain consensus algorithms for IoT systems.

(3) *The conflicts between transparency and privacy*: As an important characteristic of blockchain, transparency allows all participants to access blockchain data

and audit the transactions. However, it brings concerns on privacy issues for some IoT systems, such as e-health and smart home, where the collected sensitive user data should be confidential and are only accessible to authorized entities. Enforcing access control mechanism to some extent encounters the transparency principle of blockchain. But for some IoT applications like supply chain management, data traceability is mandatory at the cost of transaction transparency. Thus, trade-off between transparency and privacy becomes an import factor in the design of blockchain based IoT systems.

(4) *Security on IoT data and blockchain*: IoT devices are vulnerable to network attacks compared with computers and cloud-based services. Corrupted IoT data from compromised devices make the cast that data itself is not correct before sending transactions to blockchain network. Hence, the data finalized in blockchain is polluted. On the other hand, the blockchain consensus protocol can tolerate some certain level of Byzantine failure given Byzantine nodes are below a threshold. However, more compromised IoT devices also make the consensus vulnerable, so that data in the chain are not immutable. Security should be considered in terms of IoT data and blockchain network.

Considering the above challenges on incorporating blockchain technology into IoT systems, designing optimized blockchain fabrics empowered with light and efficient consensus protocols become a prospective solution. In the next section, *Microchain* is introduced as a case study to demonstrate how to design and implement a partially decentralized, scalable, and lightweight distributed ledger protocol for IoT-based applications.

9.5 Microchain: A Lightweight Blockchain Fabric for IoTs

To address challenges in integrating blockchain technologies into IoT systems, Microchain [37] was proposed by designing an efficient consensus mechanism running on a small number of validators.

9.5.1 Microchain System Architecture

The Microchain network is shown in the upper part of Fig. 9.4. The Microchain is built on top of a permissioned network, in which only registered entities are authorized to access the network, allowed to interact with other validators, and contribute to consensus protocol, such as transactions propagation, block verification, and mining. For dynasty epoch during a fixed time period, a final committee is responsible for key functions of consensus protocol, like transactions processing, blocks generation, and chain finality. A random committee formation protocol ensures that the committee election process is unpredictable. During the

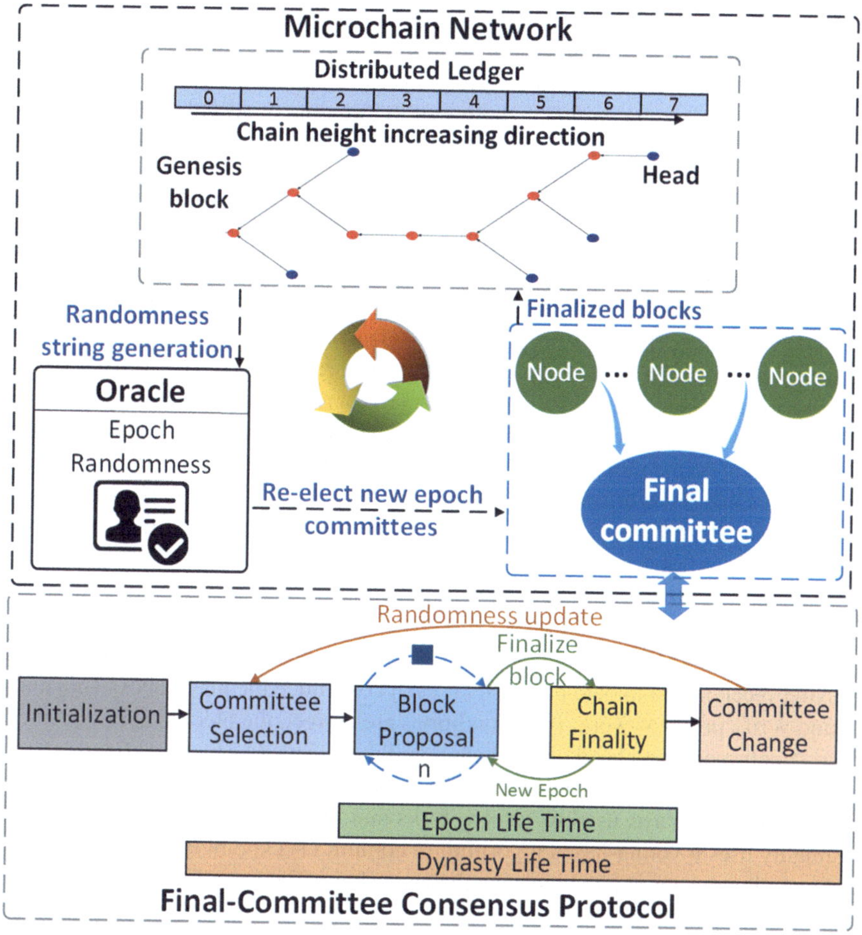

Fig. 9.4 Microchain system overview

lifetime of each dynasty, a final-committee consensus mechanism is responsible for proposing blocks and finalizing the chain history given an unbounded time delay [37]. Given the assumption that a synchronous network in which operations of processes are coordinated in rounds with bounded delay constraints, Microchain ensures *persistence* and *liveness*, which are two formal proprieties of a robust distributed ledger.

The performance and security proprieties of Microchain system rely on a light and efficient consensus protocol design. The final-committee consensus protocol is illustrated in the lower part of Fig. 9.4. The key components and workflows are described as follows:

– *Initialization*: In the initialization process, a special dynasty, which includes a group of validators specified by the administrator, acts as initial committee D_{init}

to initialize blockchain. Each validator creates a genesis block B_0 and sets the local blockchain $\mathcal{C} = B_0$ and $head = B_0$. The initial committee will work as the first dynasty of the system until the election of the next dynasty.

- *Committee Selection*: At the beginning of the lifetime of each dynasty, the final-committee formation protocol exploits a Verifiable Random Function (VRF) based cryptographic sortition scheme [8] to randomly choose a subset of validators V as the final committee according to their credit weight. The selected committee members D will be added to the current block, which is marked as the beginning block of the new dynasty epoch. The lifetime of dynasty epoch starts from committee selection and ends after dynasty change is finished.
- *Block Proposal*: The block proposal mechanism uses a pure Proof-of-Stake (PoS) protocol, called Proof-of-Credit (PoC), to generate new blocks in each block proposal run. Only validators in the current dynasty can propose a new block. The probability that a validator v_j could propose a block is associated with its credit distribution of the current dynasty (pk_j, c_j), where pk_j is the public key. Given an adjustable difficulty condition parameter ξ, a validator firstly computes a hash value $hc = \mathcal{H}(B_i, pk_j, c_j)$ in upper bounded time. Then he/she uses output of $\mathcal{TB}(hc, \xi)$ as work proof, where $\mathcal{TB}(hc, \xi)$ function outputs lower ξ bits of the hash code hc. If work proof $\mathcal{TB}(hc, \xi) \leq d_{cond}(\xi, p_j)$, $(d_{cond}(\cdot, \cdot)$ is difficulty condition function denoted as: $d_{cond}(\xi, p_j) = (2^{\xi} - 1) \cdot p_j)$, he/she generates new block B_{i+1} and broadcasts it with a valid signature to all committee members. After receiving candidate block B_{i+1}, each committee member checks signature and verify proof-of-work. If all conditions are correct, the block will be accepted as "confirmed," and head of local chain will be changed as $head = B_{i+1}$.
- *Chain Finality*: At the end of an epoch, the *head* with epoch height becomes a checkpoint that is used to resolve forks and finalize chain history. The chain finality uses a voting-based algorithm to commit checkpoint block and finalizes those already committed blocks on the main chain. The chain finality ensures that only one path, including finalized blocks, becomes the main chain, as the upper part of Fig. 9.4 shows. Therefore, the following blocks in the new epoch are only extended on such a unique main chain. The chain fork problem is prevented by resolving conflicting checkpoints and finalizing the history of the blockchain.
- *Committee Change*: At the end of the lifetime of a dynasty, the current committee members agree on a new dynasty randomness string. The epoch randomness string generation uses the RandShare mechanism to make an agreement on proposing the next epoch randomness string among members of the final committee. RandShare is a randomness protocol which is based on Publicly Verifiable Secret Sharing (PVSS) [23, 24] to ensure unbiasability, unpredictability, and availability in public randomness sharing. The proposed unbiasable and unpredictable public randomness string will be used for the committee selection process of the next dynasty lifetime.

There are two core functions in epoch lifetime of chain extension: (1) the Proof-of-Credit (PoC) protocol, which is a pure PoS mechanism, determines whether a participant is selected to propose a block given fair initial distribution of the

Table 9.1 Configuration of experimental nodes

Device	Dell Optiplex 760	Raspberry Pi 3 Model B+
CPU	3 GHz Intel Core TM (2 cores)	Broadcom ARM Cortex A53 (ARMv8), 1.4 GHz
Memory	4 GB DDR3	1 GB SDRAM
Storage	250 G HHD	32 GB (microSD card)
OS	Ubuntu 16.04	Raspbian GNU/Linux (Jessie)

credit stake to the committee members in a given epoch; and (2) a Voting-based chain finality (VCF) mechanism could protect against fork by resolving conflicting checkpoints and finalize the history of chain data.

9.5.2 Prototype Implementation and Evaluation

To verify the proposed solution, a concept-proof prototype of Microchain is implemented in Python, consisting of approximately 3000 lines of code. Flask [7] is used, which provide a micro-framework for Python application, to develop networking and web service functions. All security functions are produced by using the standard python lib cryptography [21]. The key generation and signature are implemented over RSA, and the hash function is the SHA-256. SQLite [25], which is a lightweight and embedded SQL database engine, is adopted to manage data such as node, block, and vote information.

The prototype is deployed on a physical network environment, including multiple nodes. Table 9.1 describes devices used for the experimental setup. Five validators are deployed on a desktop while other validators are distributed on sixteen Raspberry Pi (RPi) devices to emulate an IoT environment. Each validator is only deployed on one host machine.

9.5.3 Network Latency

To evaluate the network latency incurred by executing Microchain on IoT devices in terms of the number of validators in committee, validators are deployed on 16 RPi devices performing an entire round of final-committee consensus. The block size used in the test is 128 KB to reduce the influence of block sizes on network performance. Figure 9.5 shows the network delay that takes for Microchain to complete an entire round of consensus epoch cycle with the number of validators varying from 4 to 16.

The latency of committing a transaction $\mathcal{T}_{ct}$ is used for evaluating the time for all nodes of the dynasty to accept a broadcasted transaction. Since the communication

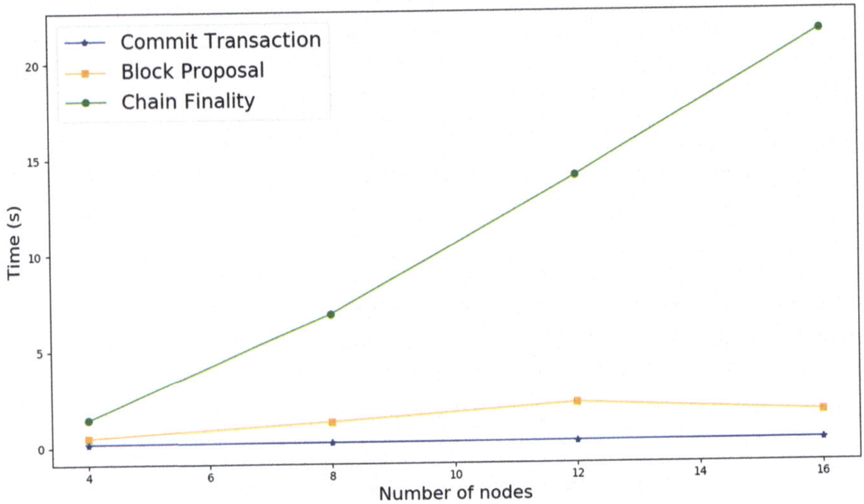

Fig. 9.5 Time latency for one round of Microchain with different node size

complexity of broadcasting transactions is $\mathcal{O}(K)$. The latency of committing transactions is a linear scale to committee size K, and it varies from 162 ms to 246 ms, as the blue line at the bottom of Fig. 9.5 shows.

The yellow line in the middle of Fig. 9.5 indicates the latency of block proposal $\mathcal{T}_{bp}$ process. It evaluates how long the candidate blocks could be generated and verified by validators. Since the block proposal opportunity is proportion to validator's credit distribution $\mathcal{D}$, which has expectation $E(\mathcal{D})$, the latency of block proposal is scale to communication complexity $\mathcal{O}(\frac{K^2}{E(\mathcal{D})})$ and varies from 0.5 s to 1.7 s.

Finally, the latency of chain finality $\mathcal{T}_{cf}$ is the time it takes the voting process for finalizing the checkpoint block to complete among all nodes. The voting-based chain finality process requires that each validator broadcasts its vote among committee members such that the total communication complexity is $\mathcal{O}(K^2)$, thus, the $\mathcal{T}_{cf}$ is greatly influenced by the committee size K. Given 16 validators in the committee, the latency could be 21.5 s, while the scenario with four nodes only introduces 1.4 s latency.

9.5.4 Throughput Evaluation

In the following experiments, five RPi devices work as validators in the committee to focus on throughput given limited influence from the committee size. To evaluate how much data could be processed during a certain period, the block confirmation time is calculated $\mathcal{T}_{bc} = (\mathcal{T}_{ct} + \mathcal{T}_{bp} + \mathcal{T}_{cf})$, which takes for Microchain to complete

Table 9.2 Latency for one epoch cycle of Microchain vs. Block sizes

Block size	512 K	1 M	2 M	4 M
Latency (Second)	8.9	12.2	17.8	54.7

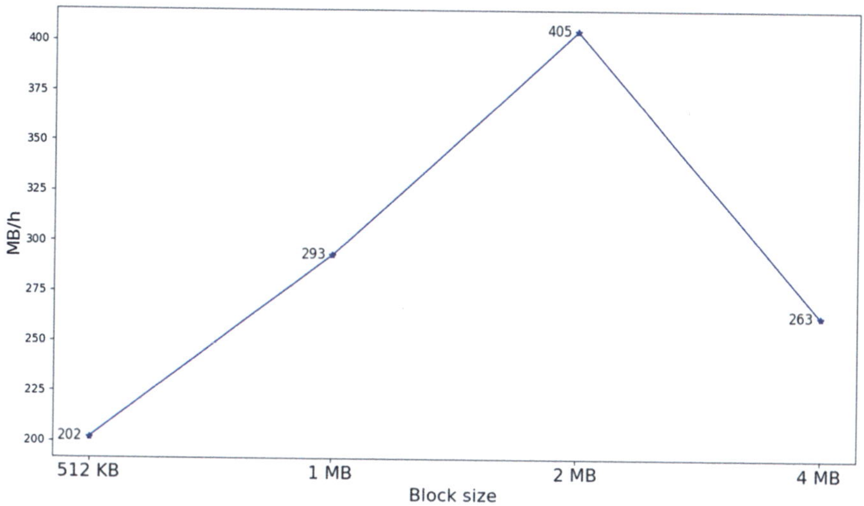

Fig. 9.6 Throughput with different block size

an entire round of final-committee consensus with a varying block size between 512 K and 4 M. With variant block sizes, corresponding time latency is obtained as shown in Table 9.2.

Figure 9.6 demonstrates throughput for one epoch cycle of Microchain given different block size. The block data throughput could be specified as $Th = \frac{Block\_size}{T_{bc}} \times 3600$ (M/h), where M/h means Mbytes per hour. Given varying block size and time delay in Table 9.2, the throughput results are calculated as: Th_{512K}=202 (M/h), $Th_{1M} = 293$ (M/h), $Th_{2M} = 405$ (M/h), and $Th_{4M} = 263$ (M/h). Given a fixed transaction size, like 1K, increasing the block size allows committing more transactions, and therefore reach a higher throughput, which maximizes the system capability.

In the test, running Microchain with 2M block size implies a theoretical maximum rate of $\frac{405\times10^3 K}{3600\times 1K} \approx 113$ (tx/s). As block size increases, however, Microchain achieves higher throughput at the cost of increased latency, and the throughput is constrained by network and system capability. For comparison, Bitcoin achieves a throughput of processing about seven transactions per second by committing a 2 MB block per 10 min.

Table 9.3 Comparative evaluation on different blockchain platforms

	Microchain	Tendermint	Ethereum
Tx committed time (s)	6.5	2.9	4.7
CPU usage (%)	7	26	100
Memory usage (MB)	27	63	1200

9.5.5 Comparative Evaluation

To make a comparative evaluation between Microchain and existing blockchain platforms, a set of experimental test cases are also conducted on two blockchain benchmarks: Tendermint [9] and Ethereum [5]. For Tendermint test network, we use 16 RPi devices as validators, while the Ethereum is implemented on a private network that includes six desktops as miners and two RPi devices as nodes. To evaluate the general performance of committing transaction on blockchain, we specify test scenario that a node launches a 2K transaction (tx) and waits until it has been committed on blockchain. We conduct 100 test runs based on test scenario and evaluate the results regarding several key performance matrices.

Table 9.3 provides a comparative evaluation by running the test scenarios on different blockchain platforms. Tendermint relies on a BFT consensus protocol to achieve deterministic finality, hence, tx committed time is almost stable (about 2.9 s) and lower than Microchain and Ethereum. However, Microchain demonstrates advantages over these benchmarks regarding resource consumption in terms of CPU and memory usage, as Table 9.3 shows. Leveraging a computation intensive PoW consensus algorithm, the mining process in Ethereum almost occupies full CPU capacity and uses about 1.2 GB memory. Therefore, it is not suitable to deploy Ethereum miners on resource constraint IoT devices.

Utilizing lightweight consensus protocols allows Microchain and Tendermint to achieve efficiency in resource consumption on host machine. Thus, they are both suitable to deploy validators on IoT devices. However, Microchain outperforms Tendermint by incurring less resources cost on edge devices, like 7% CPU usage (amount to 1/4 of Tendermint does) and 27 MB memory (amount to 1/2 of Tendermint dose).

9.6 Conclusions

Consensus is the core function of a blockchain system. This chapter introduces the basics of distributed consensus and identifies consensus goals in distributed systems. Given a comprehensive overview on blockchain consensus protocols in terms of BFT-based consensus, Nakamoto consensus and their varieties, challenges on integrating blockchain with IoT are evaluated. Finally, Microchain is introduced as a case study that demonstrates the rationale and approach for designing scalable, lightweight blockchain protocols for IoT systems.

The Microchain provides a promising distributed ledger solution to IoT application scenarios. However, there remains a number of open issues in designing blockchain for IoT in terms of security, scalability, and efficiency. Although committee selection could improve the scalability of Microchain, more investigation and test are needed to evaluate how committee selection algorithm scale to the network size. Another challenge is redesigning chain structure to address the ever-growing chain data size, which has a significant impact on computation and storage capability of IoT devices.

References

1. E. Blasch, R. Xu, Y. Chen, G. Chen, D. Shen, Blockchain methods for trusted avionics systems (2019). arXiv preprint arXiv:1910.10638
2. M. Castro, B. Liskov, Practical byzantine fault tolerance and proactive recovery. ACM Trans. Comput. Syst. (TOCS) **20**(4), 398–461 (2002)
3. M. Castro, B. Liskov, et al.: Practical byzantine fault tolerance, in *OSDI*, vol. 99 (1999), pp. 173–186
4. N. Chen, Y. Chen, Smart city surveillance at the network edge in the era of IoT: opportunities and challenges, in *Smart Cities* (Springer, Berlin, 2018), pp. 153–176
5. Ethereum Homestead Documentation. https://www.ethdocs.org/en/latest/index.html
6. I. Eyal, A.E. Gencer, E.G. Sirer, R. Van Renesse, Bitcoin-ng: a scalable blockchain protocol, in *Proceedings of the 13th* {USENIX} *Symposium on Networked Systems Design and Implementation* ({NSDI} 16) (2016), pp. 45–59
7. Flask: A Pyhon Microframework. http://flask.pocoo.org/
8. Y. Gilad, R. Hemo, S. Micali, G. Vlachos, N. Zeldovich, Algorand: scaling byzantine agreements for cryptocurrencies, in *Proceedings of the 26th Symposium on Operating Systems Principles* (ACM, New York, 2017), pp. 51–68
9. J. Kwon, Tendermint: consensus without mining. Draft v. 0.6, fall **1**, 11 (2014)
10. L. Lamport, R. Shostak, M. Pease, The byzantine generals problem. ACM Trans. Program. Lang. Syst. (TOPLAS) **4**(3), 382–401 (1982)
11. X. Lin, R. Xu, Y. Chen, J. Lum, Enhance generalized exchange economy using blockchain: a time banking case study. The IEEE Blockchain Technical Briefs (2019). https://blockchain.ieee.org/technicalbriefs/march-2019/enhance-generalized-exchange-economy-using-blockchain-a-time-banking-case-study
12. X. Lin, R. Xu, Y. Chen, J.K. Lum, A blockchain-enabled decentralized time banking for a new social value system, in *Proceedings of the 2019 IEEE Conference on Communications and Network Security (CNS)* (IEEE, New York, 2019), pp. 1–5
13. B. Liskov, J. Cowling, *Viewstamped Replication Revisited* (2012)
14. L. Luu, V. Narayanan, C. Zheng, K. Baweja, S. Gilbert, P. Saxena, A secure sharding protocol for open blockchains, in *Proceedings of the 2016 ACM SIGSAC Conference on Computer and Communications Security* (ACM, New York, 2016), pp. 17–30
15. D. Nagothu, R. Xu, S.Y. Nikouei, Y. Chen, A microservice-enabled architecture for smart surveillance using blockchain technology, in *Proceedings of the 2018 IEEE International Smart Cities Conference (ISC2)* (IEEE, New York, 2018), pp. 1–4
16. S. Nakamoto, Bitcoin: a peer-to-peer electronic cash system. Tech. rep., Manubot (2008)
17. S.Y. Nikouei, R. Xu, D. Nagothu, Y. Chen, A. Aved, E. Blasch, Real-time index authentication for event-oriented surveillance video query using blockchain, in *Proceedings of the 2018 IEEE International Smart Cities Conference (ISC2)* (IEEE, New York, 2018), pp. 1–8

18. S.Y. Nikouei, R. Xu, Y. Chen, A. Aved, E. Blasch, Decentralized smart surveillance through microservices platform, in *Sensors and Systems for Space Applications XII*, vol. 11017 (International Society for Optics and Photonics, New York, 2019), p. 110170K
19. O. Novo, Blockchain meets IoT: an architecture for scalable access management in IoT. IEEE Internet Things J. **5**(2), 1184–1195 (2018)
20. B.M. Oki, B.H. Liskov, Viewstamped replication: a new primary copy method to support highly-available distributed systems, in *Proceedings of the Seventh Annual ACM Symposium on Principles of Distributed Computing* (ACM, New York, 1988), pp. 8–17
21. pyca/cryptography documentation. https://cryptography.io/en/latest/
22. F.B. Schneider, Implementing fault-tolerant services using the state machine approach: a tutorial. ACM Comput. Surv. (CSUR) **22**(4), 299–319 (1990)
23. B. Schoenmakers, A simple publicly verifiable secret sharing scheme and its application to electronic voting, in *Annual International Cryptology Conference* (Springer, Berlin, 1999), pp. 148–164
24. M. Stadler, Publicly verifiable secret sharing, in *International Conference on the Theory and Applications of Cryptographic Techniques* (Springer, Berlin, 1996), pp. 190–199
25. SQLite. https://www.sqlite.org/index.html
26. W. Wang, D.T. Hoang, P. Hu, Z. Xiong, D. Niyato, P. Wang, Y. Wen, D.I. Kim, A survey on consensus mechanisms and mining strategy management in blockchain networks. IEEE Access **7**, 22328–22370 (2019)
27. R. Wu, B. Liu, Y. Chen, E. Blasch, H. Ling, G. Chen, A container-based elastic cloud architecture for pseudo real-time exploitation of wide area motion imagery (WAMI) stream. J. Signal Process. Syst. **88**(2), 219–231 (2017)
28. Y. Xiao, N. Zhang, J. Li, W. Lou, Y.T. Hou, Distributed consensus protocols and algorithms, in *Blockchain for Distributed Systems Security* (2019), p. 25
29. Y. Xiao, N. Zhang, W. Lou, Y.T. Hou, A survey of distributed consensus protocols for blockchain networks (2019). arXiv preprint arXiv:1904.04098
30. R. Xu, Y. Chen, E. Blasch, G. Chen, Blendcac: a blockchain-enabled decentralized capability-based access control for IoTs, in *Proceedings of the 2018 IEEE International Conference on Internet of Things (iThings) and IEEE Green Computing and Communications (GreenCom) and IEEE Cyber, Physical and Social Computing (CPSCom) and IEEE Smart Data (Smart-Data)* (IEEE, New York, 2018), pp. 1027–1034
31. R. Xu, Y. Chen, E. Blasch, G. Chen, Blendcac: a smart contract enabled decentralized capability-based access control mechanism for the IoT. Computers **7**(3), 39 (2018)
32. R. Xu, X. Lin, Q. Dong, Y. Chen, Constructing trustworthy and safe communities on a blockchain-enabled social credits system, in *Proceedings of the 15th EAI International Conference on Mobile and Ubiquitous Systems: Computing, Networking and Services* (ACM, New Yotk, 2018), pp. 449–453
33. R. Xu, S.Y. Nikouei, Y. Chen, E. Blasch, A. Aved, BlendMAS: a blockchain-enabled decentralized microservices architecture for smart public safety, in *The 2019 IEEE International Conference on Blockchain (Blockchain-2019)* (IEEE, New York, 2019), pp. 1–8
34. R. Xu, G.S. Ramachandran, Y. Chen, B. Krishnamachari, BlendSM-DDM: Blockchain-enabled secure microservices for decentralized data marketplaces, in *Proceedings of the 2019 IEEE International Smart Cities Conference (ISC2)* (IEEE, New York, 2019)
35. R. Xu, S. Chen, L. Yang, Y. Chen, G. Chen, Decentralized autonomous imaging data processing using blockchain, in *Multimodal Biomedical Imaging XIV*, vol. 10871 (International Society for Optics and Photonics, New York, 2019), p. 108710U
36. R. Xu, Y. Chen, E. Blasch, G. Chen, Exploration of blockchain-enabled decentralized capability-based access control strategy for space situation awareness. Opt. Eng. **58**, 58–16 (2019). https://doi.org/10.1117/1.oe.58.4.041609

37. R. Xu, Y. Chen, E. Blasch, G. Chen, Microchain: a hybrid consensus mechanism for lightweight distributed ledger for IoT (2019). arXiv preprint arXiv:1909.10948
38. R. Xu, Y. Chen, E. Blasch, G. Chen, A. Aved, D. Shen, Hybrid blockchain-enabled secure microservices fabric for decentralized multi-domain avionics systems, in *Sensors and Systems for Space Applications XIII*, vol. 11422. Journal of International Society for Optics and Photonics (2020), p. 114220

Chapter 10
Leveraging Blockchain and SDN for Efficient and Secure IoT Network

Nitin Shukla, Charu Gandhi, and Tanupriya Choudhury

10.1 Introduction

Today, the Internet of Things (IoT) is getting utilized in every field for its prominent applications in real-world scenarios. The IoT applications can be used in different sectors—health, education, industry, logistics, smart city, smart homes, etc. The IoT devices use a variety of sensors for sensing the environment and record the data onto storage devices connected to the IoT sensors with the help of different communication technologies. With the continuous increase of IoT devices throughout the globe, the security of these devices and data getting communicated is of much concern. As these IoT devices have low computational resources, it discourages the application of high-level security measures to safeguard the devices and the data captured by them. IoT devices are, therefore, very much vulnerable to various cyber-attacks due to the always-connected nature of these devices. An attacker may attack to take control of these devices and maliciously configure them to affect the privacy of the end-users.

Further, different IoT devices host different operating systems and link-layer communication protocols. Different operating platforms make these devices prone to a range of attacks. A variety of platforms restricts the application of a general security measure on all the devices with the same effectiveness. Additionally, a successful attack on a single device may lead to access to the complete network.

N. Shukla (✉) · C. Gandhi
Department of Computer Science and Engineering, Jaypee Institute of Information Technology, Noida, India
e-mail: nitin.shukla@jiit.ac.in; charu.gandhi@jiit.ac.in

T. Choudhury
Department of Informatics, School of Computer Science, University of Petroleum and Energy Studies (UPES), Dehradun, Uttarakhand, India

T. Choudhury et al. (eds.), *Blockchain Applications in IoT Ecosystem*, EAI/Springer Innovations in Communication and Computing,
https://doi.org/10.1007/978-3-030-65691-1_10

A large number of heterogeneous IoT devices and their interactions tend to create a very complex system. With the current infrastructure, it is quite hard to monitor the security aspect continuously. These security challenges in an IoT environment lead to the design of new networking and security framework that can deal with attacks on IoT devices.

Traditional networking infrastructure makes it hard to identify the attacks on their own due to their limited capacity for packet inspection. On the other hand, Software-Defined Networks (SDN) [23] enable packet analysis based on application, transport, network, or data-link layer. This facility can add to the security of a communication environment by identifying potential attacks. SDN also offers a comprehensive set of services to enable security in the IoT network. As SDN decouples the data plane and the control plane, it offers logically centralized control management of underlying network devices. Using this centralized control and management, we can quickly identify a security attack as early as it occurs and instruct the underlying network devices to block it by specifying dynamic flows. Several research proposals [3, 12, 13, 19–21, 31] advocated using SDN-based IoT systems to overcome security-related issues.

Additionally, blockchain is a new and cryptographically proven technology to add security to a transaction system of any kind. Blockchain is a distributed and decentralized collection of validated, verified, and permanent transaction records across a large peer-to-peer (P2P) network. These transaction records are considered as blocks and other peers verify the blocks before adding it to a previous chain. Further, it applies various public/private key-based cryptographic schemes. Therefore, it is almost impossible for an attacker to change the data once it is added to a blockchain. Further, its distributed and decentralized nature avoids a centralized attack or failure. This security feature of blockchain can quickly be adopted for securing data over an IoT network. This security feature of blockchain can easily be adopted for securing data over an IoT network. Some researchers [2, 5, 7–10, 16, 18, 25, 29, 32, 34, 36] have also suggested blockchain-based IoT frameworks for provide a secure IoT environment.

In this chapter, a secure and efficient IoT framework based on a combination of SDN and blockchain is proposed. Our proposal intends to achieve adaptability, fault tolerance, security, and reliability for an IoT infrastructure. Blockchain provides secure and decentralized services to the IoT networks. With the use of blockchain, the management of distributed IoT devices will become easier. It also facilitates additional security to the system by adding trust at meager operational cost. SDN, on the other hand, support the IoT system by providing better network management, performance, reduced latencies, and flexibility to introspect the network and packets at any instance.

Rest of the chapter is organized as follows. A brief introduction of related technologies like Blockchain and SDN is presented in Sect. 10.2. In Sect. 10.3, we discuss and analyze major security issues associated with IoT and SDN technologies. Further, the proposed secure blockchain-based software-defined IoT framework is presented and discussed in Sect. 10.4. Finally, Sect. 10.5 concludes this chapter.

10.2 Related Technologies

In this section we provide a ready reference of the technologies—*SDN* and *blockchain*, which helps in developing a secure IoT infrastructure. This ready reference presents the motivation behind utilizing these technologies for proposed framework.

10.2.1 Software-Defined Networks

The Open Networking Foundation (ONF) [15] defines SDN [23] as *Decoupling of network planes i.e., data, control and management. Data planes are responsible for forwarding of data, control plane is responsible for taking control decisions and management plane manages the network applications/policies. SDN abstracts data plane from control and management plane.*

Traditional network devices tightly couple the data, control, and management planes of the network. These devices are fabricated in such a way so that they cannot be programmed later. These devices can only be configured with the set of primitives defined by the manufacturer. One cannot change the policies of these devices as per the dynamic needs of the organization. In order to implement a new policy in the organization, or to achieve new functionality, we need to procure a new network device. On the other hand, in last few years, we have witnessed extreme growth in the development of different network based applications. Further, major portion of network accessibility now lies in the domain of wireless and mobile environments instead of wired networks. This enormous development in application layer and data-link layer further requires the services of network layer responsible for data forwarding/routing, which failed to be developed at the same pace. Thus, SDN helped the overall development of protocol stack to provide better network performance.

Traditional networking devices have tightly coupled data and control plane. SDN decouples these planes from each other. It places the intelligent control plane at a logically centralized device known as controller. A SDN controller maintains a complete topological view of the underlying infrastructure and has control of all the protocols and the policies. The decoupling of the control plane allows programmers to write modules for controller applications. These modules can manage network resources as per the dynamic needs of the organization. The control application also has a complete view of network topology, which aids in optimizing the resources and securing it. So, SDN allows network programmers to write customized applications to meet the organization's day-to-day needs.

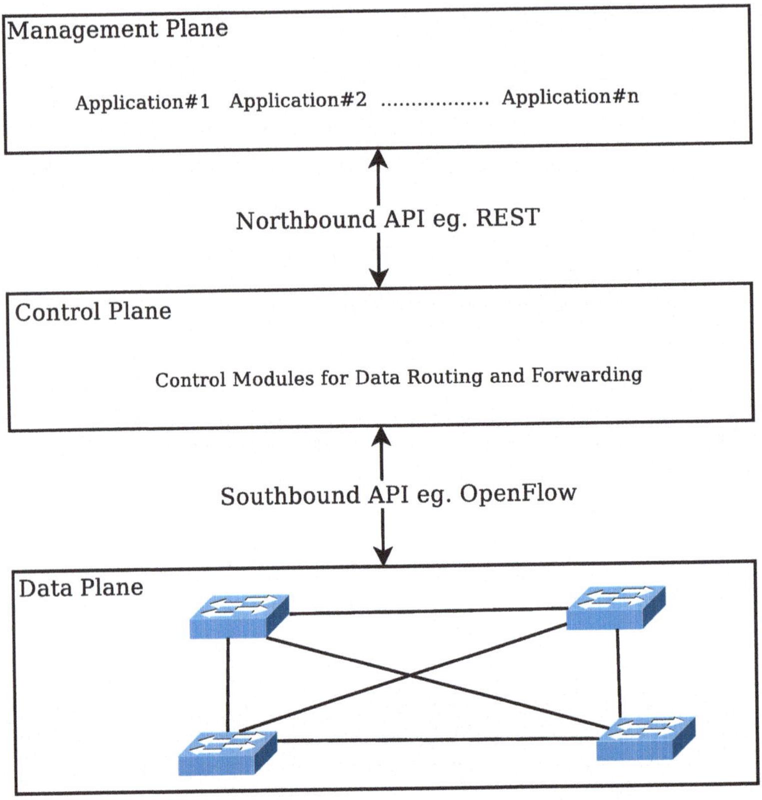

Fig. 10.1 Basic SDN architecture

10.2.1.1 SDN Architecture

Figure 10.1 depicts the underlying architecture of Software-Defined Networking. There are three layers in SDN architecture; the application layer, the control layer, and the infrastructure layer. The control layer is the heart of the architecture which comprises applications for decision-making and is placed at logically centralized device known as controller. The controller application continuously maintains and updates a topological view of the network substrate.

Different application programming interfaces (APIs) are utilized to have communication between these layers. The north-bound APIs between control and application layers assist in providing an abstracted network view to business applications. In the case of distributed controllers, the west-bound APIs are utilized to have communication between them. The most important is the south-bound API that assists switches to communicate with the controller. OpenFlow [24] by ONF is the most popular and accepted south-bound API. Most of the available controllers are written for the OpenFlow enabled switches. The SDN controllers such as NOX [30], POX [4], Floodlight [14], etc. are tested to process a minimum of 50,000 flow requests per second under different test-cases.

10.2.2 Blockchain

Blockchain [28] is a distributed and secure ledger of transactions spread across a large number of computers globally. These computers are termed as *nodes* and are connected by using a peer-to-peer network. The ledgers are used to keeping track of any kind of transaction that can be carried out between two or more parties. These transactions are recorded as *blocks*. A block in a blockchain is secured using cryptographic techniques and cannot be removed from its chain. These blocks are distributed across peers participating in creating a blockchain, and a copy of the block is available with all the peers. A participating peer is termed as *node*.

A block, as seen in Fig. 10.2, stores hash from the previous block and current transactions. The value to hash is verified at every stage for its correctness. A blockchain, as shown in Fig. 10.3, is a distributed network of cryptographic blocks containing transaction information. Every block shares a connection with its preceding and succeeding block. To perform modification of a transaction in one block involves altering the information stored in all the other blocks associated with it. Additionally, the transactions stored in a block are secured using different

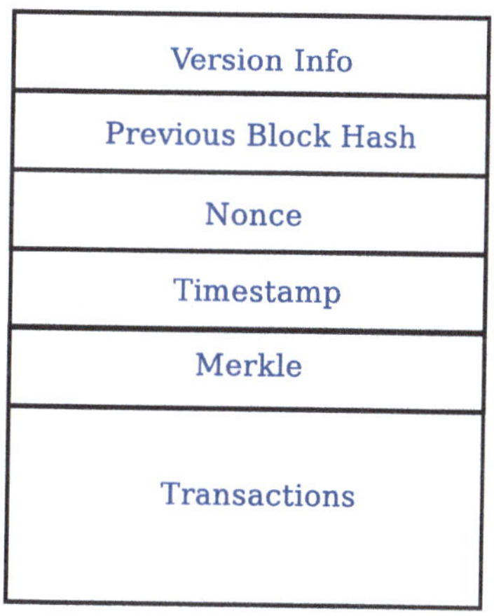

Fig. 10.2 Simple block structure

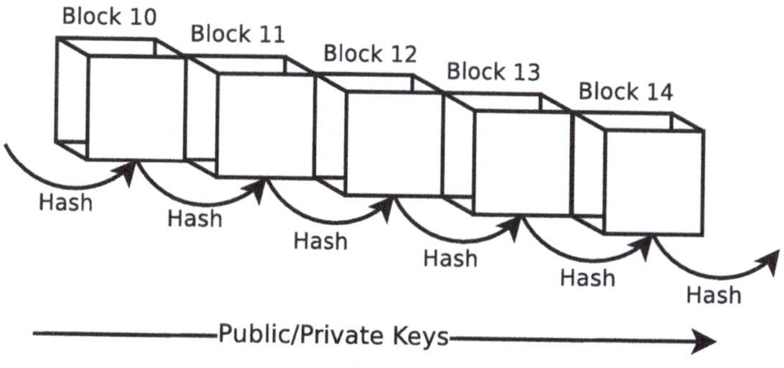

Fig. 10.3 A blockchain

cryptographic means like public/private keys and hashes. Every node participating in the blockchain holds a private key that is used to make transactions secure. Digital signatures are also used to provide authenticity to the block for stored transactions. If an attacker attempts to alter any transaction, the signatures associated with that transaction become changed, resulting in the attack's identification.

10.3 Security Issues

In this section, we discuss the potential threats compromising the security of an IoT environment. Several researches presented in [1, 22, 41] discuss various security challenges and their possible solution. Further, to provide a software-defined IoT environment, it is important to investigate security challenges associated with SDN as well. This section also provides details of possible security attacks in SDN, which need to be considered while creating a secure IoT infrastructure.

10.3.1 Security Issues in IoT

IoT architecture, in general, mainly consists of three layers—*sensing layer, network layer,* and *application layer*.

1. *Sensing Layer:* This layer is also known as the perception layer. It comprises different sensors. This layer is responsible for sensing the data from the surroundings. It then makes the sensed data available for transmission to the cloud services for further processing.
2. *Network Layer:* This layer acts as a middleware between the sensors and the data centers. The sensed data is forwarded to the application layer using wired/wireless networks.
3. *Application Layer:* The topmost layer of a high-level IoT architecture is known as the application layer. This layer provides storage services and various tools/techniques to process the data.

Different IoT layers are susceptible to different kinds of attacks. Several factors need to be considered while suggesting a security solution for IoT devices. It is, therefore, essential to identify and analyze attacks possible at different layers. Figure 10.4 presents a set of possible attacks at different layers of IoT.

The attacks possible in the IoT framework are not limited to the list provided in Fig. 10.4. We have attempted to provide a list of some of the major attacks at different layers. A brief description of these attacks is given as follows:

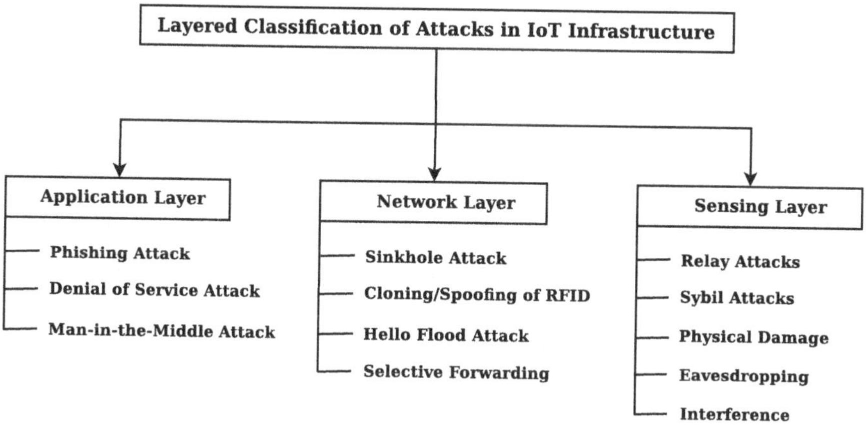

Fig. 10.4 Layered classification of attacks in IoT infrastructure

- **Sensing Layer :**
  (i) *Relay Attacks:* In this attack, the attacker uses the authentication data of a pre-approved device. This authentication data is being replayed again to gain trust in the network.
  (ii) *Sybil Attacks:* An attacker may use the credentials of other nodes in the network and may participate in the network's private communications.
  (iii) *Physical Damage:* In this attack, the attacker causes physical damage to the device by either breaking or stealing the device.
  (iv) *Eavesdropping:* The attacker listens to the communication and can collect confidential information.
  (v) *Interference:* Attacker adds noise signal to the actual data signals sent by the IoT. This causes loss of data and requires re-transmissions.
- **Network Layer :**
  (i) *Sinkhole Attack:* An attacker may use a malicious network device that manipulates the routing table to redirect all the traffic towards itself. The attacker can then change the data.
  (ii) *Cloning/Spoofing of RFID:* In the absence of authentication and trust between management plane and control plane, an impersonation attack can occur by spoiling the subsequent API messages. This may result in the stealing of confidential information of a user.
  (iii) *Hello Flood Attack:* Here, a malicious node is introduced in the network that uses high transmission frequencies to convince the other network devices that the malicious nodes are their neighbor. They then try to forward the data to the end device via that malicious node. This causes an additional delay in data forwarding.

(iv) *Selective Forwarding Attack:* A malicious network device may refuse to forward all the traffic it receives. It only allows a fraction of original packets to be further communicated.

- **Application Layer :**

  (i) *Phishing Attack:* The applications at the application layer are prone to this attack. Using this attack, an attacker can send a spam mail to a user with a link to provide the credentials. The link may appear to be similar to an actual webpage. This way, an attacker can theft the login credentials and other private information of a user.

  (ii) *Denial of Service Attack:* This attack is caused by exhausting resources of a system by forcing excessive use of its resources. The system then stops proper functioning and refuses to provide its services. This attack is also possible to devices available at other layers in the architecture.

  (iii) *Man-in-the-Middle Attack:* While communication is being carried out between two devices, the attacker pretends to be the other party to both of them. The attacker then accesses the confidential information being communicated between the devices.

10.3.2 Security Issues in SDN

In addition to security problems associated with IoT devices, we are also required to focus on security aspects in Software-Defined Networks as well. This will lead to ensure the overall security of a Software-Defined IoT Framework. SDN architecture comprises three layers/plane—*infrastructure layer or data plane, control layer or control plane,* and *application layer or management plane*. All these planes can be prone to several attacks. Further, there are various APIs that are used to enable communication between these devices, as shown in Fig. 10.5.

The APIs are categorized as:

- *South-Bound APIs:* Enable communication between data plane and control plane.
- *North-Bound APIs:* Enable communication between control plane and management plane.
- *East/West-Bound APIs:* Enable communication between control plane of controllers in a multi-controller environment.

All the communication carried out using these APIs is also susceptible to different attacks. Here, we will discuss the possible attacks for layers of SDN and its communication APIs, as shown in Fig. 10.6.

- **Attacks on Network Layers/Planes:**

  (a) *Data Plane:*

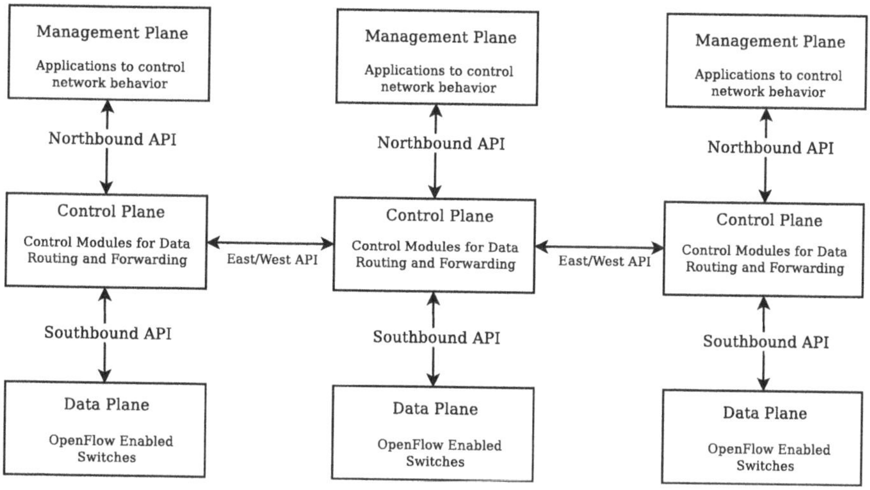

Fig. 10.5 Layers and APIs in SDN architecture

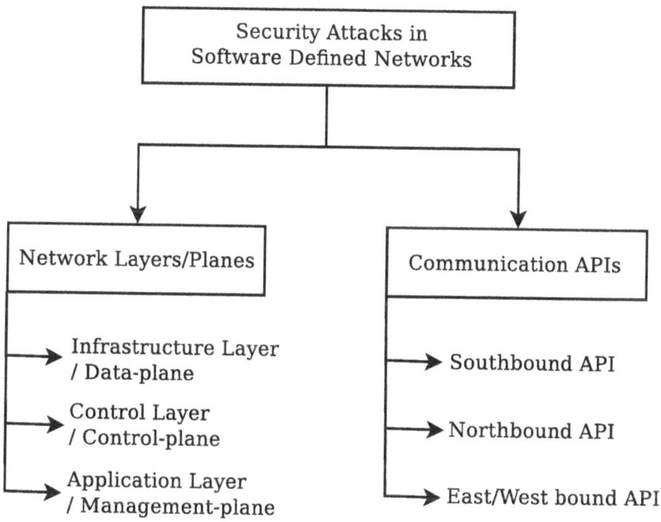

Fig. 10.6 Classification of attacks in SDN

(i) The switches available in the data plane have limited storage capacity; thus, an attacker can flood the flow tables of the switch using a DoS attack.
(ii) As the switches have a module to get connected with the controller, this connection endpoint can be attacked by an attacker if the switch does not support access control and authentication.

(iii) Insertion of manipulated flow-rule may lead to drop/wrong forwarding of the packets.
(iv) Inconsistency among flow rules may lead to incorrect packet processing.

(b) *Control Plane:*

(i) Controllers can be subjected to a DoS, which can compromise the complete SDN network. Various spoofed packets for different OpenFlow messages and malicious controller application can potentially lead to a DoS attack.
(ii) A controller is a logically centralized device that controls the functioning of SDN infrastructure. An attacker can create malicious control plane modules, thus creating inappropriate entries in the switch flow tables.
(iii) Just like any other software system, the controller(s) are also prone to hacking attacks that may lead to complete access to the SDN infrastructure.

(c) *Management Plane:*

(i) The applications used for networking monitoring and policymaking can be malicious or compromised. This can lead to taking complete control of the network and having access to the private information being shared over the network.
(ii) If an attacker can use a compromised API, he can gain control of the SDN network through the controller. If the controller does not have some form of security for the North-bound API, the attacker can create its own SDN policies and can control the whole SDN infrastructure.
(iii) Lack of application isolation can lead to inconsistent flow rules. Flow separation has the advantage of allowing for continuous network updates using different packets with different versions of the policy, thus enabling continuous flow rules.

- **Attacks on Communication APIs:**

(a) *South-bound APIs:*

(i) Lack of encryption between messages exchanged among control and data plane can result in eavesdropping and compromised south-bound exchanges.
(ii) Lack of trust and weak authentication can lead to man-in-the-middle or spoofing attacks, making it easier for attackers to analyze the flow rules and permitted traffic.
(iii) Flow rules can be modified in-between for malfunctioning of the network.

(b) *North-bound APIs:*

(i) In the absence of authentication and trust between management plane and control plane, an impersonation attack can occur by spoiling the subsequent API messages.
(ii) Improper authentication can lead the attacker to communicate malicious instructions to controller applications.

(c) *East/West-bound APIs:*

(i) Insecure endpoints of these APIs may allow the attacker to take control of the controllers.
(ii) If the APIs do not support security features, the data in communication may be modified, leading to inconsistent state exchanges among the controllers.

10.4 Blockchain-Based Secure Software-Defined IoT Framework

In this section, we provide details of our proposed *Blockchain-based secure Software-Defined IoT Framework*. This framework is a combination of three advanced technologies—*Blockchain, SDN, and IoT*. With an increase in the use of IoT devices in major sectors, it is important to ensure the security of IoT devices and communicated data. As the IoT environment, alone, could not avoid security breaches, the use of SDN and Blockchain would help to provide security and efficiency to restrict data theft and manipulation. Various researches available in [6, 11, 17, 26, 27, 33, 35, 37–40] have discussed different applications of blockchain and SDN in IoT environments to make it more secure.

Our proposed framework for a secure IoT framework is shown in Fig. 10.7. The framework comprises three layers—*sensing layer, network layer,* and *application layer*. The sensing layer here comprises different IoT sensors sensing data for different applications. At the network layer, we propose to replace traditional network devices, with blockchain-based SDN infrastructure. The topmost layer of our framework, i.e., the application layer, comprises the blockchain-based distributed cloud. We have divided our design into two stages to provide a better understanding of the proposed framework:

A. SDN as Network Layer in IoT Infrastructure

In any system, significant security issues occur due to the absence of proper authentication and authorization. Using SDN in our proposed IoT framework, we ensure that only authenticated and authorized devices can participate in data sensing

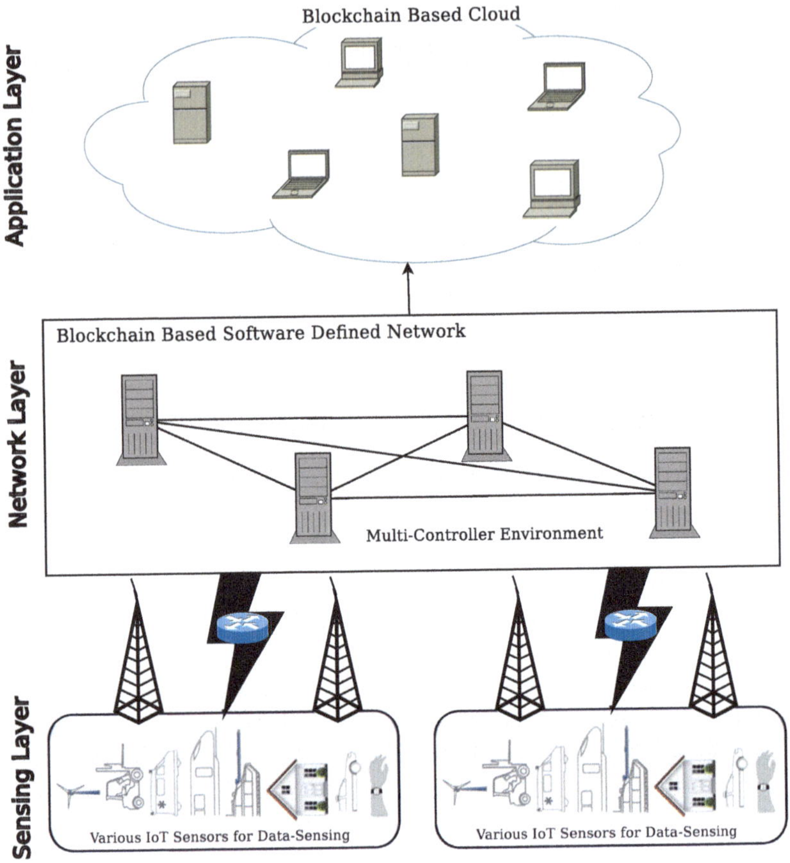

Fig. 10.7 Blockchain-based secure software-defined IoT framework

and data forwarding. SDN provides network programmability and has a global view of each component in the underlying physical network. Using this feature, we proposed to program SDN controllers to develop a list of authenticated end-devices (sensors) and network devices. As soon as the controller identifies data from an unauthenticated device, it blocks all the traffic from that device. Further, if any unauthenticated network devices participate in data forwarding, the controller restricts that device to receive or forward data. At the same time, a new device joins a network, the controller checks for its authentication status and assigns the device to allowed devices or black-list devices.

Additionally, with the help of SDN-based communication framework, the controllers can identify flow characteristics and traffic patterns. Using these patterns and statistics, we can identify the possible attack. This controller-based application can help mitigate attacks like—flooding, DoS, etc. Further, SDN also provides efficient data forwarding based on the dynamic needs of the applications. This can help in reducing communication latency and can provide better throughput to the IoT framework.

B. Use of Blockchain-Based SDN and Distributed Cloud

A blockchain can be used for it contributions towards enhancing efficiency and security in SDN and IoT infrastructures. In the network layer of our proposed model, we suggest to build a blockchain-based multi-controller environment. At application layer we propose to utilize a blockchain-based distributed cloud instead of a traditional centralized data-center. Incorporating blockchain at both the network and application layer of the proposed IoT framework leads to a secure infrastructure.

At network layer, the SDN controllers are connected to a blockchain. Blockchain is broadly for two purposes—to analyze the flow control and to ensure security to the controller applications. While analyzing the flow control it includes the whole network and detects any malicious event. After every communication request, the controller needs to maintain updated flow rules at the data plane network devices. The request contains information about the source and destination. The controller also maintains a global view of the underlying network topology and keeps updating the same as soon as a change occurs. All this information is shared with blockchains to ensure that flow information does not tamper while applying it at the data plane. Additionally, with the help of blockchain technology, details of access policies are available on each controller. When a device participates in the communication, the controller checks its database for device verification and shares it in the blockchain. Each controller compares this requirement with an access policy. When device description matches the access policy, blockchain investigates the description, sends a confirmation to the controller, and adds this transaction to the blockchain.

At the application layer, the blockchain-based distributed cloud provides secure storage, data access, and transaction monitoring. For storage, when an IoT device sends data to store in the cloud, it is sent to the blockchain via the network layer. The nodes in the blockchain then validate the permissions and hash of the previous block. The miner node then stores the data to cloud storage using a unique and random identifier assigned to the transaction. The storage checks the transaction's validity and confirms that there is space available in the cloud storage. The storage node then calculates the hash of the received transaction and compares it with the hash received by the miner. Once the hashes match, the data is stored, and a new block is generated and published to the blockchain. This block is available to all the nodes, and thus, any change in the stored data can be identified by all the nodes in the blockchain. Further, when an application needs to access the data for processing, the application needs to sign the request. As the signed transaction is available with miners, they verify the access rights of the application. After successful verification, the data access is given to the application, and the access request transaction is added as a block to another blockchain. This transaction information can act as proof of data access request made by the application. In addition to these security measures, it is always important to monitor the transactions required to configure the devices in the IoT infrastructure. Blockchain miner directly intervenes each such transaction request. The miner asks the user to send the configuration change request from the owner of the device. Each such transaction is first verified and then added to the blockchain. After that, the request to change the configuration of the device is forwarded to the device. This way, each configuration change request is recorded for security.

10.5 Conclusion

In this chapter, we investigated various security issues associated with IoT frameworks. We further proposed the use of blockchain and SDN to provide a secure and efficient IoT framework. In our framework, we have used blockchain-based SDN for data communication. Although SDN provides certain security features like authentication, authorization, packet inspection, and flow-based security, it also suffers from some implicit security issues. To overcome this aspect, we have used blockchain-based SDN. In addition to this, we also proposed a blockchain-based distributed cloud to provide storage and hosting various data processing services.

The use of blockchain restricts the attacker to change any transaction due to the size of the blockchain and its distributed nature. Blockchain also uses cryptographic techniques to build trust among the nodes and to secure the information. These features of SDN and blockchain are utilized to propose a framework that overcomes the significant security challenges of an IoT ecosystem.

References

1. F.A. Alaba, M. Othman, I.A.T. Hashem, F. Alotaibi, Internet of things security: a survey. J. Network Comput. Appl. **88**, 10–28 (2017)
2. M. Banerjee, J. Lee, K.K.R. Choo, A blockchain future for internet of things security: a position paper. Digital Commun. Networks **4**(3), 149–160 (2018)
3. P. Bull, R. Austin, E. Popov, M. Sharma, R. Watson, Flow based security for IoT devices using an SDN gateway, in *Proceedings of the 2016 IEEE 4th International Conference on Future Internet of Things and Cloud (FiCloud)* (IEEE, New York, 2016), pp. 157–163
4. P. Controller, Pox wiki
5. H.N. Dai, Z. Zheng, Y. Zhang, Blockchain for internet of things: a survey. IEEE Internet Things J. **6**(5), 8076–8094 (2019)
6. A. Derhab, M. Guerroumi, A. Gumaei, L. Maglaras, M.A. Ferrag, M. Mukherjee, F.A. Khan, Blockchain and random subspace learning-based IDS for SDN-enabled industrial IoT security. Sensors **19**(14), 3119 (2019)
7. A. Dorri, S.S. Kanhere, R. Jurdak, Towards an optimized blockchain for IoT, in *Proceedings of the 2017 IEEE/ACM Second International Conference on Internet-of-Things Design and Implementation (IoTDI)* (IEEE, New York, 2017), pp. 173–178
8. A. Dorri, S.S. Kanhere, R. Jurdak, P. Gauravaram, Blockchain for IoT security and privacy: the case study of a smart home, in *Proceedings of the 2017 IEEE International Conference*

on Pervasive Computing and Communications Workshops (PerCom Workshops) (IEEE, New York, 2017), pp. 618–623
9. A. Dorri, S.S. Kanhere, R. Jurdak, P. Gauravaram, LSB: a lightweight scalable blockchain for IoT security and anonymity. J. Parallel Distrib. Comput. **134**, 180–197 (2019)
10. A.D. Dwivedi, G. Srivastava, S. Dhar, R. Singh, A decentralized privacy-preserving healthcare blockchain for IoT. Sensors **19**(2), 326 (2019)
11. S. Faizullah, M.A. Khan, A. Alzahrani, I. Khan, Permissioned blockchain-based security for SDN in IoT cloud networks (2020). arXiv preprint arXiv:2002.00456
12. I. Farris, T. Taleb, Y. Khettab, J. Song, A survey on emerging SDN and NFV security mechanisms for IoT systems. IEEE Commun. Surv. Tutorials **21**(1), 812–837 (2018)
13. O. Flauzac, C. González, A. Hachani, F. Nolot, SDN based architecture for IoT and improvement of the security, in *Proceedings of the 2015 IEEE 29th International Conference on Advanced Information Networking and Applications Workshops* (IEEE, New York, 2015), pp. 688–693
14. P. Floodlight (2017). http://www.projectfloodlight.org. [Online; accessed 24-April-2017]
15. O.N. Foundation, Software-defined networking:the new norm for networks (2013). https://www.opennetworking.org/sdn-resources/openflow
16. K. Jaswal, N. Kashyap, M. Singla, T. Choudhury, A framework for security and protection in internet of things (IoT) devices, in *Proceedings of the 2018 International Conference on Communication, Computing and Internet of Things (IC3IoT)* (IEEE, New York, 2018), pp. 123–126
17. K. Kataoka, S. Gangwar, P. Podili, *Trust list: Internet-wide and distributed IoT traffic management using blockchain and SDN*, in *Proceedings of the 2018 IEEE 4th World Forum on Internet of Things (WF-IoT)* (IEEE, New York, 2018), pp. 296–301
18. A. Khanna, R. Anand, IoT based smart parking system, in *Proceedings of the 2016 International Conference on Internet of Things and Applications (IOTA)* (IEEE, New York, 2016), pp. 266–270
19. A. Khanna, A. Sah, T. Choudhury, Intelligent mobile edge computing: a deep learning based approach, in *International Conference on Advances in Computing and Data Sciences* (Springer, Berlin, 2020), pp. 107–116
20. D.S.R. Krishnan, S.C. Gupta, T. Choudhury, An IoT based patient health monitoring system, in *Proceedings of the 2018 International Conference on Advances in Computing and Communication Engineering (ICACCE)* (IEEE, New York, 2018), pp. 01–07
21. Y. Liu, Y. Kuang, Y. Xiao, G. Xu, SDN-based data transfer security for internet of things. IEEE Internet Things J. **5**(1), 257–268 (2017)
22. R. Mahmoud, T. Yousuf, F. Aloul, I. Zualkernan, Internet of things (IoT) security: current status, challenges and prospective measures, in *Proceedings of the 2015 10th International Conference for Internet Technology and Secured Transactions (ICITST)* (IEEE, New York, 2015), pp. 336–341
23. N. McKeown, Software-defined networking. INFOCOM Keynote Talk **17**(2), 30–32 (2009)
24. N. McKeown, T. Anderson, H. Balakrishnan, G. Parulkar, L. Peterson, J. Rexford, S. Shenker, J. Turner, Openflow: enabling innovation in campus networks. ACM SIGCOMM Comput. Commun. Rev. **38**(2), 69–74 (2008)
25. D. Miller, Blockchain and the internet of things in the industrial sector. IT Prof. **20**(3), 15–18 (2018)
26. A. Muthanna, A. Ateya, A. Khakimov, I. Gudkova, A. Abuarqoub, K. Samouylov, A. Koucheryavy, Secure and reliable IoT networks using fog computing with software-defined networking and blockchain. J. Sens. Actuator Networks **8**(1), 15 (2019)
27. A. Muthanna, A.A. Ateya, A. Khakimov, I. Gudkova, A. Abuarqoub, K. Samouylov, A. Koucheryavy, *Secure IoT Network Structure Based on Distributed fog Computing, with SDN/Blockchain* (2019)
28. M. Nofer, P. Gomber, O. Hinz, D. Schiereck, Blockchain. Bus. Inf. Syst. Eng. **59**(3), 183–187 (2017)

29. O. Novo, Blockchain meets IoT: an architecture for scalable access management in IoT. IEEE Internet Things J. **5**(2), 1184–1195 (2018)
30. Nox Controller (2017). https://github.com/noxrepo/nox
31. F. Olivier, G. Carlos, N. Florent, New security architecture for IoT network. Procedia Comput. Sci. **52**, 1028–1033 (2015)
32. A. Panarello, N. Tapas, G. Merlino, F. Longo, A. Puliafito, Blockchain and IoT integration: a systematic survey. Sensors **18**(8), 2575 (2018)
33. M. Pourvahab, G. Ekbatanifard, An efficient forensics architecture in software-defined networking-IoT using blockchain technology. IEEE Access **7**, 99573–99588 (2019)
34. Y. Rahulamathavan, R.C.W. Phan, M. Rajarajan, S. Misra, A. Kondoz, Privacy-preserving blockchain based IoT ecosystem using attribute-based encryption, in *Proceedings of the 2017 IEEE International Conference on Advanced Networks and Telecommunications Systems (ANTS)* (IEEE, New York, 2017), pp. 1–6
35. N. Rajabi, J. Qaddour, *SDioBoT: A Software-Defined Internet of Blockchains of Things Model* (2019)
36. S. Rathore, B.W. Kwon, J.H. Park, BlockSecIoTNet: Blockchain-based decentralized security architecture for IoT network. J. Network Comput. Appl. **143**, 167–177 (2019)
37. P.K. Sharma, M.Y. Chen, J.H. Park, A software defined fog node based distributed blockchain cloud architecture for IoT. IEEE Access **6**, 115–124 (2017)
38. P.K. Sharma, J.H. Park, Blockchain based hybrid network architecture for the smart city. Future Gener. Comput. Syst. **86**, 650–655 (2018)
39. C. Tselios, I. Politis, S. Kotsopoulos, Enhancing sdn security for IoT-related deployments through blockchain, in *Proceedings of the 2017 IEEE Conference on Network Function Virtualization and Software Defined Networks (NFV-SDN)* (IEEE, New York, 2017), pp. 303–308
40. A. Yazdinejad, R.M. Parizi, A. Dehghantanha, Q. Zhang, K.K.R. Choo, An energy-efficient SDN controller architecture for IoT networks with blockchain-based security. IEEE Trans. Serv. Comput. **13**(4), 625–638 (2020)
41. Z.K. Zhang, M.C.Y. Cho, C.W. Wang, C.W. Hsu, C.K. Chen, S. Shieh, IoT security: ongoing challenges and research opportunities, in *Proceedings of the 2014 IEEE 7th International Conference on Service-Oriented Computing and Applications* (IEEE, New York, 2014), pp. 230–234

Chapter 11
The Biometric Signature as a Blockchain Application

Ahmet Koltuksuz

11.1 The Definition and the Function of a Signature

A signature is a person's name or a nickname handwritten by himself or by herself as a proof of intent and identity for the authorship of the contents of a document. Traditionally, the primary function of a signature has been to create a binding in between a person and a record by perpetually attaching a person's uniquely identifiable handwritten sign as direct evidence that the document that is in part or in whole belongs to that person.

Therefore, the peculiarities of a given signature to fulfill this traditional function are that it should be authentic, unique, unforgeable, undeniable, unalterable, and not reusable and must be created in such a way that it cannot be repudiated.

The determination of the genuineness of the signature has been a problem for centuries. Even today, when a signature drawn by a person's handwriting with the use of a pen, which in such a case is known as the wet signature, is in question in court, many measures must be taken by the forensic laboratories of law enforcement bodies to determine whether the signature is authentic or not. All the metrics and endeavors related to a wet signature are bundled together under the science of graphology.

In the history of humankind, every century has witnessed some technological advancements like that of the invention of the wheel, a clock, steam engines, electricity, and electrical motors until the twentieth century. All these technological improvements have changed societies and their lifestyles tremendously, however, without much affecting the belonging proving methodologies. The signature, produced as a handwritten wet signature for proof of belonging, has always been around

A. Koltuksuz (✉)
Yasar University, Izmir, Turkey
e-mail: ahmet.koltuksuz@yasar.edu.tr

T. Choudhury et al. (eds.), *Blockchain Applications in IoT Ecosystem*,
EAI/Springer Innovations in Communication and Computing,
https://doi.org/10.1007/978-3-030-65691-1_11

for hundreds of years, that is, until the era of Information and Communication Technologies (ICTs).

The Internet, a direct result of the ICTs, has changed humankind unprecedentedly and thus very much deserved to be defined as a game-changer. Thus, the concept of a traditional handwritten wet signature has witnessed quite a few new forms that are altogether different in structure and generation than the classical one the very first time many centuries after. So, the game both for the production and forensics of signature has changed radically.

11.2 A Brief Literature Review on Digital Signatures

The digital signature is a protocol-level application of an asymmetrical cryptosystem. It is the direct result of a combination of both the hash functions and the asymmetrical encryption. In contrast to a symmetrical cryptosystem, in which there is only one key utilized for both processes of encryption and decryption, there are two separate keys involved for each operation in asymmetrical cryptosystems.

The idea belonged to Merkle [1] and Diffie and Hellman [2]. The methodology put forward by these researchers declared that the keys should be created in pairs and be utilized one by one for the processes of encryption and decryption. Moreover, in mathematical terms, it should be intractable to generate one key from the other. The encryption key which belongs to the receiver is publicly known and thus employed by anybody in the process of sending a message to that specific receiver. That way, the encryption key was renamed as the public key.

On the other hand, the legitimate receiver is the only possessor of the decryption key. Thus, as in the renaming of the encryption key, the decryption key is renamed as the secret key. Today, by this renaming convention, the cryptographic system is designated as the public key cryptography (PKC) and the related hardware and software infrastructure as the public key infrastructure (PKI).

The Institute of Electrical and Electronics Engineers (IEEE) standardized PKC as P1363-2000 [3].

While PKC is one compound of the digital signature, the hash functions are the other. A mathematical hash function of h is given as:

$$H = h(m) \tag{11.1}$$

where h is the hash function, m is the variable length message, and H is the fixed length hash value of the message. The hash value for the message can also be termed as the message digest, as the fingerprint, or as the digital fingerprint of the message. The peculiarities of a hash function are as follows: (i) the hash value should be computed in P time for any given message of m; (ii) the hash function should be a one-way function, that is, it must be computationally intractable; and (iii) there should only be one hash value for any given message which means that the hash function should be collision-free.

Table 11.1 An algorithm in pseudocode for the creation of a digital signature

Step number	Pseudocode
1	**start**
2	**read the plaintext as the message (m)**
3	**apply SHA function to (m) to obtain the 512bit hash value (H)**
4	**apply PKC to encrypt the (H) with sender's SECRET key to obtain the digital signature (DS)**
5	**append (DS) to the end of (m) to sign the message**
6	**end**

Secure Hash Algorithms (SHA) have been standardized by the US National Institute for Standards and Technology (NIST), and they produce H values with 160, 256, 384, and 512 bits, respectively [4].

11.2.1 Conventional Digital Signatures

When combined with a hashing function, as mentioned above, one of the marvelous outcomes of asymmetrical cryptosystems is a digital signature. Whether one chooses either the RSA or elliptic curve or ElGamal cryptosystem as the PKC, the digital signature can be created and be added to the document after a couple of steps. The algorithm for digital signature creation is given in Table 11.1.

Currently, one can obtain a digital signature either by having the asymmetrical keys stored on a flash memory to be utilized through the USB port of a computer or having these keys on a SIM card of a mobile phone, which in that case it is called as the mobile signature. However, each of these technologies heavily underlines a hardware dependency; thus, user reluctance has always been an issue for those systems.

11.2.2 Server Signing

Nevertheless, there is one alternative way of digital signing known as the server signing, which runs with asymmetrical keys that are stored on a networked server, and the digital signature is created by that server whenever there is a demand by the signee.

Server signing is founded upon the EU regulation on "electronic identification and trust services for electronic transactions" known as eIDAS [5]. eIDAS might be considered as one of the underlying framework regulations for server signing standard CEN/TS 41924 [6]. The server signing option frees the users form hardware dependencies. However, it is not free from the complexities of the networking hardware and software.

Although a digital signature thus obtained is mathematically proven to be secure, it is nevertheless not so easy to utilize by the signees, and unfortunately, underlying computing intractabilities are susceptible to quantum computing attacks, which seems to be the new revolutionizing technological breakthrough in the days to come along with IoT.

11.3 The Biometric Signature

Biometric authentication can be done in many ways, such as retina, voice, palm, or fingerprint recognition. Along with these, behavioral biometric verification can be used very effectively. A biometric signature is a behavioral biometric recognition that can be done by one's actual handwriting signature on – say – a tablet computer or on a cell phone using a digital pen (a stylus). Since a very conventional way of handwriting does it, it is of no surprise to find the fast acceptance of biometric signatures by banks, hospitals, companies, and various government departments all throughout the world, hence the ISO's standard 19794/7, "Biometric data interchange formats-Part 7: Signature/sign time series data" on biometric signatures [7].

The popularity of biometric signatures is continuously increasing. Recently, Páez et al. proposed an architecture for a biometric electronic document identification implemented on blockchain for enhanced security measures [8]. While Delgado-Mohatar discusses blockchain technologies for storing data in biometric templates [9], Tolosana et al. discuss the biometric signature application on smartphones not with a stylus but with an actual finger touch [10]. Moreover, Bibi et al. delineate the offline and online biometric signature verification systems by taxonomical classification models [11].

The biometric signature consists of three steps that are capturing the image, extracting the signature specific features, and comparing the signature with that of the master signature recorded earlier, respectively. After capturing the handwritten signature on a tablet or a mobile phone, 20 different features on each point of the signature (usually a signature consists of 300–350 points depending on how large the signature is) are extracted for signature recognition. Here are the typical features extracted: the normalized x coordinate, the normalized y coordinate, the pressure of the pen, the altitude angle, the azimuth angle, velocity in x coordinate, velocity in y coordinate, the absolute speed, x coordinate acceleration, y coordinate acceleration, absolute acceleration, tangential acceleration, press derivation, sine of the α, cos of the α, the α-angle between the absolute $\alpha(t)$ velocity vector and the x axis, derivation of α angle, sine of the $\alpha'(t)$, cos of the $\alpha'(t)$, and the angle between two adjacent line segments at each coordinate [12–14]. Figure 11.1 depicts a captured signature image with point number 0, and Table 11.2 shows the extracted 20 features from point number 0.

Once the extraction of these 20 different features from every 300–350 points of the handwritten biometric signature is done, then this data set ($20 \times 350 = 7000$

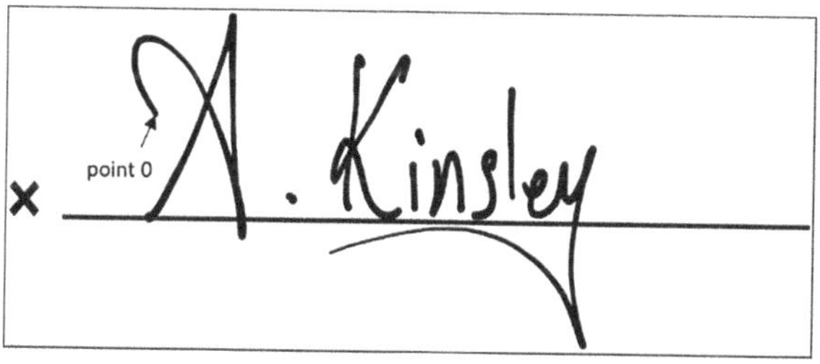

Fig. 11.1 A captured signature. Arrow shows point 0

Table 11.2 Extracted 20 features from point number 0

Feature name	Description	Feature value
x(t)	The normalized x coordinate	−2.0417045316174507
y(t)	The normalized y coordinate	0.8904726440681375
p(t)	The pressure	−0.12352253310985135
altitude(t)	The altitude angle	−0.7705902233032206
azimuth(t)	The azimuth angle	0.19341427835884628
$v_x(t)$	Velocity in x coordinate	0.2578478834273576
$v_y(t)$	Velocity in y coordinate	−0.2321773958415317
v	The absolute speed	2.047146860798148
$a_x(t)$	x coordinate acceleration	0.1153587798695471
$a_y(t)$	y coordinate acceleration	−0.21780126610015516
a(t)	The absolute acceleration	−0.64522618184873
$a_t(t)$	Tangential acceleration	−0.05318895898172835
$p'(t)$	Press derivation	0.993728388123432
$\alpha(t)$	The angle between the absolute Velocity vector and the x axis	−0.3296822027324172
$\sin\alpha(t)$	Sine of the α	−0.3296843007139452
$\cos\alpha(t)$	Cosine of the α	0.15743419748851212
$\alpha'(t)$	Derivation of α angle	0.0026385951811272353
$\sin\alpha'(t)$	Sine of the $\alpha'(t)$	0.0026385951811272353
$\cos\alpha'(t)$	Cosine of the $\alpha'(t)$	0.0026385951811272353
$\beta(t)$	The angle between two adjacent line segments at each coordinate	0.002739030665447192

specific data item in total) is dynamically compared with the original master handwritten signature data of the user which was obtained earlier. The dynamic comparing process creates a threshold value. Once and if the comparison threshold value is in the acceptance interval, then the biometric signature can be accepted, hence no forgery. Figure 11.2 shows a comparison of a genuine signature against a fraud by selected features.

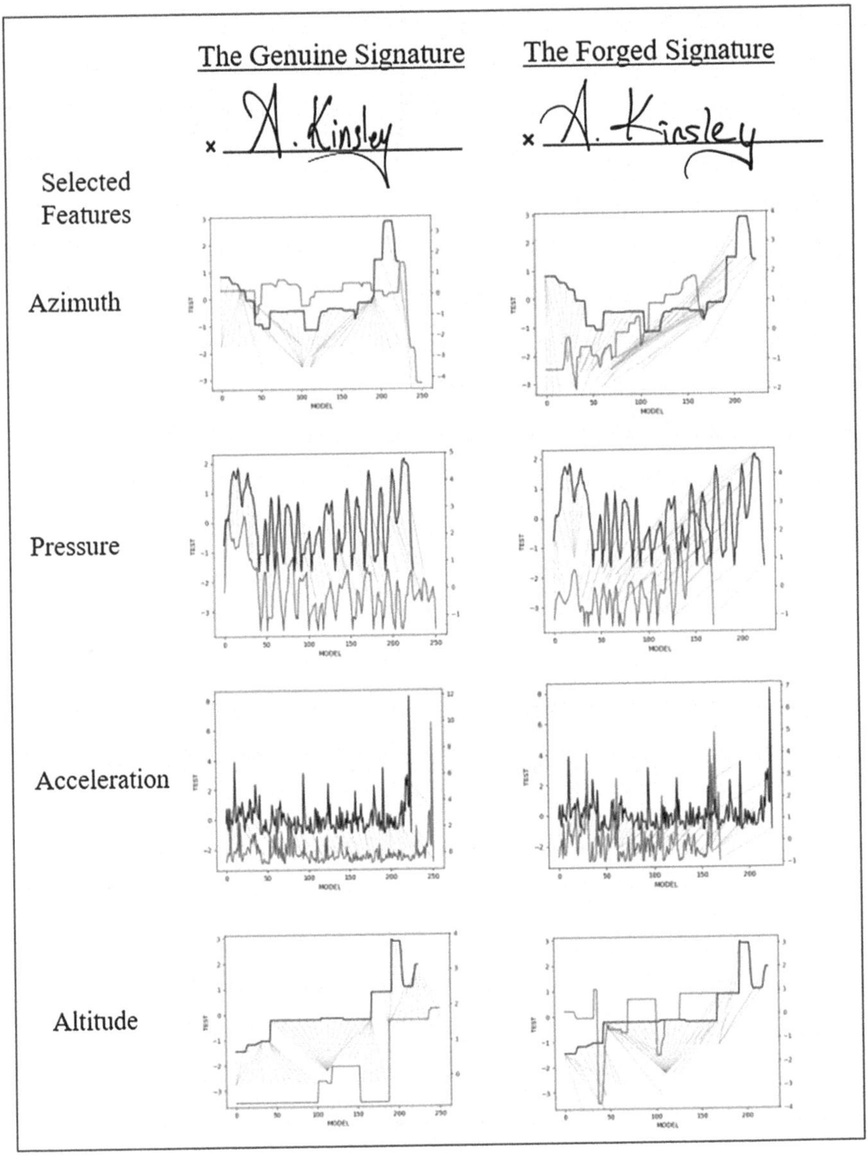

Fig. 11.2 The comparison of genuine and forged signatures by selected features only

11.4 The Biometric Signature on a Blockchain

Hardware dependency has always been a significant issue for conventional digital signing. Even though server signing was kind of an answer to that unsettled question, it has not without its networking issues. On the other hand, computational intractability, which provides the security and reliability for all these asymmetrical cryptosystem protocols, is due to our current computational model.

The advancements in the science of physics and engineering make it possible that quantum computing will be in use in a decade or so. If this will be the case, the conventional asymmetrical cryptosystems will be useless. Thus, digital signing methodologies as we know them today will be pushed aside.

While handwritten signing on a touch-sensitive screen like that of a tablet and/or a mobile phone is natural, hence the frequent and rapid acceptance by the industry, the data which is composed of the signee's signature image should still be kept under tight security. Therefore, all the information reflected as the extracted features from the points of the image must be stored along with the image of the signature itself.

The idea of utilizing the conventional cryptographic protocols to provide security for biometric data is by no means the only alternative due to the issues mentioned above.

What we propose is to have all that biometric information added to a blockchain. With a new hashing algorithm that will be developed as a quantum computing resistant, the blockchain will be one of the safest solutions to come.

As detailed in Sect. 11.3, the biometric info in the form of extracted features from all the points of the signature image provides the base for comparison. However, there must be at least five authentic signature images obtained from the signee to develop the genuine signature base with all extracted features to be kept in a blockchain. Table 11.3 shows the basic model for a blockchain.

Table 11.3 A blockchain entry for a biometric signature genuine base

Data Structure in a Blockchain


```
{
  "data": {
    "client": "2020",
    "threshold": 2.316184737819842,
    "signatures": {
      "s1": {350 items...}
      "s2": {350 items...}
      "s3": {350 items...}
      "s4": {350 items...}
      "s5": {350 items...}
    },
}
 "prev_hash":
"d498df3a5b9e4935cf965da99b2e2dbb64e876453479a05fb1c0801df3126a7
5",
  "timestamp": 1594906588.52126,
  "proof": 32,
  "index": "16"
}
```

Table 11.4 A blockchain entry for the extracted features of the first point of the first signature

Data Structure of First Points of First Signature


```
"signatures": {
        "s1": {
          "0": {
            "x": -1.988754379338032,
            "y": 0.4511260131208926,
            "p": -0.7661117399152955,
            "ax": -0.08439360522909167,
            "ay": 0.6915132498531755,
            "vx": 0.06674528445182873,
            "vy": 0.1399171613281847,
            "altitude": -1.4444335626533926,
            "azimuth": 0.8187890660248799,
            "v": 1.9903431712158168,
            "a": -0.2087518489067404,
            "att": 0.026736406092351797,
            "dvp": 0.4651733290018387,
            "alfa": -0.2052921036689894,
            "sina": -0.20529230586040206,
            "cosa": 0.3752404477999517,
            "deva": 0.11523422497388082,
            "devsina": 0.11523422497388082,
            "devcosa": 0.11523422497388082,
            "beta": -0.00020099889827516584
          },
```


Each block includes extracted features from all points of five genuine signatures along with client, threshold, previous hash, and timestamp info. Table 11.4 indicates the extracted features of the first point of the first signature.

All the details of a transaction must also be added to the block. Table 11.5 depicts the transaction details as kept in blockchain. Note that the latitude and longitude info along with the time info also stored in blockchain for the increased reliability of the whole transaction.

11.5 Conclusion: The Biometrix Project

The idea of storing the biometric information on blockchain was realized in a project called Biometrix. The issues in signing and the related biometric solutions along with a blockchain implementation outlined above were addressed in the Biometrix project. The detailed information concerning the application of Biometrix can be accessed in GitHub [15].

Table 11.5 Transaction information on blockchain

Transaction Information Data Structure


```
{
  "transaction-id": "B90CFE6E-CCD6-46A1-84BE-1CE5B861F1DE",
  "client-id": "2020",
  "end-user-id": "taylanakbas@bx.com",
  "document-id": "2C700091-D9C6-4266-ABA3-D60A07FF03B0",
  "timestamp": "16-07-2020 16:37:50",
  "signature": {
    "0": {
      "x": -2.0417045316174507,
      "y": 0.8904726440681375,
      ...
      ...
      ...
    }
    "1": {
      "x": -2.0221943900950627,
      "y": 0.8639206844542605,
      ...
      ...
      ...
    }
    ...
    ...
    ...
  },
  "score": "2.301189502775008",
  "threshold": "2.316184737819842",
  "difference": "-0.015",
  "transaction-result": "Genuine",
  "latitude": 38.44594114808755,
  "longitude": 27.202107367501608
}
```


References

1. R.C. Merkle, Secure communications over insecure channels. Commun. ACM **21**(4), 294–299 (1978)
2. W. Diffie, M.E. Hellman, New directions in cryptography. IEEE Trans. Inf. Theory **22**(6), 644–654 (1976)
3. IEEE Homepage., https://standards.ieee.org/standard/1363-2000.html. last accessed 16 July 2020
4. Secure Hash Standard (SHA), *Federal Information Processing Standards (FIPS) Publication 180–4* (2015). https://doi.org/10.6028/NIST.FIPS.180-4. August 2015

5. eIDAS: Regulation (EU) No 910/2014 of the European Parliament and of the Council of 23 July 2014 on electronic identification and trust services for electronic transactions in the internal market and repealing Directive 1999/93/EC
6. EN 419241-1:2018: *Trustworthy Systems Supporting Server Signing – Part 1: General System Security Requirements*. 15-Jul-2018
7. ISO/IEC 19794/7, *Biometric Data Interchange Formats-Part 7: Signature/Sign Time Series Data* (2007)
8. R. Páez, M. Pérez, G. Ramírez, J. Montes, L. Bouvarel, An architecture for biometric electronic identification document system based on blockchain. Future Internet **12**, 10 (2020)
9. O. Delgado-Mohatar, J. Fierrez, R. Tolosana, R. Vera-Rodriguez, *Blockchain Meets Biometrics: Concepts, Application to Template Protection, and Trends* (2020). https://arXiv.org/2003.09262 [cs.CV]
10. R. Tolosana, R. Vera-Rodriguez, R. Guest, J. Fierrez, J. Ortega-Garcia, Exploiting complexity in pen- and touch-based signature biometrics. Int. J. Doc. Anal. Recognit. (IJDAR) **23**, 129–141 (2020). https://doi.org/10.1007/s10032-020-00351-3
11. K. Bibi, S. Naz, A. Rehman, Biometric signature authentication using machine learning techniques: Current trends, challenges, and opportunities. Multimed. Tools Appl. **79**, 289–340 (2020). https://doi.org/10.1007/s11042-019-08022-0
12. O. Hurtada-Miguel, *Online Signature Verification Algorithms and Development of Signature International Standards*. Ph.D. Thesis, Universidad Carlos III de Madrid, September (2011)
13. T.Q. Ton, T. Pham Tung, Online signature verification using dynamic time wrapping and extended regression. Int. J. Adv. Res. Comput. Eng. Technol. (IJARCET) **4**(5), 1854 (2015 May)
14. T. Giorgino, Computing and visualizing dynamic time warping alignments in R: The dtw package. J. Stat. Softw. **31**(7), 1–24 (2009). https://doi.org/10.18637/jss.v031.i07
15. T. Akbaş, *Biometrix- BIOMETRIX – Artificial Intelligence Assisted Biometric Signature on Block Chain* (Engineering Graduation Project, Yasar University, Department of Computer Engineering, Izmir, 2020) https://github.com/taylanakbas/Biometrix

Chapter 12
BlockTwins: A Blockchain-Based Digital Twins Framework

Ezz El-Din Hemdan and Amged Sayed Abdelmageed Mahmoud

12.1 Introduction

Recently, digital twin has gained significant interest from academia and the industry due to its considerable impact on increasing productivity. DT is the digital replica of an actual-world physical asset, product, or a system around us. Digital twin concepts employed in previous works demonstrate two important features: (1) Every idea explores the relation between the actual system and the consequent simulated system [1], and (2) this relation is proven by producing real-time information from sensors [2]. The idea of a digital twin can be linked with other ideas such as cross-real worlds or co-spaces and mirror prototypes, which aim to, by and large, synchronize part of the physical world with its cyber representation [3, 4].

The digital twin comprises of different modules: actual device, simulated product, and communication between the actual product and virtual product. A digital twin for a car product is depicted in Fig. 12.1. The communication between the physical and the virtual product is important for preserving the vitality of digital twins. The data transmission from a virtual product to a physical product can be utilized to observe and support the execution of the actual product.

Recently, with the appearance of blockchain technology, digital twins have been redefined in its various applications in the Internet of Things. It can be used for transferring data and value onto the Internet with full transparency and security. Conservatively, to build a digital twin system, it needs a central intermediary that

E. E.-D. Hemdan (✉)
Department of Computer Science and Engineering, Faculty of Electronic Engineering, Menoufia University, Shibin Al Kawm, Egypt

A. S. A. Mahmoud
Department of Industrial Electronics and Control Engineering, Faculty of Electronic Engineering, Menoufia University, Shibin Al Kawm, Egypt

T. Choudhury et al. (eds.), *Blockchain Applications in IoT Ecosystem*, EAI/Springer Innovations in Communication and Computing, https://doi.org/10.1007/978-3-030-65691-1_12

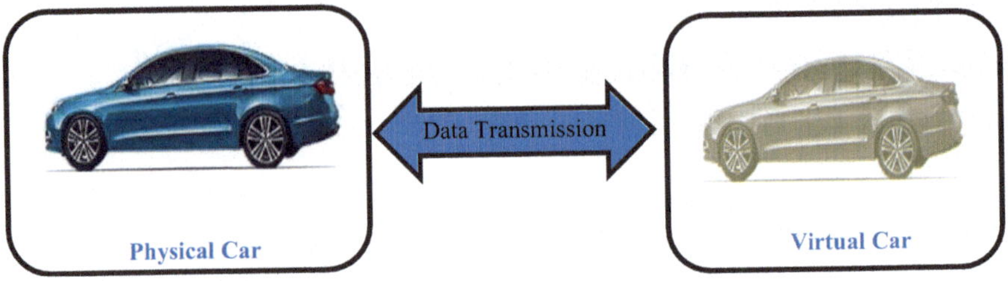

Fig. 12.1 A digital twin of a car product

is dependable in performing analytics and holding data. To generate and monitor digital twins in a secure and immutable manner, blockchain can achieve this aim. Likewise, to monitor the various stages in the construction of DTs, there is a need for a secure, reliable, robust, and consistent method.

Likewise, linking digital twin and blockchain will support businesses and brands to protect their products from being counterfeited and prevent financial losses. Hence, this work aims to propose a decentralized blockchain-based digital twins framework. Concisely, the main contributions of this work are as follows:

- Explore the basic concepts of blockchain and digital twin technologies
- Present a blockchain-based digital twin framework that assures secure and reliable traceability, convenience, and accessibility of transactions and data provenance of its creation process as well as governing and tracking connections initiated by applicants engaged in the digital twins system

The rest of this work is organized as follows: Sect. 12.2 presents the concept of the digital twins, while the blockchain terminology and its concepts are provided in Sect. 12.3. Section 12.4 provides the importance of combining the digital twins with blockchain as a perfect pair, while the proposed framework is introduced in Sect. 12.5. Finally, the conclusion of this work is provided in Sect. 12.6.

12.2 Digital Twins

Thanks to tremendous development in communication and information technology in the last decade, digital twins has become a dynamic topic recently and has been applied in different fields such as manufacturing, smart cities, biomedicine, and aerospace [5–9]. DT can be defined as a replicate or twinning the real system, product, and/or assets using a computer-based model based on collected data and information from the system [10]. Likewise, there are various definitions of DT as tabulated in Table 12.1. The advantage of DT is that it can simulate and model a simple or more complicated process ranging from vehicle parts to homes, cities, and even humans [8].

With the proposed digital twins technology by Grieves in 2002, its purpose is to simulate the system to increase productivity, optimize the operation, and reduce

Table 12.1 Definitions of digital twins

References	Definition
Grieves & Vickers [14]	"The Digital Twin is a set of virtual information constructs that fully describes a potential or actual physical manufactured product from the micro atomic level to the macro geometrical level. At its optimum, any information that could be obtained from inspecting a physical manufactured product can be obtained from its Digital Twin"
Glaessgen & Stargel [15]	"A Digital Twin is an integrated multiphysics, multiscale, probabilistic simulation of an as-built vehicle or system that uses the best available physical models, sensor updates, fleet history, etc., to mirror the life of its corresponding flying twin"
Tao, Sui [16]	"Digital twin is a real mapping of all components in the product life cycle using physical data, virtual data and interaction data between them"
Bolton [17]	"A dynamic virtual representation of a physical object or system across its lifecycle, using real-time data to enable understanding, learning and reasoning"
Söderberg [18]	"Using a digital copy of the physical system to perform real-time optimization"

the cost of the productions. The main three parts of DT are the physical system, the virtual model, and the connections between the physical and virtual [11] It is important to understand the behavior and characteristics of DT. DT has some characteristics which discriminate it from other technologies:

- Connectivity
- Reprogrammable and smart
- Digital traces
- Modularity
- Homogenization

The digital twin persistently monitors and observes data from several supplies that support to forecast product safety and recognize the defect in working condition and then send the information to the physical systems to drive prime result. Precisely, if an issue happens in one system and is perceived and regulated, then that solution and operation is not only applied in that system but likewise in other identical systems across the world to provide optimized operation and service [12]. Similarly, predictive modelling is employed in DT to prognosticate the upcoming changes in the real system such as failures in the product's life cycle [8]. Thus, DT can be installed on the device itself or the cloud or edge computing, and the data from the sensors are transferred to the virtual model [13].

There are many applications of digital twins, from the production process, aviation, and agriculture to smart city applications and healthcare systems. Any digital twin platform must be designed and built with special care because it must be resilient to malware and viruses due to the usage of IoT and cloud computing. Important data and relevant information can be damaged because of hacking. Consequently, safety and privacy should be taken very seriously, particularly when it comes to biomedical and healthcare fields [8]. Through technological developments such as blockchain, there are many ways to improve privacy and emphasis on seeking approaches that support securing digital twins' data.

12.3 Blockchain

In the last decade, a new technology called blockchain was developed by Nakamoto [19] to operate as the decentralized transaction ledger of the digital currency called Bitcoin. Blockchain is constructed by a collection of blocks connected by cryptography. The structure of the blockchain network is considered as an ordered list of blocks shown in Fig. 12.2 where each block belongs to a prior block, providing a blockchain. When a block has been generated and connected to the blockchain, the operations in that block cannot be altered or returned [20].

The blockchain core is the coordination process that certifies that all compromise nodes on the network agree on a single global state of the blockchain. A blockchain network typically consists of data producers, consensus nodes, and data pool. When data producers want to write the data on the blockchain, they first submit their data to the data pool as presented in Fig. 12.3. Then the data will be collected by

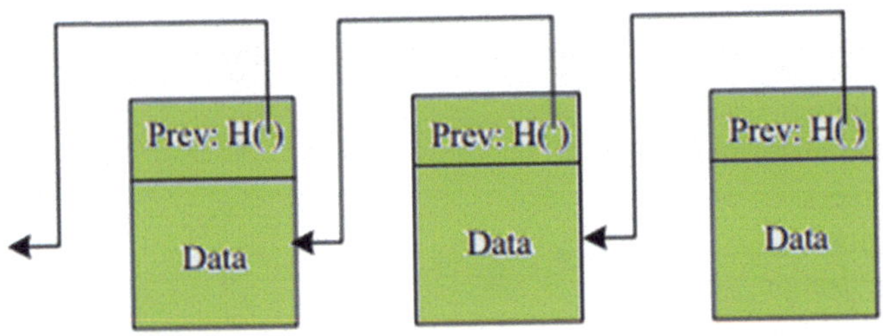

Fig. 12.2 Blockchain connected network [20]

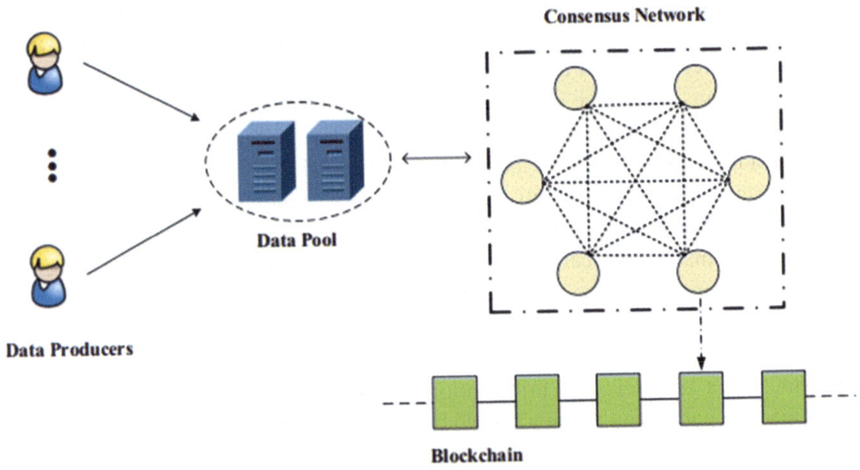

Fig. 12.3 The working process of the blockchain network [20]

compromise points in the consensus network from the data pool. After validating the collected information, the consensus protocol is run by the consensus node, and the bookkeeping node will be picked. The bookkeeping node shall submit the data to the blockchain [20].

The blockchain system is a decentralized and public digital ledger where every engaged history cannot be changed retroactively, without the change of all subsequent blocks. There are several types of blockchain as tabulated in Table 12.2. Currently, blockchain is used in different fields, such as transportation, healthcare, electronic voting, logistics, and so on.

12.4 Blockchain and Digital Twins Pair

Digital twins and blockchain can be leveraged together for their security features and assist businesses to thwart instances of fraud and duplication of their products and services. In the last days, businesses always have been counterfeiting their products. Technology is universal and is advancing at a quick step. Therefore, it has become much easier for fraudsters to create replicas and sell it to unsuspecting customers. These fraudsters not only cause financial losses for reputable brands but may also cause permanent reputational losses. The combination of digital twin and blockchain can provide us with a solution to prevent frauds and help businesses to maintain the authenticity of their offerings.

In 2020, the approximated IoT devices are about over 20 billion. These devices will be able to support millions of digital twins. Digital twins will form one of the fundamental pillars of the digitization of physical objects. Blockchain technology, on the other hand, with its decentralized framework, will bring in transparency, further strengthening the security of the digital data. The concept of combining digital twin and blockchain can be applied in various applications such as in logistics and the medical field. The benefits of using blockchain for digital twins are shown in Fig. 12.4.

12.5 Blockchain-Based Digital Twins Framework

In this section, we describe our proposed blockchain-based digital twins framework called BlockTwins to secure a digital twins system. Transactions would not have to rely on third-party verifications to ensure the security of each transaction in the twins' system during the communication between the virtual and physical assets. Instead, each transaction would be timestamped and then hashed into an ongoing chain of hash-based proof of work. This can prevent any external malicious tampering and modifications by criminals and illegitimate users.

In the industrial control system and manufacturing, the product life cycle contains the strategy, production, servicing, and so on, in which tremendously various

Table 12.2 Types of blockchain systems

Type	Features					
	Access to data	***Network expansion***	***Proof of transaction***	***Identifiability***	***Transaction speed***	Transaction maker
Private	Authorized users	Very easy	Central agency	Possible to identify	Fast	Only authorized users
Public	Anyone	Difficult	Verification algorithm	Anonymity	Slow	Everybody
Consortium	Authorized users	Easy	Previously agreed rule	Possible to identify	Fast	Only authorized users

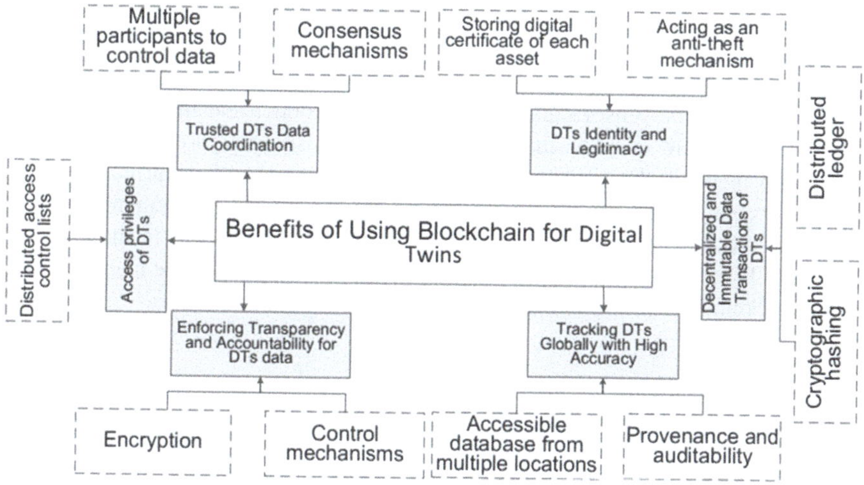

Fig. 12.4 Benefits of using blockchain for digital twins [21]

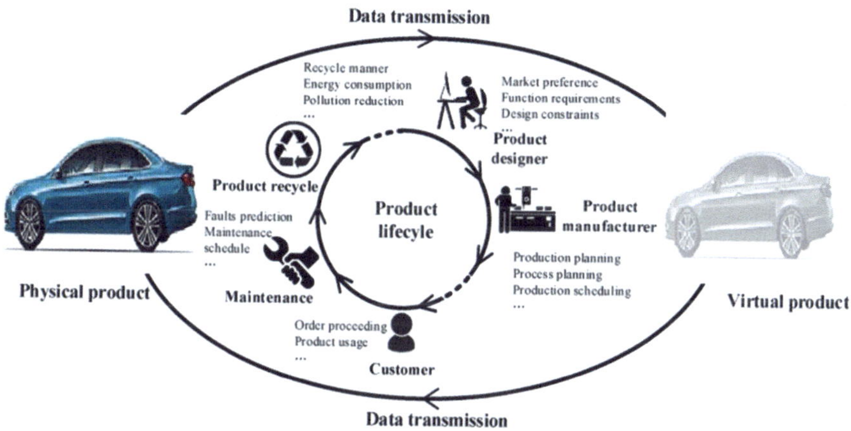

Fig. 12.5 Product life cycle management based on a digital twin

product data are produced and termed as product life cycle data. It is a complex task that should be focused on complicated details in each cycle. The product life cycle management is required to guarantee that all the processes within the product design phases are under control. The advent of DT supports a method to supervise every action of the device inside the whole life cycle and improve the performance of the device based on the virtual model of the digital twin, as demonstrated in Fig. 12.5. Thus, blockchain can be used to handle the problems in the data management of digital twins for all phases of device production securely and efficiently. These problems involve data storage, data sharing, data access, and data authenticity.

To paradigm the proposed system, it is required to build a blockchain network to connect all components within the product life cycle. Every activity of the DT of the product between contenders is logged by the transaction. The transaction will also record the sensor data between the real product and the digital model. All transactions are saved in the linked blocks by hashing algorithms along with timestamp involves the entire processes to mark the time of occurrence. These blocks are connected to establish a blockchain-based product management network. Likewise, blockchain is used to handle the main phases involved in the creation process of DTs as shown in Fig. 12.6. The proposed product life cycle management-based blockchain is shown in Fig. 12.7.

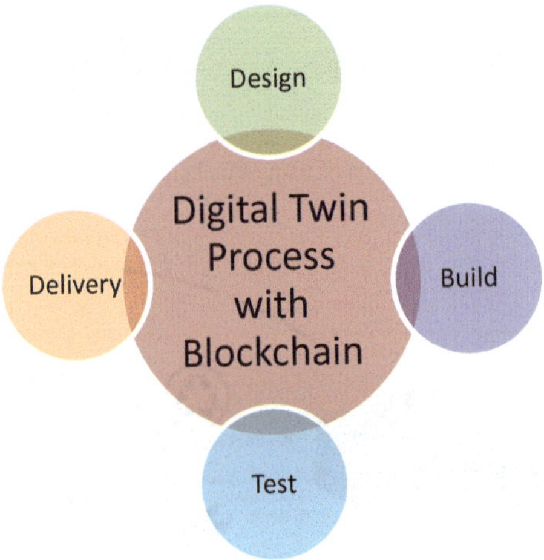

Fig. 12.6 Blockchain as the managing entity for the DT creation process

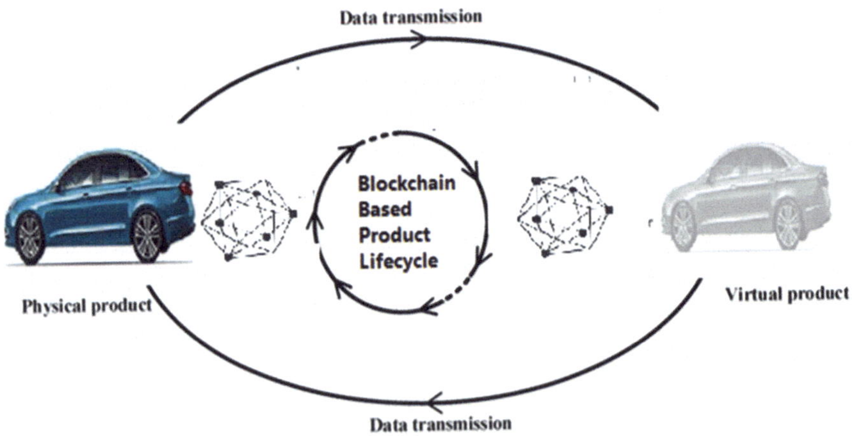

Fig. 12.7 Blockchain-based product life cycle management

12.6 Conclusion

The concept of digital twins can be redefined with the advent of blockchain. It became a decisive technology to aid the IoT-based digital twin's applications for transferring data and value onto the Internet with full transparency, accessibility, trusted traceability, and immutability of transactions and data provenance. Therefore, this work presents a framework of blockchain-based digital twins.

References

1. M.B. Chhetri, S. Krishnaswamy, S.W. Loke, Smart virtual counterparts for learning communities, in *International Conference on Web Information Systems Engineering*, (Springer, Berlin, Heidelberg, 2004, November), pp. 125–134
2. G. Bacchiega, *Creating an Embedded Digital Twin: Monitor, Understand and Predict Device Health Failure*. Inn4mech-Mechatronics and Industry, 4
3. S.W. Loke, S. Smanchat, S. Ling, M. Indrawan, Formal mirror models: An approach to just-in-time reasoning for device ecologies. Int. J. Smart Home **2**(1), 15–32 (2008)
4. S.W. Loke, B.S. Thai, T. Torabi, K. Chan, D. Deng, W. Rahayu, A. Stocker, The La Trobe e-sanctuary: Building a cross-reality wildlife sanctuary, in *2015 International Conference on Intelligent Environments*, (IEEE, Prague, Czech Republic, 2015, July), pp. 168–171
5. A. Fuller, Z. Fan, C. Day, C. Barlow, Digital twin: Enabling technologies, challenges and open research. IEEE Access **8**, 108952–108971 (2020)
6. D. Jones, C. Snider, A. Nassehi, J. Yon, B. Hicks, Characterising the digital twin: A systematic literature review. CIRP J. Manuf. Sci. Technol. **29**, 36 (2020)
7. F. Tao, F. Sui, A. Liu, Q. Qi, M. Zhang, B. Song, et al., Digital twin-driven product design framework. Int. J. Prod. Res. **57**(12), 3935–3953 (2019)
8. B.R. Barricelli, E. Casiraghi, D. Fogli, A survey on digital twin: Definitions, characteristics, applications, and design implications. IEEE Access **7**, 167653–167671 (2019)
9. Y. Zheng, S. Yang, H. Cheng, An application framework of digital twin and its case study. J. Ambient. Intell. Humaniz. Comput. **10**(3), 1141–1153 (2019)
10. S. Boschert, R. Rosen, Digital twin—The simulation aspect, in *Mechatronic Futures*, (Springer, Cham, 2016), pp. 59–74
11. Q. Qi, F. Tao, Digital twin and big data towards smart manufacturing and industry 4.0: 360-degree comparison. IEEE Access **6**, 3585–3593 (2018)
12. M. Grieves, Digital twin: Manufacturing excellence through virtual factory replication. White Pap. **1**, 1–7 (2014)
13. S. Boschert, C. Heinrich, R. Rosen, Next generation digital twin, in *Proceedings of the TMCE*, (Committee of TMCE, Las Palmas de Gran Canaria, 2018, May), pp. 209–217
14. M. Grieves, J. Vickers, Digital twin: Mitigating unpredictable, undesirable emergent behavior in complex systems, in *Transdisciplinary Perspectives on Complex Systems*, (Springer, Cham, 2017), pp. 85–113
15. E. Glaessgen, D. Stargel, The digital twin paradigm for future NASA and US Air Force vehicles, in *53rd AIAA/ASME/ASCE/AHS/ASC Structures, Structural Dynamics and Materials Conference 20th AIAA/ASME/AHS Adaptive Structures Conference 14th AIAA*, (Red Hook, Curran, 2012, April), p. 1818
16. F. Tao, F. Sui, A. Liu, Q. Qi, M. Zhang, B. Song, Z. Guo, S.C.-Y. Lu, A.Y.C. Nee, Digital twin-driven product design framework. Int. J. Prod. Res. **57**(12), 3935–3953 (2019)
17. R.N. Bolton, J.R. McColl-Kennedy, L. Cheung, A. Gallan, C. Orsingher, L. Witell, M. Zaki, Customer experience challenges: Bringing together digital, physical and social realms. J. Serv. Manag. **29**, 776 (2018)

18. R. Söderberg, K. Wärmefjord, J.S. Carlson, L. Lindkvist, Toward a digital twin for real-time geometry assurance in individualized production. CIRP Ann. **66**(1), 137–140 (2017)
19. S. Nakamoto, *Bitcoin: A Peer-to-Peer Electronic Cash System* (Manubot, 2019)
20. H. Wang, Y. Song, Secure cloud-based EHR system using attribute-based cryptosystem and blockchain. J. Med. Syst. **42**(8), 152 (2018)
21. I. Yaqoob, K. Salah, M. Uddin, R. Jayaraman, M. Omar, M. Imran, Blockchain for digital twins: Recent advances and future research challenges. IEEE Netw. **34**, 290 (2020)

Chapter 13
Blockchain Technologies for Securing IoT Infrastructure: IoT-Blockchain Architectonics

Mobasshir Mahbub

13.1 Introduction

Recent developments in ICT have facilitated traditional computerized industries' transition into the intelligent market, which features data-driven supervision and decision-making [1]. The IoT plays a significant part in linking the real industrial world with the cyber area of computer networks, creating a cyber-physical structure (CPS), during this shift of paradigm. IoT may serve a large variety of commercial uses, such as production chain, distribution, livestock, and service industries. IoT is targeted at increasing manufacturing and operating efficiencies, decreasing system downtime, and enhancing production efficiency.

There is an increasing enthusiasm in Blockchain technologies to ensure the privacy, confidentiality, and resolve challenges of 50 billion IoT devices by 2020. Reinventing network-able IoT gadgets are limited memory-capable, low power, and lightweight; therefore, much of the resources they produce must be utilized, and the flexibility required in maintaining the privacy and security of the user must be effectively met. Numerous systems that challenge each other also rely on the unified infrastructure. The complexity of scalability and one-to-many existence of traffic are thus not inherently appropriate for IoT networks. Many of the current data-sharing networks between separate nodes often expose unnecessary or incomplete details of maintaining consumer privacy and security. In these instances, IoT needs a lightweight, flexible, and distributed framework to ensure privacy and security [2]. To address these difficulties, this work has combined IoT with Blockchain

M. Mahbub (✉)
Department of Electrical and Electronic Engineering, Ahsanullah University of Science and Technology, Dhaka, Bangladesh

Department of Electronics and Communications Engineering, East West University, Dhaka, Bangladesh

T. Choudhury et al. (eds.), *Blockchain Applications in IoT Ecosystem*,
EAI/Springer Innovations in Communication and Computing,
https://doi.org/10.1007/978-3-030-65691-1_13

technologies in a scalable, decentralized, lightweight, and private manner. In addition to addressing such decentralized, scalable, and private behaviors, Blockchain technology often offers a way to merge all IoT gadgets into a shared Blockchain-aware infrastructure [3]. In the year 2008, S. Nakamoto disclosed a document on Bitcoin which proposed a novel, decentralized, and trustless infrastructure of digital currency. All transactions are processed and constantly checked on a central ledger utilizing public-key encryptions. The growth of IoT infrastructures is influenced by the aggregation of data, data management, and cloud data processing [4]. Another issue is to pick a suitable system for the transmission of data from heterogeneous IoT gadgets. In essence, Blockchain technology is a brilliant addition to the infrastructure of IoT with enhanced scalability, anonymity, stability, durability, and interoperability.

The forthcoming sections of the chapter are structured as follows. A review of current works and literature is given in Sect. 13.2. Section 13.3 represents an epitome of IoT. The introductory illustration of Blockchain is given in Sect. 13.4. Section 13.5 describes the security issues in IoT. The convergence of Blockchain and IoT is elaborated in Sect. 13.6. In Sect. 13.7, open research issues are discussed. Finally, the chapter concludes with Sect. 13.8.

13.2 Relevant Literature

The use of Blockchain technology into IoT has gained wide acceptance and popularity as being evidenced by the rapid growth research literature since 2011.

H. F. Atlam et al. [5] provided an overview of Blockchain integration with IoT to highlight advantages and challenges. T. A. Syed et al. [6] concluded that there were numerous issues to address and argued that before further Blockchain studies were overviewed, architectural aspects and components, consensus protocols, challenges, and future IoT guidelines were critical. M. Banerjee et al. [7] generally described aspects of IoT security enhancement technology of Blockchain. As a key to addressing these questions, E. L. C. Macedo et al. [8] classified IoT-related security problems, analyzed IoT security standards, and identified intrinsic aspects of intelligent contracts. A systematic review of the current literature in 2019 was conducted by F. Casino et al. [9]. A recent survey by M. Conoscenti et al. [10] has been closely linked with our work. They discussed Blockchain technology and IoT integration which includes categorizing and analyzing current trends in research in various areas and specifying usage patterns with challenges and directions for the future.

The article [11] offers in specific a systematic literature analysis on IoT-Blockchain, in which a variety of usage cases are classified. The research of [12] provides an IoT protection study and discusses the promise of Blockchain technology. S. K. Lo et al. [13] explored the scope and development challenges of Blockchain incorporation with IoT. The [14] research provides a study of the implementation dimension of combining Blockchain and IoT.

K. Salah et al. [15] addressed the issues and opportunities of study and different technologies for Blockchain-centered IoT. Moreover, this work has presented an inclusive review of current research and development activities in various sections of the infrastructure of IoT that can demonstrate significant impacts on Blockchains. The aspects of IoT security and privacy dependent on Blockchain are investigated in [16]. But most articles or works are constrained by the following:

1. Most of them have not presented a detailed overview of IoT and relevant features of Blockchain.
2. Limitations of details of integration scope of Blockchain in IoT.
3. No generalized framework for Blockchain-aware IoT is given.

Contributions of this Work From the context of prior work, this work aimed to (i) provide a technical overview of IoT and Blockchain, (ii) present a comprehensive study of the prospective of Blockchain integration in IoT, and (iii) offer in-depth discussions on technological challenges of Blockchain-based IoT security. In brief, the significant contribution of this work has been illustrated as follows:

1. Initially, there is a short overview of IoT, and a description of the main features of IoT is given.
2. An outline of Blockchain technology is then presented with a description of core functionality and a review of Blockchain concepts.
3. Typical vulnerabilities and privacy issues in the IoT network.
4. Motivations for the incorporation of Blockchain technologies in IoT and application mechanisms for Blockchain-IoT.
5. Overview of the latest Blockchain-aware approaches for IoT.
6. Furthermore, this work discusses Blockchain-aware implementations of IoT infrastructure and describes research and development issues.

13.3 An Overview of IoT

Throughout the section, the chapter will discuss the IoT briefly, and the challenges of IoT are summarized.

13.3.1 Introduction to the Internet of Things

The current market is shifting from the traditional IT market to the intelligent field, powered by recent advancements of IoT and data analytic technologies. During this growth, IoT performs a crucial task in connecting the industrial-physical world and the computer network cyberspace, while data analytics is able to assist with secret values derived from massive data of IoT to make adequate decisions.

IoT is a system of intelligent artifacts (e.g., things) with a variety of industrial facilities. The following layered subsystems represent a standard IoT framework.

Perception Layer A wide range of IoT gadgets includes controllers, sensors and actuators, RFID, NFC, smart meters, and other wired or wireless gadgets. Such gadgets can experience and gather surrounding environmental data. In the meantime, some of the gadgets can perform certain tasks for sensing the environment.

Communications Layer Different wireless and/or wired devices, including controllers, sensor devices, actuators, and different other gadgets, may be able to connect to a network via gateways, wireless access points, and cellular base stations. Such networks are enabled by numerous networking protocols such as Bluetooth, NFC, WirelessHART, 6LoWPAN, Sigfox, NB-IoT, LoRa, and Ethernet (Industrial) [17].

Applications IoT may be utilized massively for a variety of industrial applications. Usual applications of IoT technologies include the production, industrial production chain, livestock, intelligent grid, medical sector, and vehicles.

13.3.2 Challenges of the IoT

The IoT ensures the interconnection of different things (intelligent objects) connected to different electronic and mechanical sensors, actuators, and software that can sense and gather physical data and act on the surrounding environment. IoT's particular features present a variety of challenges in the research sector.

Interoperability Interoperability involves the ability to exchange, utilize information, and collaborate between IoT systems (both hardware and software). Due to the decentralization and heterogeneity of IoT systems, data sharing between various industries, supervisory bodies, and IoT ecosystems has been challenging. As a consequence, it is challenging to obtain interoperability.

Resource Constraints of IoT Devices Actuators, sensors, RFIDs, and smart meters are such IoT gadgets subjected to restricted resources, including computing, storage, and power. For instance, passive tags of RFID have no battery power and hence have to depend on RFID readers or the ambient environment for the collection of energy. Moreover, the limitations of resources also lead to IoT devices being vulnerable to infectious attacks.

Security and Privacy Vulnerability The term privacy is intended to ensure the proper utilization of IoT ecosystem generated data while private information of the client is not revealed without the consent of the client. Due to the complexity, heterogeneity, and decentralized nature of IoT, it will be difficult to ensure data privacy in IoT infrastructure. Besides, the integration of IoT gadgets with cloud computing has become a trend because cloud computing empowers IoT with additional storage and computing capacities. The confidentiality of IoT infrastructure data may, however, be impaired by the use of the external cloud server [2] as well.

Recent ICT advancements may transcend some inherent constraints of IoT. For instance, communications assisted by the ambient backscatter technique can enable IoT nodes to obtain additional energy from the environment. Multi-access edge computing (MEC) may expand the capabilities of IoT nodes by offloading computing-intensive tasks to edge computing servers. Furthermore, recent developments in Blockchain-aware security technology offer significant resolutions to challenges like weak interoperability, confidentiality, and several other privacy issues. Furthermore, the heterogeneous infrastructure of IoT is benefitted by Blockchain.

13.4 Features of Blockchain

For the P2P (peer-to-peer) network and a shared agreement to establish "distributed ledger" coordination [18], Blockchain-aware infrastructures are a combination of cryptographic technologies, PKI, and economic modeling. Generally, Blockchain technology is a dispersed data structure and is referred to in its usefulness as a "distributed ledger" for recording transactions in a network. As cryptocurrency is one of the authentication features in Blockchains, the "distributed ledger" can be used in networks where variable data sharing occurs. Both active peers retain similar versions of the ledger in a peer-to-peer Blockchain network. In the Blockchain, new entries are supplemented by mutual agreement among peers containing details about the transactions.

It is necessary to develop an understanding of how Blockchain technologies and how the Blockchains accomplish decentralization, to realize the potential applications of Blockchain in IoT.

13.4.1 Salient Features of Blockchains

The main features which make Blockchain technology something that could fundamentally reshape many industries are discussed below.

Decentralization Data transmission is authenticated and approved by trustworthy central third-party organizations in consolidated network infrastructures. This involves costs in terms of maintenance of the centralized servers and bottlenecks in performance. Two nodes may exchange information with one another in Blockchain-aware infrastructures without needing to trust the central authority to manage records or execute tasks.

Immutability Because all new entries in Blockchain infrastructure have been agreed on by peers through decentralized procedures, Blockchain resists censorship and is almost impossible to manipulate. Also, all the previously held records in Blockchain are immutable, and an assailant would have to compromise multiple nodes of the Blockchain network to alter any records. Any alterations in the substance of Blockchains are otherwise detected easily.

Fault Tolerance Every pair of Blockchains contains the same replicates of the records. Every failure or information leakage occurring in the Blockchain framework can be detected by decentralized accord, and information thefts can be eliminated utilizing the replication of peers in Blockchain.

13.4.2 Block and Blockchain

A Block is generally a data structure that records the transaction. In particular, it consists of a header (Block-header) and numerous transaction records. For example, in the Bitcoin Blockchain framework, the Block-header preserves the hash number of the previous Block, Merkle root, timestamp, nonce, and other required information. Blocks can be connected together, thus forming a Blockchain based on the previous hash value. It is worth mentioning a similar structure in terms of other Blockchains. Block hash enables each Block to be recognized, and the previous Block's hash ensures the prevention of alteration in the consistency of Blocks by modifying the header of the forthcoming Block. In other words, if an opponent tries to manipulate a Block, he also has to manipulate all the Blocks afterward which exist. The records for transactions are organized into a Merkle tree in Bitcoin for any node. The Merkle tree is a tree formed by hashes in which each leaf (node) has a data Block hash and each non-leaf node has a cryptography-based hash of the child node labels. The root of the Merkle tree of each Block is extracted from all the transaction hashes found in the Block. Therefore, by altering the header, no transactions can be changed.

13.4.3 Transactions and Digital Signatures

The peers of Blockchain required public-private key pairs to make transactions in cryptocurrency or a simple exchange of data. Peers use the private keys to register transfers and to give them the Blockchain address of the receiver. These addresses are determined through the calculation of a cryptographic hash of public key of users. The SHA-256 algorithm is utilized to determine client addresses [19], for instance, in Bitcoin. Encoding and encryption of the obfuscate Blockchain public keys of peers are essential. There are no serial tokens in cryptocurrency implementations, rather an elementary number of secured tokens relevant to the addresses usually entailed in the Blockchain's primary periods. The transactions keep track of ownership of the tokens through the addition or subtraction of the tokens associated with each participant's address imitating the Genesis Block. In outsider cryptocurrency implementations, transfers do not allocate secured tokens and instead entail the interchange of encrypted data utilizing digital (secured) signatures.

13.4.4 Cryptography

For two key reasons, Blockchains have to utilize cryptography. Firstly, anonymity is safeguarded. Because of Blockchain transparency, each network node will see the entire list, which may contribute to secrecy leakage. The second is the security of the properties of the customer. Because digital objects cannot readily prove their properties as tangible resources, the ownership of digital assets must be proved through crypto-techniques (for instance, digital signature).

Technically, asymmetrical encryption and hash function were the cryptographic variables used in Blockchain. Asymmetric encryption is also known as shared cryptography, which includes a distributed public key and holds a secret private key. The hash function will create a fixed-length message digest for various inputs. In the central system, only to show ownership of the wealth, one has to communicate with the accounting center. The evidence can be satisfied as long as the core is aware of its existence. In a decentralized environment, though, the entire network must demonstrate its identity and guarantee the integrity of its identity through the popular identification of several nodes. Centralized authentication of identity is no longer applicable. There's a problem with proving your own identity without revealing your password. The answer is given by asymmetric encryption. The user can maintain the private key secrecy but spread the public key to the network as a whole. The identity of the user can be proven with the pair of keys so that the digital property can prove itself. A pseudonym is important to maintain confidentiality to address the privacy problem.

13.4.5 Blockchain Architecture

Instead of merely linking Blocks to the network, Blockchain technology contains many layers. The Blockchain interconnects interest; on the other hand, the Internet interconnects content. Blockchain technology can be split into four layers: data, network, consensus, and application layer. Each layer performs its function in this architecture.

Data Layer The data layer performs the task of data structure design, data management, and data storage. Access to data and performance are important factors to consider. The primary knowledge for Blocks and transfers is placed on the fundamental data system such as Blockchain. Various Blockchains take different data organization and data storage strategies. Blockchain, for example, adopts a Merkle tree, while Ethereum uses the Merkle-Patricia tree, to organize and store transaction information. To conserve disk space, old Blocks must always be taken into consideration when stubbing the branches of the Merkle tree. As regards data storage, Ethereum and Bitcoin exploit level DB to enhance the performance of data access.

Network Layer A P2P network composed of miners and users autonomously maintains and manages Blockchain. There is no hub in the P2P network, and each node may still join or leave the network. Every node can enter every other node without any restrictions. The P2P network in Blockchain will enable the tolerance of node failure to some degree. All of the nodes in a network preserve characteristics such as fair treatment, autonomy, etc. They can relay information for mining and transfers, check for new nodes, etc. In the case of an attack, contact and authentication should be protected.

Consensus Layer The consensus layer in Blockchain is crucial as it is managed through a P2P network, in which each node has its perspective. Developing a consensus is a non-negligible role to obtain, and this goal has been achieved through a large number of consensus algorithms. These algorithms or processes can be classified as PoS, PoW, and variants and BFT and variants.

Application Layer This layer seeks to expand Blockchain's ability and make Blockchain applications easier for a researcher to build. This layer is made to make Blockchain more performance-efficient by the smart contract, the side chain, and other techniques. In 1996, Nick Szabo created a smart contract, a computer protocol to automate contract execution in Blockchain. The smart (intelligent) contract was an IT protocol. Intelligent contracts can resolve any discrepancy between the parties to the contract. The side chain is a system that makes possible the usage of digital properties in one network in a different network. While they may be transferred to digital properties between the initial chains and the side chains, it is isolated. The rest would not be harmed if just one chain breaks. Blockchain-as-a-Service (BaaS) is an emerging term which computes fast like lightning networks [20].

General Applications Blockchain was introduced in different fields, such as fintech, payment, insurance, government, etc. since its proposal in 2008. It can be mixed with several other trendy innovations including artificial intelligence, big files, and quantum computing, as it is an integral technology.

13.5 Security Issues in IoT Infrastructure

13.5.1 Issues

Interference Severe interruption or interference may occur in gadgets due to different reasons. Jamming is one of them which may affect wireless gadgets by sending poor radio signals to deteriorate networks to interfere with lawful users. Sleep destitution assault triggers sensing gadgets to stay alert and induces battery diminution (such as 6LoWPAN). Deployment mobility and distribution of gadgets generate significant vulnerabilities.

Insecure Configuration Uncertain IoT gadgets and software configuration and initialization indicate the security of the entire IoT ecosystem that might be penetrable to exterior users. This circumstance is often enkindled by human mistakes, scarcity of information, and the utilization of vincible versions of the software. A major barrier to the convergence of hardware or equipment and software and a large pool of interconnected systems contributes to inadequate servicing and dangerous setups. Mirai malware illustrates the threat incident. DDoS is released on ten million IoT computers.

Spoofing Spoofing attacks are targeted at security mechanisms in IoT system vulnerability. Owing to the scarce capital and the vast number of IoT apps, advanced systems are challenging to implement. Sybil is a case that falsifies identity or impersonates a valid identity through inflicting.

Physical Security IoT applications are used in multiple contexts, where physical risks are specific. The system loss triggered by earthquakes, vandalism, burglary, etc., for example, represents the efficiency of operations. Open, unprotected devices like the installation of malicious software by their physical deployment interfaces are easier to compromise.

Resource Depletion Restricted IoT system services contribute to resource depletion vulnerabilities. For instance, IEE E 802.15.4 standard devices permit the transmission of tiny frame-sized packets that allow packet fragments to be retrieved. Like the conventional TCP/IP stack, this condition may allow malicious fragments to be replayed, duplicated, or completed. These attacks contribute to heavy memory utilization and the exhaustion of resources. Services delivered by these devices are therefore stopped.

Identity and Authentication To be uniquely recognizable, IoT devices rely upon IPv6 which provides adequate addresses. Pre-transmission identity discovery such as address resolution should be carried out securely. Identity attacks involve spoofing, disappointment, Sybil attack, and impersonation attack. The result ranges from denying services to complete loss of control. The key administration for device authentication is extensively practiced. Effectiveness and scarcity of IoT resources are key concerns. The protection technologies for Datagram Transport

Level Safety (DTLS), Secured Socket Layer (SSL), the VPN, and the IPSec include large overheads and need to be configured for IoT. Resource-restricted systems may use poor security methods to expose vulnerabilities. The hijacking of a session in IoT is an example that becomes easier.

Privacy Concerning privacy in this framework, apps may contain private details or even protect the lives of the people. To manage complicated algorithms, including the processing of a vast volume of IoT data, IoT resources can be offered by computers, servers, or clouds. Sending details to a single organization from computers poses a chance of disclosing classified information.

13.5.2 Security Threats

Key Attack It happens by the manipulation of the weakness of private key leakage. The LNSC protocol [21] provides a shared authorization mechanism for electric vehicles and charging batteries to cope with this threat. To this end, various private temporary keys of electric vehicles, batteries, and operators are used for each session agreement, and therefore elliptical curve encryption is used to calculate the hash (Hash Function).

Replay Attack The purpose of this assault is to spoof two parties' identities, intercept their data packets, and transfer them without alteration to their destinations. The key could leak, and the replay attack could be initiated by leveraging the weakness if the Blockchain produces a private key with minimal randomness in the signature phase. To resist this attack, LNSC [21] uses the idea to calculate hash functions using temporary private keys for each session agreement and elliptical curve cryptography.

Sybil Attack An opponent creates several fake identities under this attack. The opponents will achieve a significant influence within the system, i.e., increase/decrease the credibility of other agents through the success of other interactions over the network. TrustChain [22] substitutes for proof work with a mechanism for establishing transaction validity and integrity. Transaction agents could decide not to add a transaction to their local chain because of Blockchain architecture. TrustChain solves this by building an unchanging contact chain for each user. TrustChain measures the trust of agents in an electronic network utilizing previous transactions as inputs to Sybil-resist [23].

Tampering Attack When adversaries can take advantage of a private key leakage or 51% loophole, after signature, they can misrepresent Bitcoin transactions, sums, and other details. In fact, through human behaviors in everyday transactions, it is possible to connect Bitcoin accounts with identity. W. Yin et al. [24] use a public-key crypto-system, consistent with the existing Bitcoin system, to prevent such an attack. They propose the addition of the homomorphic Paillier encryption system to cover the amount of the plaintext in transactions, with the Commitment Proof checking for the encrypted amounts.

Man-in-the-Middle Attack Through leveraging certain weaknesses like private key leakage and 51% weakness, an intruder through spoofing the identity of two parties may secretly communicate and even change the contact between such parties, who think they are interacting directly, but in reality, the entire conversation is under the control of the intruder. BSein guarantees safely shared authentication to resist this assault. In work [21] LNSC provides reciprocal authentication with provisional private keys and the elliptical curve cryptography for calculating the hash functions for each meeting agreement.

Cryptanalytic Attacks These kinds of attacks are aiming at cracking the code and revealing their buttons. In [24] Blockchain is tested for a quantum attack. This attack aims to resolve the digital logarithm of the elliptic curve, i.e., the private key is derived from the public elliptical curve. This enables an opponent to sign unauthorized transactions and establish a valid user signature.

DDoS/DoS Attack It needs the transmission of a significant number of requests for network failure. X. Li et al. [25] propose a mixing arrangement to disguise the transfers between coins. To this end, they use an Elliptic Curve Digital Signature Algorithm (ECDSA) which is a ring-based signature approach. By restricting consumer behavior, the security of the mixing facility against the DoS assault is accomplished. In Bitcoin, resilience against DoS is achieved by limiting Block sizing, checking for transaction inputs for the maximum number of signatures, and using multi-receiver coding to provide authorized participants with confidentiality.

13.6 Integration of Blockchain Technology and IoT

Some of the case studies have found that enterprises are running new businesses over IoT using Blockchain technology. For example, C. Qu et al. [26] prescribed a hyper-graphic Blockchain method used in a network regarding smart home, focusing on the application system's storage capacity and performance enhancement. In Blockchain-based edge-IoT networks, B. W. Nyamtiga et al. [27] implemented a quick and safe payment solution. The off-chain methodologies including the lightning-aware network were utilized for the protection of systems against dual-flow (double-spending) attacks. For authentication of IoT devices, Blockchain technology and physical un-clonable function (PUF) have been combined. They also gained from numerous authentication variables, continuous authentication, and paths of origin. X. Xu et al. [28] analyzed a Blockchain technology-based offer analytics software framework, utilizing an intelligent contracting tool to reduce the IoT overhead of data collection as transaction protocol, respectively. Intelligent storage contracts could be trusted as they execute themselves and are not controlled by the users. The technology used by U. Javaid et al. [29] for the DDoS of IoT devices is used in Blockchain. An alternative named Ethereum having intelligent contracts was developed and deployed to replace centralized supervision. The case study of Blockchain technology securing a smart and intelligent home was introduced by P. Cui et al. [30]. An intelligent home has been equipped with a "miner" who handles internal and external communications. To control and audit the communication, the miner used Blockchain. Some attempt to narrow the range by emphasizing Blockchain's importance, opportunities, and challenges in certain IoT fields. Sadouskaya's thesis described the priorities and threats to the logistics and supply chain usage of Blockchain technology and analyzed recent application adoptions in different start-up ventures. J. J. Sikorski et al. [31] suggested Blockchain platform implementations in the sense of the chemical industry linked to the automated machine-to-machine (M2M) interactions. A. Dorri et al. [32] presented challenges to safety and privacy and explained how Blockchain technology can enhance vehicle ecosystem security. NXXTECH claimed to establish Nxxtech as a fully scalable and usable Blockchain platform with an emphasis on corporate structures. It created for businesses a scalable, versatile, and powerful toolkit for implementing decentralized technologies through business and sectors.

Miraz and Ali (2018) assessed Blockchain development deployment for improved IoT protection but focused on an overall overview of possible Blockchain advantages, including storage and networking data redundancy.

Nonetheless, technological information or test results to show its capabilities were not published. Alphand developed Blockchain technology and IoT as an IoTChain incorporating the design and acceptance system for the Constrained Environment (ACE) Objective Safety Framework for IoT (OSCAR). The Exchange to Exchange (E2E) approach was given to maintain a safe approved connection to IoT products. In a private Ethereum network, IoTChain was deployed and checked.

13.7 Research Issues

Although there are many incentives to update Blockchain and IoT, there are also obstacles that must be overcome before the promise of Blockchain-IoT can be released in its entirety. Within this segment, we describe some main problems as Blockchain is integrated into IoT and explore possible solutions.

13.7.1 Resource Constraints

Blockchain technology sometimes requires extensive computational power and higher level of energy because of their decentralized consensus algorithms. In Bitcoin, for example, power consumption in PoW is high. Therefore, low-power IoT devices cannot accomplish consensus processes with massive energy consumption. On the other side, the massive Blockchain data reflects that Blockchain technologies cannot be completely implemented by resource-constrained IoT infrastructure. For instance, by the end of 2018, Bitcoin's Blockchain size will be nearly 185 GB. The entire Blockchain on each IoT device cannot be stored fully. In the meantime, the large amount of real-time data produced by the IoT ecosystem almost exacerbates that status quo. IoT-aware Blockchains are primarily built for stabilized networks which are not achievable in the IoT ecosystem due to node malfunction (for instance, battery depletion), weak network compatibility with IoT nodes, and the unreliable network.

13.7.2 Security Vulnerability

While integrating Blockchain technology into IoT improves IoT protection through encryption and digital signature, safety remains a major concern for Blockchain-aware IoT, given its vulnerabilities.

On the other hand, there is a growing trend in the deployment and scalability of wireless communications in an industrial ecosystem. But IoT also suffers from safety breaches, like passive eavesdropping, jamming attacks, and replay attacks, in open-service wireless media. Moreover, because of IoT gadgets' resource limitations, IoT may not be feasible with conventional sophisticated heavy cryptographic algorithms. Furthermore, it's also difficult to orchestrate keys in distributed environments (which are essential for cryptographic algorithms).

13.7.3 Privacy Leakage

Blockchain has certain mechanisms to preserve the confidentiality of Blockchain transaction records. For instance, transactions in Bitcoin are carried out via IP addresses instead of the individual names of users, guaranteeing some anonymity. To maintain the privacy of users, one-time accounts are created in Bitcoin. But such protective systems are not sufficiently robust. For instance, researches show that the client's pseudonyms might be cracked through the study of multiple transactions that are connected to a common user. Furthermore, the complete stocking of Blockchain transaction data may also generate security threats as indicated in [33].

13.7.4 Scalability of Blockchain-Aware IoT

The scalability concern of Blockchains also restricts the wide use of a massive Blockchain-aware IoT ecosystem. Blockchain's scalability can be determined by the number or measure of IoT hubs and the count of concomitant workloads by transaction per second. Some of the Blockchain ecosystems have poor performance.

For example, researches demonstrate that in each second, only seven transactions may take place in Bitcoin. On the contrary, in each second, almost 2000 transactions can be processed at VISA, and per second 170 transactions are performed by PayPal. In brief, the occupant Blockchain may not be appropriate for the services having a massive pool of transactions, especially for IoT environment.

13.8 Conclusion

IoT facilities face many obstacles, including complexity, lack of interoperability, infrastructure constraints, anonymity, and health vulnerability. The latest emergence of Blockchain technology provides a solution to the problems of improved interoperability, anonymity, protection, reliability, and traceability. The chapter discussed the incorporation of Blockchain technologies in IoT. The work provided a thorough overview of Blockchain-aware Blockchain. In particular, the chapter first briefly introduces the IoT ecosystem and Blockchain technology and then talked about Blockchain-aware IoT's opportunities and elaborates on the architecture of Blockchain-aware IoT. After that, certain issues for further research and development of Blockchain-aware IoT were outlined. Finally, the chapter concluded with the conclusion elaborating briefly on the work done. This work will be highly supportive of its audience and researchers to get the necessary insight into Blockchain-aware IoT and will be supportive to perform further researches and enhancements.

References

1. D.S. Park, Future computing with IoT and cloud computing. J. Supercomput. **74**, 6401–6407 (2018). https://doi.org/10.1007/s11227-018-2652-7
2. J. Li, X. Liao, N. Puech, Security and privacy in IoT communication. Ann. Telecommun. **74**, 373–374 (2019). https://doi.org/10.1007/s12243-019-00718-6
3. D. Pavithran, K. Shaalan, J.N. Al-Karaki, et al., Towards building a blockchain framework for IoT. Clust. Comput. **23**, 2089 (2020). https://doi.org/10.1007/s10586-020-03059-5
4. E. Adi, A. Anwar, Z. Baig, et al., Machine learning and data analytics for the IoT. Neural Comput. & Applic. **32**, 16205 (2020). https://doi.org/10.1007/s00521-020-04874-y
5. H.F. Atlam et al., Blockchain with Internet of Things: Benefits, challenges, and future directions. Int. J. Intell. Syst. Appl. **10**(6), 40–48 (2018)
6. T.A. Syed, A. Alzahrani, S. Jan, M.S. Siddiqui, A. Nadeem, T. Alghamdi, A comparative analysis of blockchain architecture and its applications: Problems and recommendations. IEEE Access **7**, 176838–176869 (2019). https://doi.org/10.1109/ACCESS.2019.2957660
7. M. Banerjee, J. Lee, K.R. Choo, A blockchain future for internet of things security: A position paper. Digit. Commun. Netw. **4**(3), 149–160 (2018). https://doi.org/10.1016/j.dcan.2017.10.006
8. E.L.C. Macedo et al., On the security aspects of Internet of Things: A systematic literature review. J. Commun. Netw. **21**(5), 444–457 (2019). https://doi.org/10.1109/JCN.2019.000048

9. F. Casino, T.K. Dasaklis, C. Patsakis, A systematic literature review of blockchain-based applications: Current status, classification and open issues. Telematics Inform. **36**, 55–81 (2019). https://doi.org/10.1016/j.tele.2018.11.006
10. M. Conoscenti, A. Vetrò, J.C. De Martin, Blockchain for the Internet of Things: A systematic literature review, in *2016 IEEE/ACS 13th International Conference of Computer Systems and Applications (AICCSA)*, (2016), pp. 1–6. https://doi.org/10.1109/AICCSA.2016.7945805
11. A. Dorri et al., Blockchain for IoT security and privacy: The case study of a smart home, in *2017 IEEE International Conference on Pervasive Computing and Communications Workshops (PerCom Workshops), Kona, HI*, (2017), pp. 618–623. https://doi.org/10.1109/PERCOMW.2017.7917634
12. M.S. Ali et al., Applications of Blockchains in the Internet of Things: A comprehensive survey. IEEE Commun. Surv. Tutorials **21**(2), 1676–1717 (2019). https://doi.org/10.1109/COMST.2018.2886932
13. S.K. Lo et al., Analysis of blockchain solutions for IoT: A systematic literature review. IEEE Access **7**, 58822–58835 (2019). https://doi.org/10.1109/ACCESS.2019.2914675
14. W. Viriyasitavat et al., Blockchain and Internet of Things for modern business process in digital economy—The state of the art. IEEE Trans. Comput. Soc. Syst. **6**(6), 1420–1432 (2019). https://doi.org/10.1109/TCSS.2019.2919325
15. K. Salah et al., Blockchain for AI: Review and open research challenges. IEEE Access **7**, 10127–10149 (2019). https://doi.org/10.1109/ACCESS.2018.2890507
16. H. Dai, Z. Zheng, Y. Zhang, Blockchain for Internet of Things: A survey. IEEE Internet Things J. **6**(5), 8076–8094 (2019). https://doi.org/10.1109/JIOT.2019.2920987
17. A.A. Zaidan et al., A survey on communication components for IoT-based technologies in smart homes. Telecommun. Syst. **69**, 1–25 (2018). https://doi.org/10.1007/s11235-018-0430-8
18. E. Palm, U. Bodin, O. Schelén, Approaching non-disruptive distributed ledger technologies via the exchange network architecture. IEEE Access **8**, 12379–12393 (2020). https://doi.org/10.1109/ACCESS.2020.2964220
19. S. Ghimire, H. Selvaraj, A survey on bitcoin cryptocurrency and its mining, in *2018 26th International Conference on Systems Engineering (ICSEng), Sydney, Australia*, (2018), pp. 1–6. https://doi.org/10.1109/ICSENG.2018.8638208
20. W. Zheng et al., NutBaaS: A blockchain-as-a-service platform. IEEE Access **7**, 134422–134433 (2019). https://doi.org/10.1109/ACCESS.2019.2941905
21. V. Rusu, Verifying an ATM protocol using a combination of formal techniques: A preliminary version of this paper has appeared in, LNCS 2805, 223–243. The URL contains PVS specifications and proofs for the case study. Comput. J. **49**(6), 710–730 (2006). https://doi.org/10.1093/comjnl/bxl039
22. S. Malik et al., TrustChain: Trust management in blockchain and IoT supported supply chains, in *2019 IEEE International Conference on Blockchain (Blockchain), Atlanta, GA, USA*, (2019), pp. 184–193. https://doi.org/10.1109/Blockchain.2019.00032
23. Y. Hu, Y. Xiong, W. Huang, X. Bao, KeyChain: Blockchain-based key distribution, in *2018 4th International Conference on Big Data Computing and Communications (BIGCOM), Chicago, IL*, (2018), pp. 126–131. https://doi.org/10.1109/BIGCOM.2018.00027
24. W. Yin, Q. Wen, W. Li, H. Zhang, Z. Jin, An anti-quantum transaction authentication approach in blockchain. IEEE Access **6**, 5393–5401 (2018). https://doi.org/10.1109/ACCESS.2017.2788411
25. X. Li, Y. Mei, J. Gong, F. Xiang, Z. Sun, A blockchain privacy protection scheme based on ring signature. IEEE Access **8**, 76765–76772 (2020). https://doi.org/10.1109/ACCESS.2020.2987831
26. C. Qu et al., A hypergraph-based blockchain model and application in Internet of Things-enabled smart homes. MDPI – Sens. **18**(9), 2784 (2018). https://doi.org/10.3390/s18092784
27. B.W. Nyamtiga et al., Blockchain-based secure storage management with edge computing for IoT. MDPI – Electron. **8**(8), 828 (2019). https://doi.org/10.3390/electronics8080828
28. X. Xu et al., A novel blockchain framework for industrial IoT edge computing. MDPI – Sens. **20**(7), 2061 (2020). https://doi.org/10.3390/s20072061

29. U. Javaid et al., Mitigating IoT device based DDoS attacks using blockchain, in *CryBlock'18: Proceedings of the 1st Workshop on Cryptocurrencies and Blockchains for Distributed Systems*, (2018), pp. 71–76. https://doi.org/10.1145/3211933.3211946
30. P. Cui, U. Guin, A. Skjellum, et al., Blockchain in IoT: Current trends, challenges, and future roadmap. J. Hardw. Syst. Secur. **3**, 338–364 (2019). https://doi.org/10.1007/s41635-019-00079-5
31. J.J. Sikorski et al., Blockchain technology in the chemical industry: Machine-to-machine electricity market. Appl. Energy **195**(C), 234–246 (2017). https://doi.org/10.1016/j.apenergy.2017.03.039
32. A. Dorri et al., BlockChain: A distributed solution to automotive security and privacy. IEEE Commun. Mag. **55**(12), 119–125 (2017). https://doi.org/10.1109/MCOM.2017.1700879
33. E.F. Jesus et al., A survey of how to use blockchain to secure Internet of Things and the stalker attack. *Hindawi – Secur. Commun. Netw.*, Article ID 9675050 (2018). https://doi.org/10.1155/2018/9675050

Chapter 14
qIoTAgriChain: IoT Blockchain Traceability Using Queueing Model in Smart Agriculture

Sudhansu Shekhar Patra, Chinmaya Misra, Kamakhya Narain Singh, Mahendra Kumar Gourisaria, Subham Choudhury, and Suresh Sahu

14.1 Introduction

Currently, worldwide food safety is a serious topic for discussion and needs everybody's attention. For the food safety measures, there is a need for a trusted system called food traceability system which may track as well as monitor the entire system of span from the food production by the farmers which include cultivation, processing, logistics, warehousing to selling. After 2 months of lockdown in India, due to COVID-19, on May 12, 2020, a total of Rs 20 lakh crores has been declared as relief package infused to the economy by the Prime Minister of India. As declared by the Govenment of India on 15th May 2020, the PM announced a package of one lakh crore (out of 20 lakh crore) in the third tranche of Agri-Infrastructural fund. India is the first in the world to produce milk, jute, and cereals and second to produce sugarcane. In India 85% of farmers are small- and medium-sized, and they generally don't get the value of their products and labor. But now since the government is ready to develop the agriculture infrastructure as well as the supply chain management, a great demand for secured agricultural logistic chain is seen in India.

S. S. Patra (✉) · C. Misra · K. N. Singh
School of Computer Applications, KIIT Deemed to be University, Bhubaneswar, Odisha, India

M. K. Gourisaria
School of Computer Engineering, KIIT Deemed to be University, Bhubaneswar, Odisha, India

S. Choudhury
Department of Computer Science & Engineering, SRM Institute of Science and Technology, Chennai, Tamil Nadu, India

S. Sahu
Vice President, Delivery and Professional Services, Ignitiv Inc., Cupertino, CA, USA

T. Choudhury et al. (eds.), *Blockchain Applications in IoT Ecosystem*, EAI/Springer Innovations in Communication and Computing,
https://doi.org/10.1007/978-3-030-65691-1_14

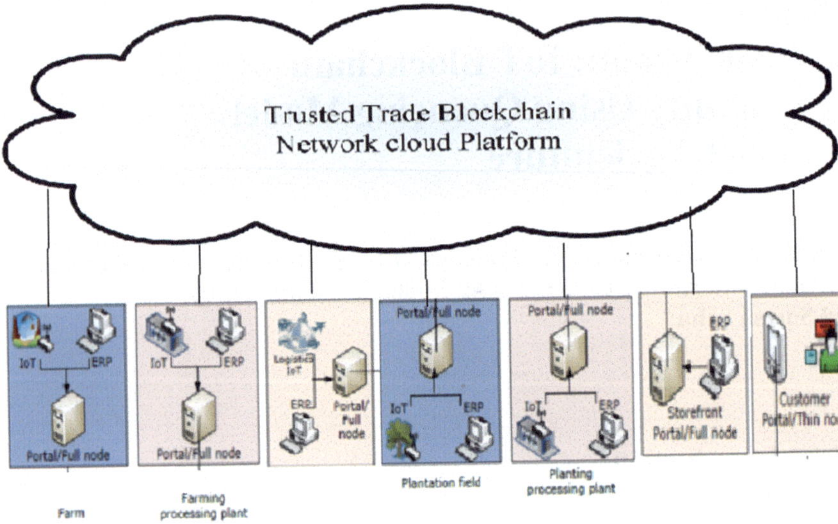

Fig. 14.1 IoT and blockchain-based smart agriculture system

In this chapter a model is designed which is based on blockchain technology as well as IoT which involves entire stakeholders of a smart and automated agricultural ecosystem, though there may not be good trust among the parties. Figure 14.1 shows the IoT and Blockchain-based smart agriculture system. The IoT sensors are used to send the data to the system without the direct intervention of humans; Fig. 14.2 shows the parties or stakeholders involved in a blockchain system. Through the innovative technology called smart agriculture, farming activities can be carried out through less manpower by maximizing the utilization of feasible and usable resources. It solves the difficulty of shortfall of farmers, improves agricultural production, resists risks, and helps small, medium, and weak farmers in joining a large-scale network through intelligent digital translation. The efficient adoption of IoT sensors, cloud infrastructures, gateways, etc. helps the stakeholders of the ecosystem to control the agricultural production and other activities through mobile platforms or computer platforms which puts more wisdom to the traditional agriculture. The smart agriculture system uses a large number of IoT devices such as sensors and actuators enclosed into physical items and transmits data to the IoT network. The data may go to a cloud server for converting the insights into action. In the entire transaction process, security measures are the main issues that have obstructed possibilities of large-scale deployment with IT. Security vulnerabilities with IoT devices are always present because they are an easy and possible target for DDoS attacks. Scalability is another issue with the IoT networks. Since the network connected through the IoT network grows with the number of devices, the authentication, authorization, and connection with different nodes in the network will be becoming a bottleneck.

Fig. 14.2 Stakeholders in the blockchain network

Blockchain, an emerging technology, also known as distributed ledger technology (DLT), has the potential in helping IoT security with scalability challenges. The development of blockchain as well as IoT is on the docket for many organizations with many implementations, research directions, solutions for different use cases, and initiatives in diverse areas of the society. It is going to be a technology game changer since it has many unique capabilities and advantages. At its core, the blockchain technology, or better DLT, shares the digital ledger between untrusted participants in the system, all residing on the Internet. Transactions are verified, validated, and recorded in the ledger and cannot further be updated or removed and are entered as well as shared by a community of participants. In between the intervals, the data is entered into the chain, known as blocks. The blocks are time-stamped, and the transactions in it along with the orders are verified (Figs. 14.3 and 14.4).

14.1.1 Application of IoT in Smart Agriculture

With the IoT environment by the use of multiple sensors, farmers are now equipped with collected data to get a better return on investment. The humidity, temperature,

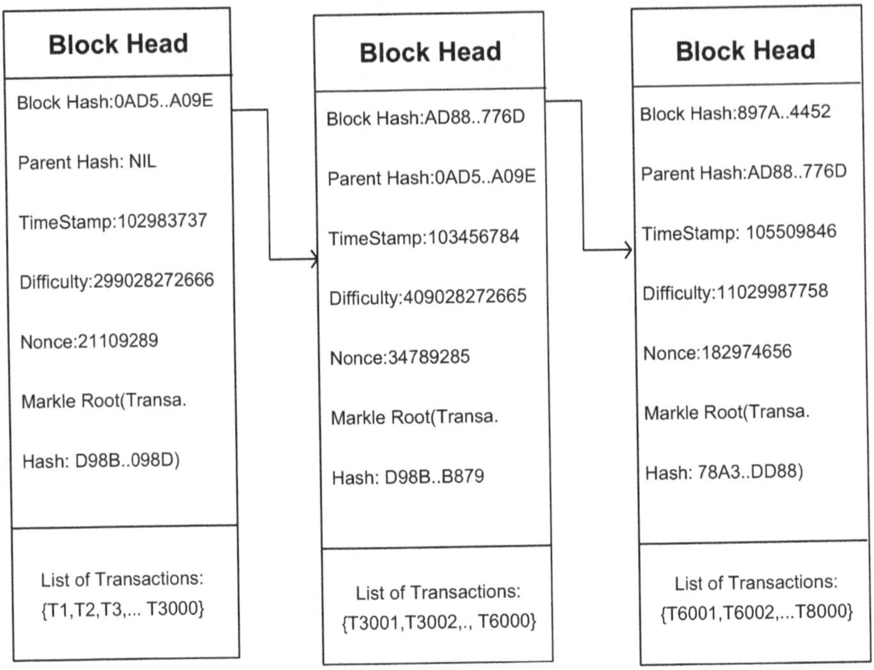

Fig. 14.3 Blockchain data structure

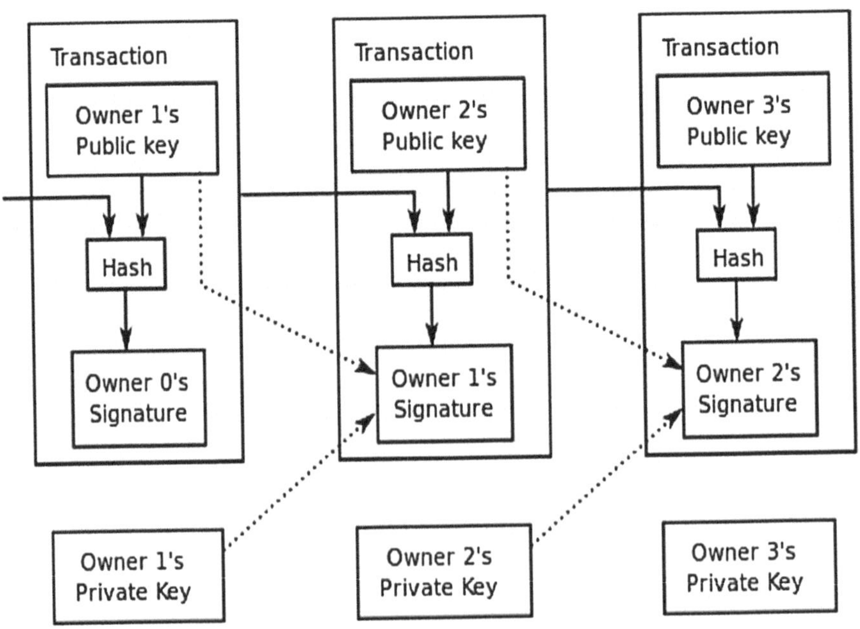

Fig. 14.4 Blockchain transactions

salt level, and other soil parameters are gathered and measured by the used sensors. The following are the benefits the farmers are getting with the use of IoT in agriculture:

- *Improved efficiency*: Today, there is a need for growth in the agricultural industry. And there is a race to fill that need. Farmers are trying to grow more crops even in the deteriorating land, the fluctuating weather conditions, and the declining available land. IoT enabled technologies to help farmers get informed decisions and monitor their products in real time. They are now able to predict issues and avoid mis-happenings and take the necessary steps. IoT implementation in the field of agriculture introduces automation and helps in demand basis irrigation and robot harvesting with fertilizing.
- *Expansion*: Now, 70% of the population are living in urban areas. IoT-grounded greenhouse system helps to deliver fresh fruits as well as vegetables with the supply chain. The smart agricultural system allows fruits and vegetables basically in the supermarket and facilitates to reach everyone's house.
- *Reduced resources*: Implementation of IoT helps in the optimization of resource use such as water, land, electricity, etc. Diverse sensors collected data helps in the proper distribution of resources by the farmers.
- *Cleaner process*: Through IoT-based precision farming, farmers are capable of saving water and energy, thereby promoting green farming and reducing consumption of pesticides and fertilizer. Through this, end users can get organic products as compared to traditional farming.
- *Agility*: IoT in agriculture brings agility in the process. Through this, farmers can quickly act to substantial changes in humidity or extreme weather changes and condition of crops and soil status of fields.
- *Quality product*: Through data-driven agriculture, end users can get a good-quality product in time as farmers understand the correlation between the conditions of the firm and crop quality, thereby recreating firm conditions and increasing product quality.

14.1.2 Blockchain Technologies in Agriculture

- *Tracking of procurement*: In the agriculture sector, the challenge is the tracking and the payment of the delivered foods. In the current scenario, a third party is involved in the total process. The buyer has an agent who takes care of the payment, and the seller has an agent who verifies the safe delivery of the goods. Since there are multiple agents involved in the process, it costs more to the system and makes the process time-consuming. Using blockchain, the distributed ledger makes the whole process simpler.
- *Crop production:* IoT sensors fetch critical information of the soil, water, and fertilizer and send it to the blockchain. Depending on the saved data in the blockchain, smart contracts may trigger to take necessary action.

- *Finance for agriculture:* Blockchain can make the finance process for agricultural needs more transparent and accessible. It is able to bring transparency to the agribusiness industry.
- *Insurance*: Low-premium agriculture insurance schemes provide social protection to farmers affected by natural calamities. But still its adoption in rural area is very low. The payout process and claim validation are time-consuming. Insurance on smart contracts will speed up the validation process and facilitate instant payout.
- *Land registration*: Incorruptible ledger of land records can be managed by blockchain technology. If it is linked with a digital ID, then the land records can be safely kept even in the time of natural calamities or war.
- *Transparency in supply chain*: Blockchain provides immutable record from the production to the retail store. This gives the consumers trust in the product, and the producers get more value of their product as there is no agent in the process.
- *Climate or green bonds*: For funding the agricultural projects the green bonds has been created. As the bond value increases, there is a need for effective tracking and verification to increase the trust of the investors. Blockchain offers carbon credit and trading.

The rest of the chapter is organized as follows: Section 14.2 depicts the related work in blockchain-based smart agriculture and the studies on queueing network in solving the different use cases related to smart agriculture. Section 14.3 describes the smart agriculture and the AgriChain, Sect. 14.4 describes the proposed model and queueing model, Sect. 14.5 measures the various performance of the system, Sect. 14.6 validates the system with various numerical illustrations and finally, and Sect. 14.7 concludes the chapter.

14.2 Related Works

Smart agriculture is the implementation of advances, for example, GPS, Big Data, cloud infrastructure, Internet of Things (IoT), machine learning, and artificial intelligence (AI) with the conventional horticulture [1, 2]. The utilization of advanced agricultural platform based on IOT, through countless detecting hubs in the objective zones, for example, farmland, nurseries, backwoods gardens, and pastures, can gather data on horticultural reproducing or planting progressively. Such data like temperature, electrical conductivity, dampness, gas concentration, light, soil dampness, and creation of pictures during producing, preparing, transportation, and deals process are collected into the cloud-based focal control framework for study or examination utilizing AI algorithm steps. Horticultural production workforce can break down ecological enormous information (or Big Data) [3] through checking irritations like pests, insects, and maladies and different hazard factors, with the goal focused on agricultural materials set up; different execution gears can be prepared as required to perform temperature control, darkening, and ventilation, just as different activities to accomplish savvy control for the developing conditions of farming. Smart agriculture is a creative method of doing cultivating exercises by decreasing human endeavors and by making the most extreme use of the accessible assets. It can take care of the issue of lack of laborers, improve the capacity of horticultural creation to oppose dangers, and help little, feeble ranchers to deliver large-scale nexus and an insightful change. The utilization of sensors, passages, cloud servers, and so forth to control farming production through versatile stages or PC platforms will make customary horticulture more "shrewd." Despite of exact recognition, control and dynamic administration from a wide perspective, smart farming incorporates agricultural web-based business, food detectability, hostile to falsifying, horticultural recreation of travel industry and agricultural data administration [4].

In [5], the authors proposed Material Conscious and Information Network (MCIN) based on smart agriculture. In [6], authors studied how IoT enhances agriculture and did a comprehensive review of the framework, considerations, and implications in implementation. In [7], authors reviewed many use cases related to agricultural applications with the help of IoT sensor and cloud computing infrastructure as the main pillar. This survey helps in understanding different technologies useful for smart agriculture and building stabilized smart agriculture [14].

Blockchain systems can be modeled as queuing systems with bulk service, which had studied in [10]. The Markovian bulk arrival $M^x/M/1$ and bulk service $M/M^x/1$ are provided in [13]. Many systems following bulk arrival as well as bulk service were studied thoroughly in [11, 12] and had applied to many use cases [8, 9], but blockchain systems are not modeled to date. Pass et al. studied the blockchain mechanism in the asynchronous system and verified the consistency of the applied mechanism [15] .

14.3 Smart Agriculture and AgriChain

14.3.1 Smart Agriculture

Smart agriculture is built on collective technologies which collaborate Big Data, IoT, cloud ecosystem, AI, and GPS, incorporating them into traditional agriculture. By the use of the sensor nodes, a huge volume of data is collected in real time from greenhouses, agricultural land, etc. used for agricultural breeding or planting. Various information such as light, humidity, temperature, soil moisture, and atmospheric condition and crop production images during time of production, transportation, processing, and sales are collected in cloud systems for future study and analyzed with the help of ML algorithms.

Smart agriculture is a modern technique for performing activities through fewer farmers and maximizing the resources available. With smart agriculture, agricultural activities are done with less number of agricultural workers, and production ability is increased by resisting risks and helping small farmers perform intelligent transformation enabling them to join to the large-scale network. The traditional platform of agriculture has transformed into smart agriculture with the help of IoT sensors, cloud infrastructures, gateways, etc. which controls the agricultural output with a mobile app or software application. Also control as well as decision-making in smart agriculture incorporates agricultural e-commerce, food safety, anti-counterfeiting, tourism, and information services.

14.3.2 AgriChain

The transmission of price signals is weak leading to over and under production by farmers. Since ASC management is weak in India, the reachability to mandis is very weak. Too many middlemen in the supply chain leads to artificial price rise and huge differences between the price the famer gets and the final consumer pay. There is also presence of asymmetric information (usually the middleman has more information than both farmers and consumers regarding policies, suppliers, and stocks available)

AgriChain incorporates blockchain-based dynamic and consistent solution with a traditional, well-established agricultural supply chain. The model is secure that unites as well as transfers information among the supply chain stakeholders.

AgriChain manages the agricultural commodities which lucidly move across the globe in a lucid way by the use of a trustworthy platform for easing the transactions. The proposed model builds networks among key stakeholders which amends the time to market, reduces supply chain costs, and improves efficiency.

The AgriChain model is built on the following major components.

14.3.2.1 The Platform

AgriChain, the proposed model, brings all stakeholders in the agricultural supply chain in a common platform allowing them to take and use better-informed decisions, eliminates unnecessary paperwork and dockets, and reduces supply chain inefficiency and risk all on one easy-to-use platform.

14.3.2.2 Stock Management

Agricultural stock management makes agricultural production detail accessible to all the partners and farmers. Every inload and outload data is being managed by recording the stock levels. Flour mills, stockfeed manufacturers, farmers, or any end user get to benefit from this design.

14.3.2.3 Contracting

Contract farming is a mechanism through which the coordination between farmers and agribusiness firms has been increased. It helps in creating and managing commodity contracts, freight contracts, etc. Since all the delivery is recorded, clients can follow contract progress at every stage of the process. The farming contracts are classified into three categories: contracts based on Market, contracts based on resources, and contracts based on production management.

14.3.2.4 Supply Chain Tracking

GPS innovation allows us to track each heap at each phase along the supply chain from enclosure to the end client.

14.3.2.5 Automation in Logistics

This is the procedure through which computer software programs or automated machines are used to improve the process and efficiency of logistic operations in the agricultural supply chain.

14.3.2.6 Broker Integration

A direct line of connection is established between the producers and agents for the smooth conductance of managing stocks and supply chain information.

14.3.2.7 Goods Receivables

There is a need for automation in the inbound and outbound deliveries at the receivable sites. Match every truck movement and the goods receivable at the sites to the correct order every time.

14.3.2.8 Traceability

With the stamping of date and time at each stage through the supply chain, the incoming and outsourcing of stock can be traced.

14.3.2.9 Position Reporting

The accurate timely position reports help in making contracting decisions.

Figure 14.5 shows a schema of a distinct blockchain network in AgriChain.

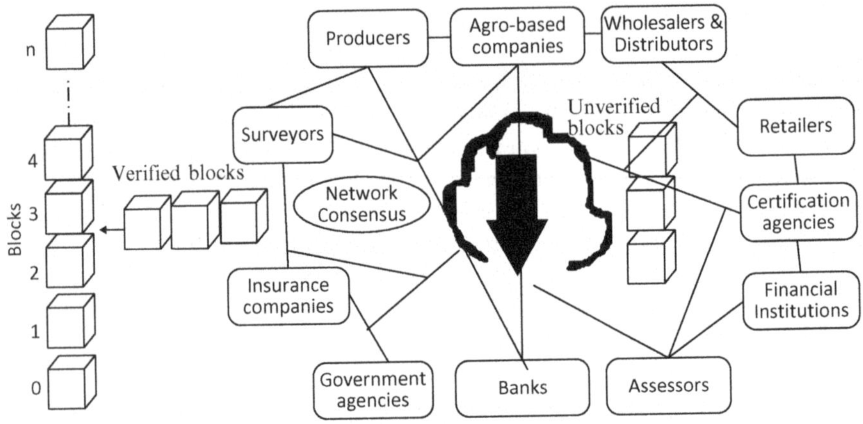

Fig. 14.5 Agriculture supply chain with blockchain technology

14.4 Model Description and Queueing Model

Considering the functioning of the blockchain system, the system can be modeled as a blockchain queue, where the generation process of blocks and the process of building the blocks are considered as the two stages where the batch service works. During the block generation stage, the confirmation of a block is done with the help of a computational problem-solving by a miner out of the miners using cryptographic hash algorithmic procedure, named mining. The nodes who contest to solve the problem are termed as miners. The miner who is the winner in the competition will be given some reward having some constant values as well as the charge of the transactions, and still has the authority for appending any newly arrival block into the system. A block is the collection of transactions, metadata having the timestamp of the newly added block, previous block, none given by the mining winner. In our model, the service time is considered as the sum of generation of blocks and block-building process.

Some of the model descriptions are defined below.

Arrival Process The transactions enter to the blockchain model with the Poisson process. The arrival rate to the system is λ. Each transaction enters into the system and waits in the queue where there is an infinite waiting room in the system. The lower left corner of Fig. 14.6 shows the arrival process of the transactions.

Service Process Every transaction once entered into the system is queued up in a buffer that is infinite in size. During the first stage of service, known as the block creation process, the transactions are successfully mined into blocks. A set of transactions are taken in a block (let it be b), a mining winner added a nonce to the block. Finally, during the second stage of the service, the blockchain is pegged

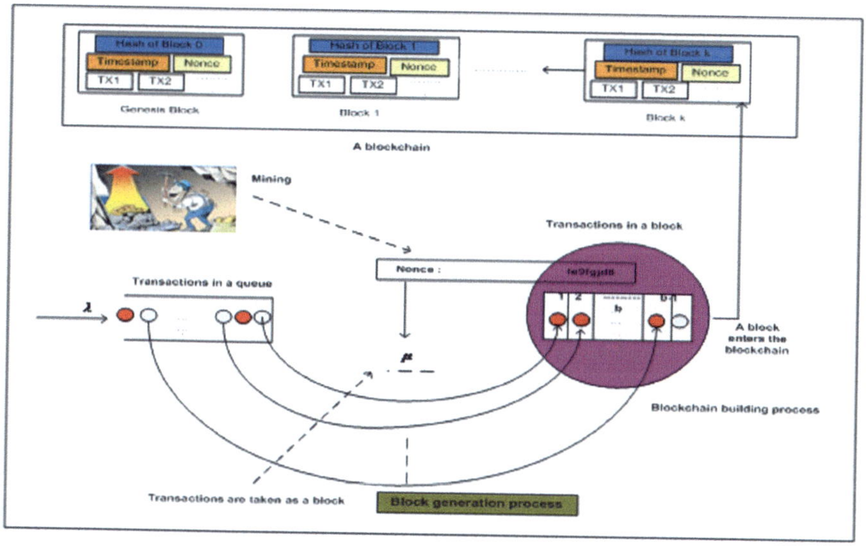

Fig. 14.6 Queueing model for the blockchain-based AgriChain system

with a block having a group of transactions. The service rate is denoted by $\mu$and is shown in the lower part of Fig. 14.6.

Discipline of Block Generation A block consists of many transactions, but a maximum of b transactions can contain inside a block. The transactions inside the block do not necessarily follow FCFS w.r.t their arrivals. Some transactions which arrived late into the queue may be mined into the block first. But for simplification, we have considered FCFS in our system model. The block generation process is depicted in the lower center of Fig. 14.6.

Maximum Block Size The maximum block size is considered to be limited, which helps in avoiding spam attacks. There are at most b transactions that can be taken in a block. If there are more than b transactions waiting in the waiting room, only at a maximum of b transactions are considered for the new block generation.

Independence Every random variable mentioned above will be independent of one another.

We model the system as a single server where transactions join the system, are processed in bulk, and enter into the system in Poisson distribution; the service time follows exponential distribution, in which whenever the server finds several transactions waiting, it begins service on b of them or the whole queue, whichever is less. Let π_n be the steady-state probability that n units are waiting in the queue ($n = 0, 1, 2, \ldots$) and the server is busy, and let π_{00}be the probability that there is no processing of blocks (the service station is empty).

The steady-state transition equations are

$$(\lambda + \mu)\,\pi_n = \lambda\pi_{n-1} + \mu\pi_{n+b}, n \geq 1 \tag{14.1}$$

$$(\lambda + \mu)\,\pi_0 = \lambda\pi_{00} + \mu\sum_{j=1}^{b}\pi_j \tag{14.2}$$

$$\lambda\pi_{00} = \mu\pi_0 \tag{14.3}$$

$$\pi\,(\tau) = \sum_{n=0}^{\infty}\pi_n\tau^n, \quad |\tau| \leq 1 \tag{14.4}$$

From (14.1) and (14.2), using (14.3), we have

$$\left(\lambda + \mu - \lambda\tau - \mu\tau^{-b}\right)\pi\,(\tau) = \mu\sum_{n=0}^{b}\pi_n\left(1 - \tau^{n-b}\right) \tag{14.5}$$

$$\pi\,(\tau) = \frac{\mu\sum_{n=0}^{b-1}\pi_n\left(\tau^n - \tau^b\right)}{\lambda\tau^{b+1} - (\lambda + \mu)\,\tau^b + \mu} \tag{14.6}$$

Now, let us take

$$f\,(\tau) = -\,(\lambda + \mu)\,\tau^b, g\,(\tau) = \lambda\tau^{b+1} + \mu \tag{14.7}$$

and take a closed contour C given by $|\tau| = 1 + \delta$, where δ is small. Then, we have on C

$$\begin{aligned} |\,f\,(\tau)| = \left|(\lambda + \mu)\,.\tau^b\right| = |\lambda + \mu|\,.\left|\tau^b\right| = (\lambda + \mu)\,(1 + \delta)^b \\ = (\lambda + \mu)\,(1 + b\delta) + O\left(\delta^2\right) \end{aligned} \tag{14.8}$$

and

$$\begin{aligned} |\, g(\tau)| &= \left|\lambda \tau^{b+1} + \mu\right| \le |\lambda| . \left|\tau^{b+1}\right| + |\mu| \\ &= \lambda \{1 + (b+1)\delta\} + \mu + O\left(\delta^2\right) \\ &= \lambda + \mu + \lambda (b+1)\delta + O\left(\delta^2\right) \end{aligned} \tag{14.9}$$

$$|\, g(\tau)| < |\, f(\tau)| \text{ on } C \tag{14.10}$$

if

$$\lambda\delta < \mu b\delta \tag{14.11}$$

i.e., if

$$\rho = \frac{\lambda}{b\mu} < 1 \tag{14.12}$$

which is the condition for the steady state to exist. When this condition is satisfied, using Rouche's theorem, we can write that the denominator of the right-hand member of (14.6), which is $f(\tau) + g(\tau)$, has equal 0's in the contour C as $f(\tau)$. But $f(\tau)$ has b zeros inside C, so the denominator $f(\tau) + g(\tau)$ also has b zeros inside the contour C as $f(\tau)$. But $f(\tau)$ has b zeros inside C, so the denominator $f(\tau) + g(\tau)$ also has b zeros inside C, one of which is τ=1, and therefore the remaining $b - 1$ zeros are within the circle of radius equal to one $|\tau| = 1$. Hence the denominator has only one zero outside $|\tau| = 1$; let us call it z_0.

Now corresponding to each zero z_i of expression, the expression should have a factor $(\tau_{-}\tau_i)$. Therefore, b of the factors of the denominator on the R.H.S of (14.6) should cancel with the b similar factors of the numerator. Also the numerator is of degree b, and the denominator is of degree $b + 1$. So, on cancelling the like factors, (14.6) can be written as

$$\pi(\tau) = \frac{A}{\tau_0 - \tau} = A \sum_{n=0}^{\infty} \frac{\tau^n}{\tau_0^{n+1}} \tag{14.13}$$

where A is a constant. The expression given in (14.13) is valid since clearly $\left|\frac{\tau}{\tau_0}\right| < 1$. Noting that the coefficient of τ^n in $\pi(\tau)$ is π_n, we have, on setting τ=0,

$$A = \pi_0 \tau_0. \tag{14.14}$$

Thus, (14.13) becomes

$$\pi(\tau) = \pi_0 \sum_{n=0}^{\infty} \left(\frac{\tau}{\tau_0}\right)^n \tag{14.15}$$

Therefore, we have $\pi_n = \pi_0 \tau_0^{-n}, \qquad n = 0, 1, 2, \ldots$ (14.16)

Also, from (14.3),

$$\pi_0 = \left(\frac{\lambda}{\mu}\right) \pi_{00} = b\rho\pi_{00} \tag{14.17}$$

Thus, (14.16) yields

$$\pi_n = b\rho\pi_{00}\tau_0^{-n}, \qquad n = 0, 1, 2, \ldots. \tag{14.18}$$

Finally, π_{00} can be evaluated from the normalizing condition.

$$\pi_{00} + \sum_{n=0}^{\infty} \pi_n = 1 \tag{14.19}$$

which after simplification gives

$$\pi_{00} = \frac{\tau_0 - 1}{(1 + b\rho)\,\tau_0 - 1} \tag{14.20}$$

14.5 Performance Evaluation

The mean number of transactions in the blockchain system waiting in the buffer is

$$L_q = \sum_{n=1}^{\infty} n\pi_n = \pi_0 \left(\frac{1}{\tau_0} + \frac{2}{\tau_0^2} + \frac{3}{\tau_0^3} + \ldots.\right)$$

$$= \frac{\pi_0 \tau_0}{(\tau_0 - 1)^2} \tag{14.21}$$

Using (14.17) and (14.20), we have

$$L_q = \frac{b\rho\tau_0}{(\tau_0 - 1)\{(1 + b\rho)\tau_0 - 1\}} \tag{14.22}$$

Note that (14.21) could also be obtained by differentiating (14.13) at τ=1.

It may be observed that z_0 is the root, outside the unit circle $|\tau| = 1$, of the equation

$$\lambda\tau^{b+1} - (\lambda + \mu)\tau^b + \mu = 0 \tag{14.23}$$

which can be written

$$\frac{1}{\tau} = \frac{\lambda}{\lambda + \mu - \mu\tau^{-b}} \tag{14.24}$$

Setting $\tau = \frac{1}{\tau}$, we see that $\frac{1}{\tau_0}$ is the root. Within the unit circle of the equation

$$x = \frac{\lambda}{\lambda + \mu - \mu x^b} = \overline{a}\left(\mu - \mu x^b\right) \tag{14.25}$$

where $\overline{a}$ is the L.T. of the inter-arrival time density function, since we have

$$\overline{a}(s) = \frac{\lambda}{\lambda + s} \tag{14.26}$$

Particular Case: M/M/1. Setting $b = 1$, it is clear that $z_0 = \frac{\mu}{\lambda} = \frac{1}{\rho}$, and we can verify the various results of M/M/1 from the foregoing. In particular, from (14.18) and (14.20),

$$\pi_{00} = 1 - \rho \tag{14.27}$$

$$\pi_n = \pi\,(n + 1 \text{ in system}) = (1 - \rho)\,\rho^{n+1}, n = 0, 1, 2, \ldots \tag{14.28}$$

and from Eq. (14.22),

$$L_q = \frac{\rho^2}{1 - \rho} \tag{14.29}$$

Queueing Time

Since the waiting time of the transactions is taken in for service in batches of b transactions of the whole queue, whichever is less, a transaction who finds $(ib + j)$ transactions are waiting in the queue $(0 < =j < =b - 1)$ has to wait until $(i + 1)$ service times are over – i for the transactions preceded by one for the residual service of the batch being served. Therefore, the queueing time density function $f_q(w)$ is given by

$$f_q(w)dw = \sum_{i=0}^{\infty}\sum_{j=0}^{b-1} \pi_{ib+j}\frac{(\mu w)^i}{i!}e^{-\mu w}\mu dw \tag{14.30}$$

$$\pi\ (w = 0) = \pi_{00} \tag{14.31}$$

Substituting from (14.16) in (14.30), we have

$$\begin{aligned} f_q(w)dw &= \sum_{i=0}^{\infty}\sum_{j=0}^{b-1} \pi_0\tau_0^{-ib-j}\frac{(\mu w)^i}{i!}\mu e^{-\mu w}dw \\ &= \pi_0 \sum_{j=0}^{b-1}\tau_0^{-j}\sum_{i=0}^{\infty}\frac{\left(\mu w\tau_0^{-b}\right)^i}{i!}\mu e^{-\mu w}dw \end{aligned}$$

$$= \pi_0\left(\frac{1-\tau_0^{-bi}}{1-\tau_0^{-1}}\right).\mu \exp\left[-\mu w\left(1-\tau_0^{-b}\right)\right] \tag{14.32}$$

Thus, the mean queueing time is

$$W_q = \int_0^{\infty} w.f_q(w)dw$$

$$= \pi_0\left(\frac{1-\tau_0^{-b}}{1-\tau_0^{-1}}\right).\mu\frac{1}{\mu^2\left(1-\tau_0^{-b}\right)^2}$$

$$= \frac{\pi_0}{\mu\left(1-\tau_0^{-1}\right)\left(1-\tau_0^{-b}\right)} \tag{14.33}$$

Now, $\tau_0$is a root of (14.23); therefore, we have

$$\tau_0^{-b} = (\lambda + \mu - \lambda\tau_0)/\mu = 1 + b\rho\,(1-\tau_0) \tag{14.34}$$

Hence, (14.33) simplifies to

$$W_q = \frac{\pi_0\tau_0}{\mu\,(\tau_0 - 1).b\rho\,(\tau_0 - 1)}$$

$$= \frac{1}{\mu b \rho} \cdot \frac{\pi_0 \tau_0}{(\tau_0 - 1)^2}$$

$$= \frac{L_q}{\lambda} \tag{14.35}$$

using Eqs. (14.12) and (14.21). Thus we see that Little's formula is applied to this system.

14.6 Numerical Results

To demonstrate the applicability of the above blockchain-based queuing Agrichain, several numerical results have been carried out, and only a few of them are presented here. Table 14.1 shows the average waiting time of the transactions in the waiting queue, and Table 14.2 shows the variance of the waiting time in the waiting queue for different arrival rates (λ) and different block sizes (b). The mean waiting time decreases in the queue when b increases. The mean waiting time also increases when λdecreases.

Figure 14.7 shows the effect of N Vs P_{block}. The effect of queue length on mean system length (L), mean queue length (L_q), and mean waiting time (W) of

Table 14.1 Mean waiting time of the transactions in the waiting queue

	$\lambda = 14$			$\lambda = 12$			$\lambda = 10$		
b	$\mu = 4$	$\mu = 6$	$\mu = 8$	$\mu = 4$	$\mu = 6$	$\mu = 8$	$\mu = 4$	$\mu = 6$	$\mu = 8$
1	9.1781	7.2726	6.3933	8.5345	6.7567	6.1323	7.5954	6.1394	5.6455
2	8.2727	6.2727	5.4303	7.4677	5.7567	5.1034	6.9594	5.4594	4.9404
3	7.2727	5.7717	5.1344	6.7656	5.1345	4.8293	6.1939	4.9409	4.1324
6	6.1526	4.7263	4.1424	6.1526	4.7263	3.6777	6.1526	4.7263	3.1454
12	5.3536	4.1737	3.5524	5.3536	4.1738	3.1336	5.3537	4.1737	2.6645
24	4.3636	3.6723	3.1323	4.3636	3.6723	2.8654	4.3636	3.6723	2.3252

Table 14.2 Variance of waiting time of the transactions in the waiting queue

	$\lambda = 14$			$\lambda = 12$			$\lambda = 10$		
b	$\mu = 4$	$\mu = 6$	$\mu = 8$	$\mu = 4$	$\mu = 6$	$\mu = 8$	$\mu = 4$	$\mu = 6$	$\mu = 8$
1	80.4838	48.8488	35.4748	63.3838	35.4884	35.3030	49.4949	35.3893	25.8939
2	63,4474	35.7637	25.9383	48.3737	25.4738	25.2303	36.5949	24.5945	24.3838
3	48.4939	25.1332	24.7374	36.6545	24.4543	17.8238	35.3893	24.3939	15.9393
6	35.7737	17.5764	16.5949	35.4554	15.5534	15.3939	35.8484	15.3939	9.12322
12	24.7373	16.3646	10.4894	25.4434	16.4455	9.39393	25.3893	14.3939	8.38393
24	16.1332	15.5747	9.30340	15.5445	11.5434	8.93933	15.3483	10.3930	5.29393

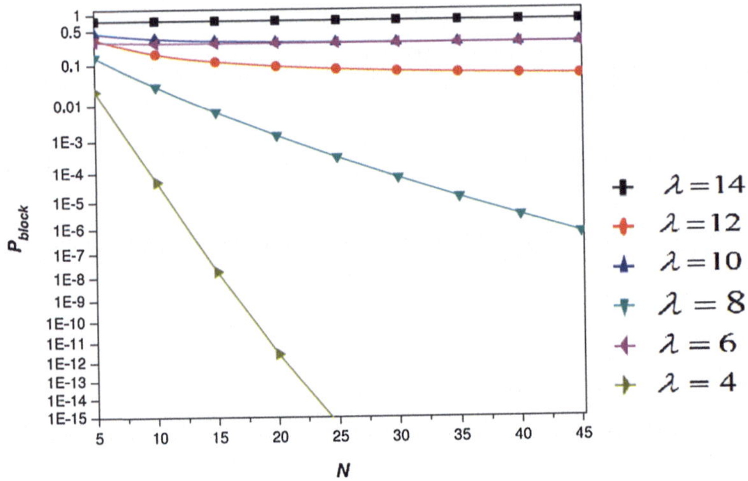

Fig. 14.7 Effect on N Vs P_{block}

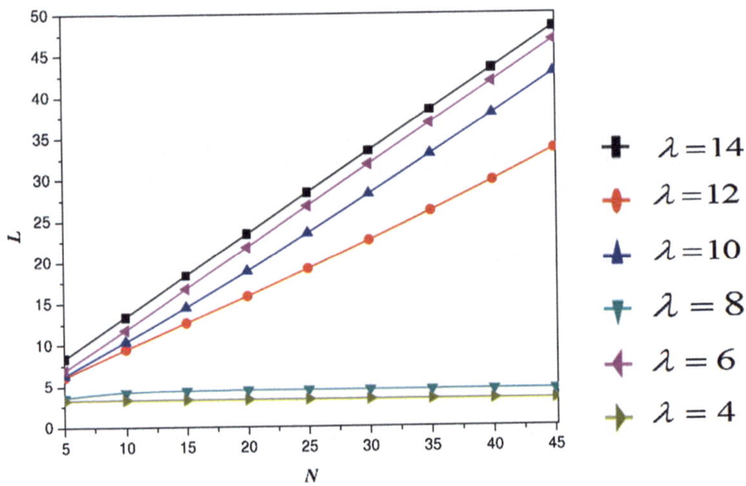

Fig. 14.8 Effect on N on L

a transaction in the system and in the buffer is presented in Figs. 14.8, 14.9, 14.10, and 14.11, respectively, for a fixed b = 5.

14.7 Conclusion

Blockchain and distributed ledger technology initially being started with the cryptocurrencies and in the later stage, many research communities and the industries

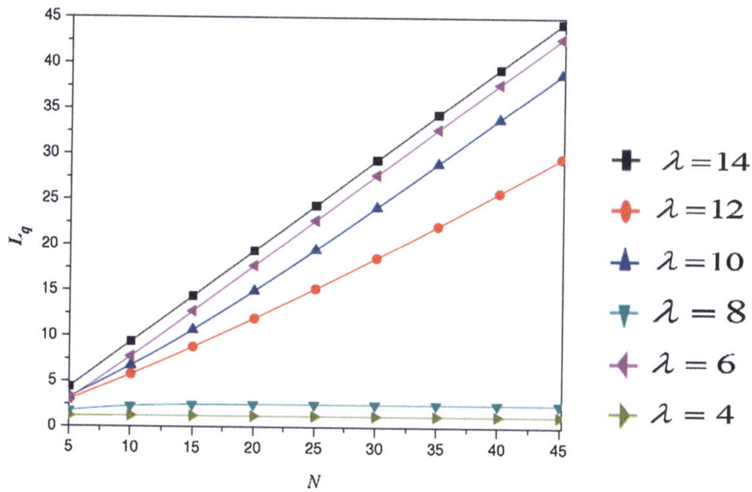

Fig. 14.9 Effect of N Vs L_q

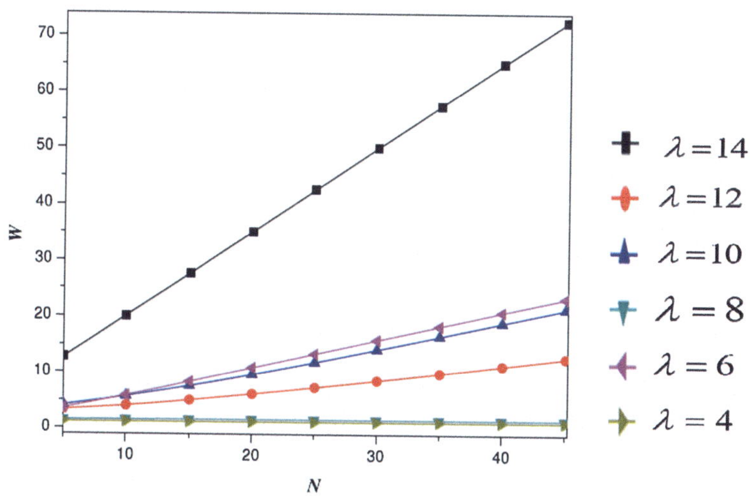

Fig. 14.10 Effect of N on W

started evaluating the technology that can apply to many use cases. After several studies, it has been found that it can apply to many use cases, and many mathematical models have also been built to prove the authenticity of the applications. Many use cases have the same fundamental question of whether the use cases are going to give the required performance measures. For more understanding, we have implemented a queueing model, the $M/M^x/1$ model, which evaluates the various performance measures of the smart agriculture system. The future research direction can include the different parameter studies and impact on different distributions.

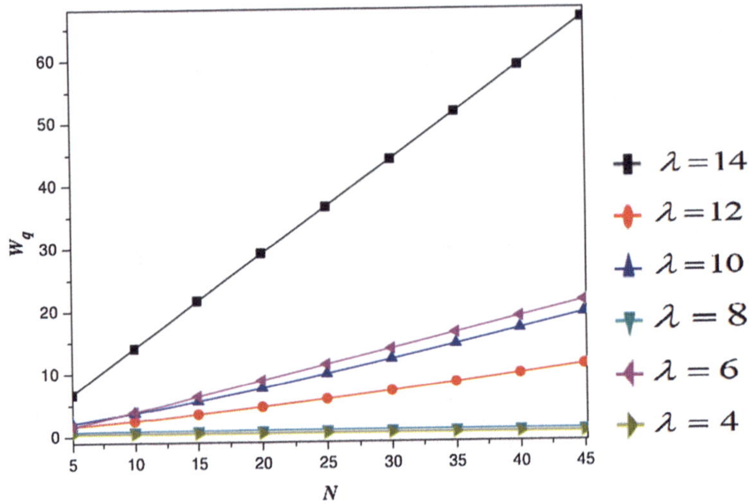

Fig. 14.11 Effect on N on W_q

References

1. O. Ahumada, J.R. Villalobos, Application of planning models in the agri-food supply chain: A review. Eur. J. Oper. Res. **196**(1), 1–20 (2009)
2. H. Allaoui, Y. Guo, A. Choudhary, J. Bloemhof, Sustainable agro-food supply chain design using two-stage hybrid multi-objective decision-making approach. Comput. Oper. Res. **89**, 369–384 (2018)
3. J. Angelis, E. Ribeiro da Silva, Blockchain adoption: A value driver perspective. Bus. Horiz. **62**, 307 (2019). https://doi.org/10.1016/j.bushor.2018.12.001
4. M. Atzori, *Blockchain Technology and Decentralized Governance: Is the State Still Necessary? As of 14 March 2019* (2015). http://www.theblockchain.com/docs/BlockchainTechnologyandDecentralizedGovernance:IstheStateStillNecessary.pdf
5. M.M. Aung, Y.S. Chang, Traceability in a food supply chain: Safety and quality perspectives. Food Control **39**, 172–184 (2014)
6. O. Bermeo-Almeida, M. Cardenas-Rodriguez, T. Samaniego-Cobo, E. Ferruzola-Gómez, R. Cabezas-Cabezas, W. Bazán-Vera, Blockchain in agriculture: A systematic literature review, in *International Conference on Technologies and Innovation*, (Springer, Cham, 2018), pp. 44–56
7. Q. Cao, D.G. Schniederjans, M. Schniederjans, Establishing the use of cloud computing in supply chain management. Oper. Manag. Res. **10**(1–2), 47–63 (2017)
8. B.B. Flynn, X. Koufteros, G. Lu, On theory in supply chain uncertainty and its implications for supply chain integration. J. Supply Chain Manag. **52**(3), 3–27 (2016)
9. X. Gu, Y. Chai, Y. Liu, J. Shen, Y. Huang, Y. Nan, A MCIN-based architecture of smart agriculture. Int. J. Crowd Sci. **1**(3), 237–248 (2017). https://doi.org/10.1108/IJCS-08-2017-0017(2017)
10. I.W. Kabak, Blocking and delays in M (x)/M/c bulk arrival queueing systems. Manag. Sci. **17**(1), 112–115 (1970)
11. K. Kaur, The agriculture internet of things: A review of the concepts and implications of implementation. Int. J. Recent Trends Eng. Res. (IJRTER) **02**, 04 (2016)

12. M.S. Mekala, P. Viswanathan, A survey: Smart agriculture IoT with cloud computing, in *Proceeding of the 2017 International Conference on Microelectronic Devices, Circuits and Systems (ICMDCS'17)*, (2017), pp. 1–7
13. J.F. Shortle, J.M. Thompson, D. Gross, C.M. Harris, *Fundamentals of Queueing Theory*, vol 399 (John Wiley & Sons, Newark, 2018)
14. X. Li, P. Jiang, T. Chen, X. Luo, Q. Wen, A survey on the security of blockchain systems. Futur. Gener. Comput. Syst. **107**, 841–853 (2020)
15. M.C. Vuran, A. Salam, R. Wong, S. Irmak, *Internet of Underground Things: Sensing and Communications on the Field for Precision Agriculture* In 2018 IEEE 4th World Forum on Internet of Things (WF-IoT) (pp. 586–591). IEEE (2018)

Chapter 15
Smart Farming: Securing Farmers Using Block Chain Technology and IOT

P. Praveen, Mohammed Ali Shaik, T. Sampath Kumar, and Tanupriya Choudhury

15.1 Introduction

Agriculture is one of the most fundamental human activities in India as the quality and amount of plant fabrication are depleted drastically due to various types of diseases that occur on cash crops. The process of identification and classification of the diseases and its types is considered to be a vital task. Due to this fact, it is very important to ensure that measures are taken to detect and mitigate any diseases on plants. This leads to huge growth in plant health that leads to economic growth of the farmer and India as a whole. Some of the plant diseases are considered to be a maximum threat in the agricultural sector, in turn reducing plant life span and drastically reducing naked-eye observation for identification and classification of various diseases. In circumstances where the crop is lost due to some disease, farmers commit suicides because of unpayable debt as per most surveys conducted [1–3].

The process that is adapted will identify various leaf diseases that leads to ambiguity in identifying the similarity with distinct visual properties that include shape, size, and color. These properties are used to evaluate the expert system based on user input. The initial phase in fighting against leaf diseases is to consider the adequate properties for recognizing the presence or absence of disease. Utilization of computers in the agricultural sector is remarkable as they are utilized in performing distinct scientific works where most are focused in the identification of diseases through foliar symptoms in various crash crops cultivated [4].

P. Praveen (✉) · M. A. Shaik · T. S. Kumar
SR University, Warangal, Telangana, India

T. Choudhury
Department of Informatics, School of Computer Science, University of Petroleum and Energy Studies (UPES), Dehradun, Uttarakhand, India

T. Choudhury et al. (eds.), *Blockchain Applications in IoT Ecosystem*,
EAI/Springer Innovations in Communication and Computing,
https://doi.org/10.1007/978-3-030-65691-1_15

The basic and significant "cash crops in India" based on good climatic conditions will support crop, and at the same time, most of the farmers prefer to perform cultivation of any regarded premier cash crop. These diseases are thoroughly analyzed as the process requires a maximum amount of time. Since few decades ago, the analysis shows that there is convincible growth in loss of yield, and this leads to drastic fall of crop productivity [5].

In present-day scenario, India has been observed a maximum drastic increase of farmers' suicides which has never happened before in the history of mankind compared with the number of suicides in any other part of the world. Based on the surveys being conducted, farmer suicides have been reported in most countries but not as many as in India. Untold misery and suffering as manifestation from farmers is the greatest human tragedy. It is neither greater nor dearer when compared with the nearer and dearer ones in one's own life because human life is the precious one and is being kept to an end unnaturally by committing suicides [6].

This chapter proposes a blockchain and IoT-based agricultural product tracking system to track the entire process of agricultural product life span which can greatly enhance the consumer confidence in food and improve the functioning of brand protection. The optimized solution to these problems is suggested in two different situations: firstly, to afford the protection of data as various organizations engage in agricultural "food supply chain" by the help of blockchain and, secondly, the addition of IoT technology to blockchain so that complete product life cycle may be monitored, avoiding poor quality and expiry of food. Research studies have proven that IoT technologies supply effective resolution to a diversity of issues related to agriculture, and in our case blockchain in collaboration with IoT will open a new corridor for agricultural food supply chain where all stakeholders (farmers, suppliers, distributors, retailers, and consumers) will make transparent transactions and trustworthy environment will be created for them without the help of mediators [7, 8].

15.2 Literature Survey

The blockchain is a ledger [5] comprises of most of the agents that obtain revolutionary stored information based on the procedure that produce and carry out [9] by acquiring the artifact or serve the collective ledger which administrates various contributing parties naturally using the peer-to-peer network [10]. All the evidences needs to be cross confirmed by network initially that includes to block chain by verifying various updates for verifying the data by pursuing the agreement of performing the decision as per the protocol specifically based on the assessment of various parties who are concerned to agree modifications of record as it will escort to alter various successive data records [7]. Almost it is not possible to update the data record in blockchain as it observes by viewing the block chain as a distributed ledger by storing the transactions rely between various parties proficiently which is supportable in a stable manner [8].

The blockchain technology tends to allocate "peer-to-peer (P2P) transactions" by considering various transparent areas without having the requirement of an intermediary aspect (similar to crypt occurrences) as the middleman in farming sector [11]. While eliminating the aspect of the "central authority" based on the modification of technology the aspect which is trusted and granted rather than authorizing and trusting by implementing cryptanalysis over P2P architecture for restoring the trust level in between various producers and customers by which the cost gets reduced drastically ("Agrifood market") [12]. The "blockchain technology" provides us with the methodology to quickly identify transactions that lie or are initiated by distinct types of customers which may further lead to reporting fraud by imposing smart contracts to track and implement the "Supply Chain Management" as the technology provides distinct levels of quality of food which is the major concern of any customer, which is easily handled by "Block chain Technology" has the capability to provide transparency in storing the various aspects by facilitating datasets with products value chain that initiates from creation to consumption based on data driven facilities and provides distinct feasible solutions that makes farming to be smarter and reduce the level of vulnerability [13–15].

15.3 Technologies Used in Agriculture

The process of adapting change and innovating is very important in any field and in farming too as in contemporary agriculture and to the "food processing industry" [2]. It has its own issues and challenges such as reduction of cost and increase in price which leads to more profits or inverse is loss as the cost merely depends on transportation, attaining supplies and labor cost as these factors show impact on price a customer preferences changes based on price and quality which has to be resolved in agri farming effectively.

The process of automating the major aspects of "Smart Farming" as it is based on utilization of technology for increasing the cultivation and to increase return on investment by automating the crop lifecycle by imparting robots or drones or custom machinery required to meet needs of a farm or cultivation by which a farmer has to work smart and not hard. Most of the technologies are under development stage or are still in the testing state. In accordance with this, there are some food processing companies who still follow the same old or traditional methodologies which have to adopt to the modern technologies (Fig. 15.1) [16].

15.3.1 Internet of Things (IoT) Technology

To connect objects with a network for information exchange and communication, IoT technology is used. IoT is capable of making billions of interconnected devices that are also termed smart objects [17, 18]. These smart objects are proficient to

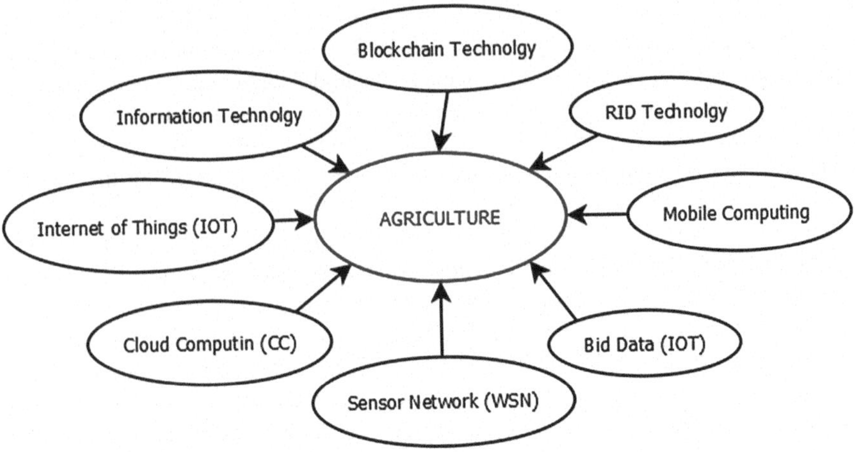

Fig. 15.1 Technologies presently used in agriculture

collect environmental information and communicate with other systems through the Internet [13]. Applications that are developed in IOT facilitates various devices by verifying distinct domains will be controlled [24]. Some of the examples are: "Home Appliances, Health Monitoring, Smart Home, Smart Cities, Smart Agriculture, etc." [7]. "IoT applications have unique importance throughout the lifespan of the agriculture sector, such as cultivate yields, irrigation, harvesting and post harvesting, crop storage, processing, transportation, and sales and for agriculture applications there are a variety of specialized sensors are available, for instance, soil moisture sensor, humidity, Leaf moisture, solar emissions, Infrared radiations, Rain predictor, etc. [19–21] In the scenario of IoT, sensors can be installed in different fields like greenhouses, seed storages, cold storages, agriculture machinery, transportation system, and livestock; and their data can be stored in the cloud for monitoring and control" shown in Fig. 15.2 [15].

15.3.2 Wireless Sensor Networks (WSNs)

For sensing and analyzing the various different parameters that are required in the agriculture domain, WSN technologies are available. To utilize sensors in agriculture, many applications have been developed. The best option available in between cyberspace and real world is established by designing sensors which connects agriculture with IOT is using Sensor Networks. WSN are cheap devices and capable to work in specific environment sand work for a long period without battery replacement [14].

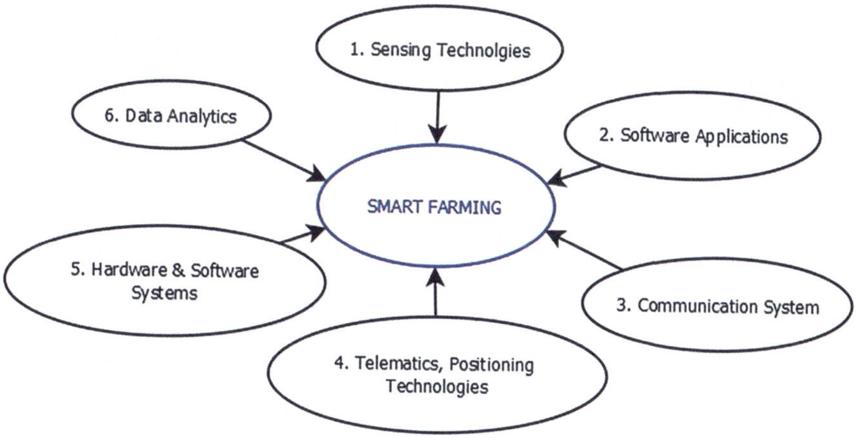

Fig. 15.2 IOT-based smart farming

15.3.3 Cloud Computing(CC)

CC is the provision of system or IT infrastructure through the use of the Internet providing share resources at a cheap cost. The service provider (SP) offers different services and platforms at a low cost to store and share agricultural data through used cloud computing [22].

15.3.4 Big Data

Big data refers to a huge quantity of data gathered from different channels for extended periods of time like data collected from sensors, social networking, and business data. Big data has many challenges like capturing, storage, investigation, and research. To cut the production cost, big data is useful in the agriculture domain for maintaining supply chain management of agricultural products [6].

15.3.5 Mobile Computing

Mobile computing has low information sharing cost and is easily available and widely being used in different sectors including agriculture. The mobile based systems are being used for sending time to time sessional update to farmers about farming to make timely decisions [23].

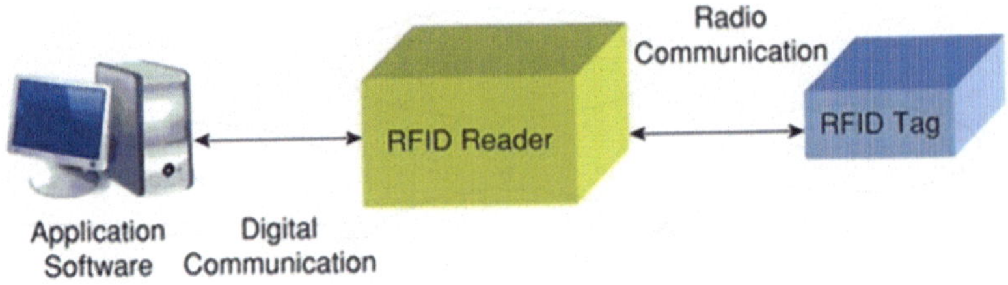

Fig. 15.3 RFID technology in agriculture

15.3.6 RFID Technology in Agriculture

For identifying various animals, we use radio frequency identification (RFID) called livestock; this process is tagged in technology for enhancing various corps; this process is adopted widely for obtaining economic results that are acceptable. Application of RFID is used in agriculture to keep track of levels of available food supply or its livestock or the level of farming precision attained along with the cold supply chain analysis as shown in Fig. 15.3.

15.3.7 Agricultural Food Supply Chain Management

In a Food supply chain network various stakeholders such as "Input supplier, Food producer (Farmer), Food processing units" or the food byproducts that are yielded at the time of processing the food or selling them to customer by verifying the process of attaining the desired quality and rate of food production by which a farmer gets profit by handing the crop to right destination at a given time span by making use of "government support where a farmer can sell his crop at any place and at his desired price without selling it to the middle man who controls the access of markets and this leads to the farmer to get poor prices". In the present-day scenario, the middle man is acquiring all the sources of profits from farmers as a farmer will not get maximum support price, as in the present context almost all farmers are selling the crop for an amount which is lesser than the minimum support price. The middleman can be eliminated as illustrated in Fig. 15.4.

15.3.8 Technologies Used in the Agricultural Food Supply Chain

The food-supply chain has attracted several important and advanced technologies in the implementation of processes like artificial intelligence and advanced analytics. Internet of Thing (IoT), autonomous mobile reboots and autonomous vehicle, "Virtual personal assistants (VPAs)", "Robotic process automation (RPAs)", "Electronic

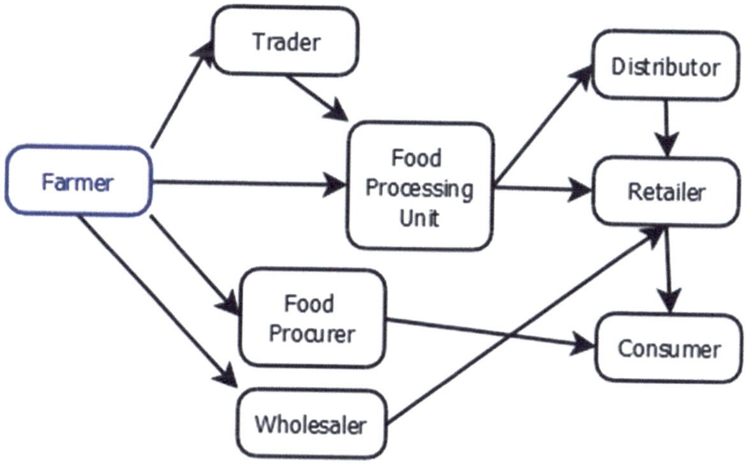

Fig. 15.4 Agricultural food supply chain

Data Interchange (EDI)", "Collaborative Planning Forecasting and Replenishment (CPFR)". Through the use of these technologies, paperwork has been reduced, product traceability has been improved, and bullwhip effect in the food supply chain has also been controlled. Related research studies have proven that technology can improve product documentation of food quality, food safety, wrapping, and software development. Most often, product documentation is created and transformed on paper or through bar codes or through RFID tags as it comprises an electronic system that tends to adopt the latest technology that possesses the capability to attract various consumers to buy more goods. "Amazon" has embraced the latest technologies which make use of components such as "camera, sensors, and sophisticated AI software" which is required to calculate buying capabilities of customers through the "Amazon app" using a smartphone to improve supply planning and overall logistic operations using the IT-based "Food Tracking System" as proposed.

The existing technology in the mostly used case scenario for implementation of food chain comprises of centralized database with distinct product features handed over to distributor from the direct seller and is mainly appropriate in performing the "centralized retail supply chain" management, as the sellers possess their own distribution system for handling effective logistics to obtain the product information related to farm is kept in a dataset to be made ready for processing. For the transformation of the existing technique of data collection, data distribution, and data safety, it is necessary to make an end-to-end tracking system based on information technology system or on blockchain technology.

15.3.9 Blockchain Technology

The power of Bitcoin improved significantly all over the world, and China produced two-thirds of those Bitcoins with blockchain technology. According to blockchain experts, it revolutionizes our daily life. Block chain is the decentralized ledger that comprises cryptocurrency stored in the form of transactions that occur in the form of a system as the digital currency by adding aspects of cryptography and block chain for empowering the existence of primary aspects of digital currency which is decentralized and denoted as "Bitcoin."

The general opinion about the block chain is that it was made for currency only. However, it can be useful to other zones by applying a decentralized operational system. Based on this background, we decided to utilize its potentials in the "agricultural food supply chain."

Blockchain is an emerging technology and presently getting the attention of many industries like finance, healthcare, education, food, and management. The main reason that blockchain is getting attention is its unique features operated only by a trusted intermediary in a decentralized method, without the help of authentication system, and capable to achieve the same goal with the same volume of dependability. Blockchain opened new pathways and introduced trustless networks because with blockchain, you can make a transaction without trust on other parties. The function of mediators has been eliminated, and transactions have become faster between different stakeholders. The security of the information may also be ensured through the use of cryptography.

Presently, companies involved in food supply chain are facing many challenges such as delay and defaults in the distribution of goods, food origin tracing, and high workforce to meet the desired demand of all stores. To address these issues, companies have digitalized their procedures to facilitate stakeholders and expand their business in the imparting supply chain which is further digitized as it tends to enhance the associated risk that comes through various attacks over the databases, and malicious users are somehow capable of updating or stealing or deleting the data. Especially in agribusiness, attack of hackers on data may cause serious issues, but the blockchain platform can provide a secure solution to these problems with decentralized, automatic, and trusted data and transportation management as shown in Fig. 15.5.

15.3.10 Control with IoT

Checking the quality of crops, plants, and animals is also a very significant process for all farmers, and IoT can extend support and make its contribution here. Quality of soil, irrigation activities, pests and diseases, and many farm-related activities can be monitored and controlled through appropriate IoT software installed in smartphones, computers, or tablets.

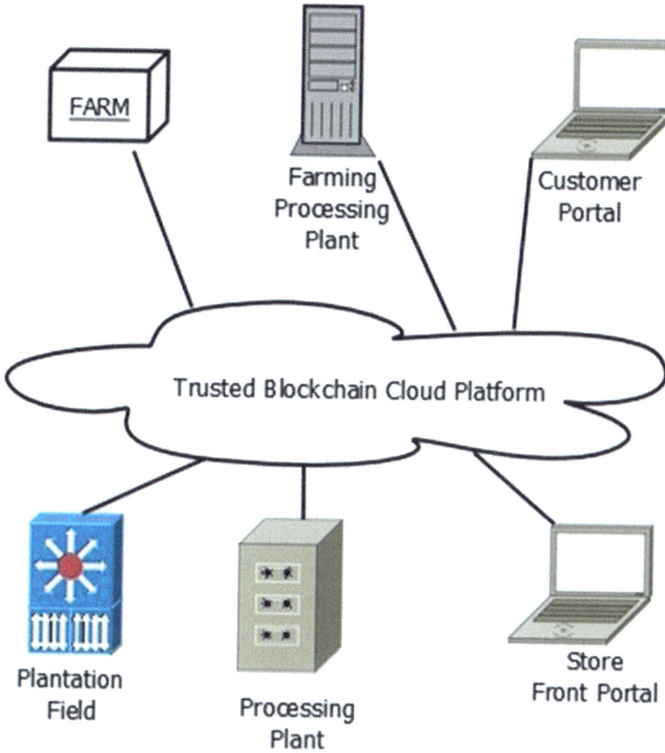

Fig. 15.5 Blockchain and IoT-based smart agriculture ecosystem

Therefore, IoT in combination with blockchain will make the agricultural food supply chain more reliable and fruitful by monitoring the whole process.

Food origin tracking. When purchasing fruits or mutton in a superstore, can you confidently say that you bought a safe food? Although store representatives will show you all relevant documents and certifications, you are not sure how the food was stored earlier and whether it contains any harmful bacteria or not. Blockchain can enable customers to find out everything about each product.

Permanent record keeping. Blockchain has the ability to keep a record on a permanent basis for future correspondence. Researchers designed an "AgriBlockIoT" [24] a decentralized, Block chain based probable resolution for agricultural based "Food Supply Chain Management" as it is capable to be flawlessly assimilate as the IoT strategy that creates and consumes digital data in accordance of the chain which is clearly defined based on the classical aspects based on vertical domain with the attributes such as named farm to fork that is designed and installed. It is further achieved to trace two distinct "blockchain implementations" which access the performance by deploying various parameters such as "latency, CPU, and network usage" that are highlighted by some of the vital pros and cons elaborately.

15.3.11 *Proposed IoT with Blockchain Smart Farming Model*

It has been proven from the presented research work that IoT combined with block chain technology can play a tremendous role in smart agriculture and food supply chain and all stakeholders can get many advantages without getting help from a trusted third party. In this chapter, we tend to propose a smart model which is based on IoT and blockchain architecture to perform smart farming activities through innovative ways as shown in Fig. 15.6.

Our smart model has three parts: IoT, blockchain, and retail market. IoT part is related to data generated through the use of sensors arranged on the farm. Data will be generated through IoT devices and will be recorded in the system, for instance, production information will be recorded during the production stage including essential information and production log information such as product name, origin, etc. and later product growing information will also be recorded at multiple times, and all stakeholders will have access to see this information. The second part is related to data storage, consensus, encryption, decryption, and verification function which will be performed by blockchain. It will run smart contracts to execute the corresponding logic at specific points in time which will increase scalability, simplify the process, and reduce cost. The third part is related to the retail market; after completion of the production, process goods will be delivered to successful bidders (distributors, retailers).

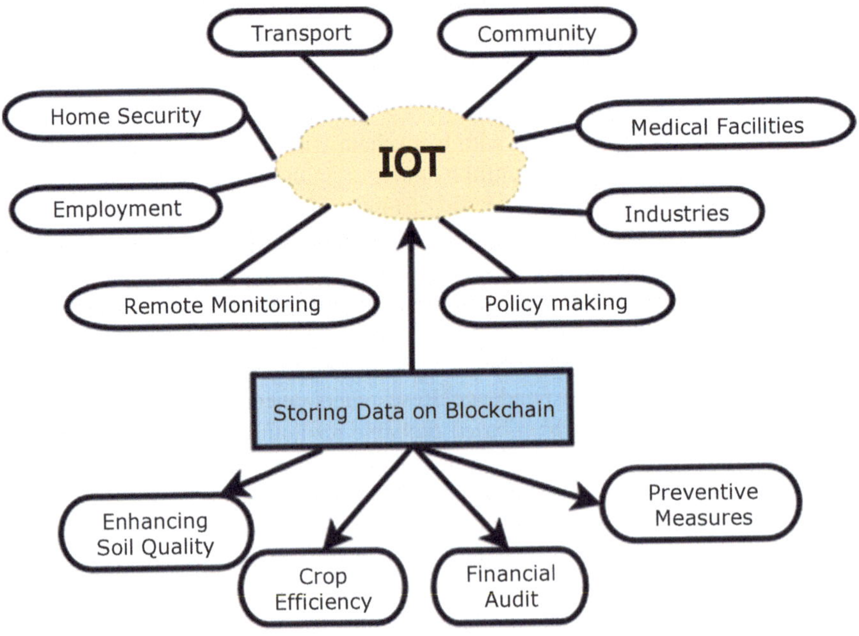

Fig. 15.6 Block diagram of IoT with blockchain smart farming model

15.4 IoT Function

15.4.1 Steps involved in IoT function

Step 1: IoT devices will monitor crop health and generate information to provide support to farmers for making a timely decision related to crop growth, and collected information will be saved on the blockchain.

Step 2: To get more insight information, machine learning is used and will provide more in-depth information like crop yield prediction, crop growth factor, and demand forecasting and recommendation to improve crop quality. Farmers can also get help from machine learning algorithms to make improvement in the irrigation system. Data collected through machine learning will be saved on the blockchain to empower stakeholders like farmers, investors, innovators, and retailers to get access fairly.

Step 3: The data that are collected by implementing machine learning will be arranged using "Interplanetary File System (IPFS)" on the blockchain over a decentralized server for avoiding the authority control and to reduce the risk of data hacking. The available system is stored for obtaining the information over a centralized server by which we can avoid the aspects of hacking, and this contrasts the generated block chain to define distinct rules. The function of Smart contracts is to facilitate specific stakeholders to exchange data stored on the blockchain; at the same time, information will be shown to each agriculture market member which will provide a unified platform to improve efficiency.

15.4.2 Food Supply Chain Process

Step 1: In smart model, IoT devices are used to provide important information related to the crop. Then data taken will be kept in the blockchain by IPFS.

Step 2: When the crops are fully grown, companies dealing with food processing will get access to the bidding platform to start bidding. Once the bidding process is completed all the crops are delivered to plants through vehicles by enabling the IoT aspects to keep and maintain the desired temperature. It will be necessary to validate through smart contracts based on which all the crops will be processed as the companies will store all the information obtained in each step over the blockchain, as this information will be accessible to all the stakeholders to confirm when the food is produced with good or low quality. Blockchain will make sure that the desired criteria have been met at each step.

Step 3: After processing of food items, wholesalers and retailers will have access on the bidding platform to offer a bid for the product they want; after completion of the bidding process, food products will be dispersed to successful bidders through vehicles enabled by IoT to maintain the desired temperature again.

Blockchain will track the whole process throughout the supply chain, which will help food businessmen to conduct food recollections or inquiries rapidly.

Step 4: Blockchain will maintain all data, from initial stage to distribution stage, for current or future check of all related information like consignment numbers, food handling, date of expiration, temperature at which food was kept, and other relevant information.

15.4.3 Improve Food Traceability

No one in the world can surely say that he has bought a good food to eat; here our smart model with blockchain infrastructure can solve the problem of consumers by providing them access to know where and how their food initially originated and how it reached to them. Traceability is another fabulous feature of our smart model which will allow farmers to record the present situation of their yield and must be capable of tracking the whole process done from planting to harvesting and further storing and delivery of the crop. In this way, food frauds will be reduced, and farmers will be paid fairly secondly through tracking system, and other stakeholders in the supply chain will also be able to track the whole process, and hence trust will be developed among all parties.

15.4.4 Improved Farmers' Productivity

Presently, majority of the farmers depend on different agriculture-related software to record their data, and they have no common platform, due to which they put on a lot of efforts and bear cost. Our smart model will permit farmers to record all information on a single platform, and everyone can easily access according to his need.

15.4.5 Fair Mode of Payment

Numbers of problems currently exist that make it hard for the farmers to acquire payment for their crops like payment through wire transfer which often takes a substantial amount to transfer money due to which farmers' profitability may decrease. In our proposed smart model, smart contracts based on blockchain will ensure payment to farmers through a fast and automatic way without being charged. Farmers will be able to get paid for their produce immediately after delivery. Another feature of this model is smart contracts through which the role of middlemen has been eliminated; farmers often face issues to put up for sale goods they produced in the marketplace at a feasible price: they often need help from middlemen whom they have to pay extra amount, or they can be cheated by mediators. Through smart contracts, farmers will interact straightaway with most of the retailers through whom there is a possibility that they will be able to get a fair price for their products.

15.5 Conclusion and Future Work

Nowadays, food supply chain companies are concentrating to find the actual food source and track the whole process of food production from food origin to end consumer which is one of the most challenging tasks for them. In this research work, we made an attempt to address this issue by providing a solution to this problem by creating a smart model based on blockchain and IoT technologies with our own understanding grounded on background literature. We have proved in our research work that blockchain in combination with IoT can be more beneficial to track the whole process of food. In addition to that, a proposed system will offer better consumer self-assurance which will reveal in sales and consumer pleasure. In our future work, we will develop a software for a proposed model for practical implementation.

References

1. P.E. Colombo, E. Patterson, L.S. Elinder, A.K. Lindroos, U. Sonesson, N. Darmon, A. Parlesak, *Optimizing School Food Supply: Integrating Environmental, Health, Economic, and Cultural Dimensions of Diet Sustainability with Linear Programming* (2019)
2. R. Casado-Vara, J. Prieto, F. De la Prieta, J.M. Corchado, How blockchain improves the supply chain: Case study alimentary supply chain. Proc. Comput. Sci. **134**, 393–398 (2018)
3. W. Chen, G. Feng, C. Zhang, P. Liu, W. Ren, N. Cao, J. Ding, Development and application of big data platform for garlic industry chain. Comput. Mater. Contin. **58**(1), 229–248 (2019)
4. Y.C. Choe, J. Park, M. Chung, J. Moon, Effect of the food traceability system for building trust: Price premium and buying behavior. Inf. Syst. Front. **11**(2), 167–179 (2009)
5. T.K. Dasaklis, F. Casino, C. Patsakis, Defining granularity levels for supply chain traceability based on IoT and blockchain, in *Proceedings of the International Conference on Omni-Layer Intelligent Systems*, (ACM, 2019), pp. 184–190
6. P. Praveen, B. Rama, An optimized clustering method to create clusters efficiently. J. Mech. Contin Math Sci, ISSN (Online): 2454-7190. **15**(1), 339–348 (2020, January). ISSN (Print): 0973-8975. https://doi.org/10.26782/jmcms.2020.01.00027
7. B.F. Glunz, W.R. Pearson, A.F. Munoz, *Method and system for creating 3D models from 2D data for building information modeling (BIM)*. U.S. Patent 9,817,922, issued 14 Nov 2017
8. B. Rama, P. Praveen, H. Sinha, T. Choudhury, A study on causal rule discovery with PC algorithm, in *2017 International Conference on Infocom Technologies and Unmanned Systems (Trends and Future Directions) (ICTUS), Dubai*, (2017), pp. 616–621. https://doi.org/10.1109/ICTUS.2017.8286083
9. A. Parikh, M.S. Raval, C. Parmar, S. Chaudhary, Disease detection and severity estimation in cotton plant from unconstrained images, in *2016 IEEE International Conference on Data Science and Advanced Analytics (DSAA)*, (IEEE, 2016), pp. 594–601

10. A.A. Sarangdhar, V. Pawar, Machine learning regression technique for plant leaf disease detection and controlling using IoT, in *2017 International Conference of Electronics Communication and Aerospace Technology (ICECA)*, vol. 2, (IEEE, 2017), pp. 449–454
11. S. Patel, I.U. Sayyed, Impact of information technology in agriculture sector. Int. J. Food Agric. Vet. Sci. **4**(2), 17–22 (2014)
12. V. Nedovic, A. Kalusevic, V. Manojlovic, S. Levic, B. Bugarski, An overview of encapsulation technologies for food applications. Proc. Food Sci. **1**, 1806–1815 (2011)
13. L. Ruiz-Garcia, L. Lunadei, The role of RFID in agriculture: Applications, limitations and challenges. Comput. Electron. Agric. **79**(1), 42–50 (2011)
14. D. Sharma, A.P. Bhondekar, A. Ojha, A.K. Shukla, C. Ghanshyam, A technical assessment of IoT for Indian agriculture sector. Int. J. Comput. Appl. (2016)
15. S. Han, H. Yang, Understanding adoption of intelligent personal assistants: A parasocial relationship perspective. Ind. Manag. Data Syst. **118**(3), 618–636 (2018)
16. G. Perboli, S. Musso, M. Rosano, Blockchain in logistics and supply chain: A lean approach for designing real-world use cases. IEEE Access **6**, 62018–62028 (2018)
17. T. Choudhury, A. Gupta, S. Pradhan, P. Kumar, Y.S. Rathore, Privacy and security of cloud-based Internet of Things (IoT), in *2017 3rd International Conference on Computational Intelligence and Networks (CINE)*, (2017), pp. 40–45
18. Z. Ajazmoharkan, T. Choudhury, S.C. Gupta, G. Raj, Internet of Things and its applications in E-learning, in *3rd IEEE International Conference On*, (2017). https://doi.org/10.1109/CIACT.2017.7977333
19. A. Khanna, A. Sah, T. Choudhury, Intelligent mobile edge computing: A deep learning based approach. Commun. Comput. Inf. Sci. **1244**(CCIS) (2020). https://doi.org/10.1007/978-981-15-6634-9_11
20. A. Khanna, R. Goyal, M. Verma, D. Joshi, Intelligent traffic management system for smart cities. Commun. Comput. Inf. Sci. **958** (2019). https://doi.org/10.1007/978-981-13-3804-5_12
21. M. Khurana, T. Choudhury, P. Malik, A review on network security challenges and the internet of things (IoT), in *Proceedings of the 4th International Conference on Contemporary Computing and Informatics, IC3I*, (2019). https://doi.org/10.1109/IC3I46837.2019.9055675
22. A. Kamilaris, A. Fonts, F.X. Prenafeta-Boldú, *The Rise of the Blockchain Technology in Agriculture and Food Supply Chain* (2018)
23. D. Ivanov, A. Tsipoulanidis, J. Schönberger, Operations and supply chain strategy, in *Global Supply Chain and Operations Management*, (Springer, Cham, 2019), pp. 81–110
24. P. Praveen, C. Jayanth Babu, *Big Data Clustering: Applying Conventional Data Mining Techniques in Big Data Environment*. Innovations in Computer Science and Engineering, Lecture Notes in Networks and Systems, vol. 74 (Springer, Singapore, 2019), ISSN 2367-3370, https://doi.org/10.1007/978-981-13-7082-3_58

Chapter 16
An Efficient Methodology for Avoiding Threats in Smart Homes with Low Power Consumption in IoT Environment Using Blockchain Technology

Vejendla Lakshman Narayana, Arepalli Peda Gopi, and R. S. M. Patibandla

16.1 Introduction

The Internet of Things are digital computer systems and are linked and may also be referred to as wired or smart gadgets. Such systems deliver, in specific, enhanced interconnectivity. Mostly IoT devices are embedded in the network and the interconnection of such devices would allow automation in all fields that help us to create smart cities, smart grids, smart homes, etc. These devices basically collect data from other devices and automate the flow of data between other devices. Throughout the Internet of Things (IoT), a human may have some biosensor installed on his body, some business, etc. Machine-to-machine connectivity can be achieved conveniently with the aid of these tools. Essentially, IoT applications are the next wave of embedded network sensor devices.

The wireless network sensor systems were not as well suited as the IoT devices. When WSNs progressed, they were gradually fitted for the introduction of innovations such as cellular technology, microservices, microelectrochemical networks, etc. Such improvements permitted the data flow between unstructured machines created by the data. The study of data transfer from various computers has rendered things simpler. Due to IoT systems, most computers have the potential to connect with two or more apps. Such tools also help to develop innovative successful market models that can strengthen all company processes. Deployment of such tools often decreases costs and risks as they grow and become more common. According to one report, almost 20.8 billion computers would be in operation by 2020 [1]. Both

V. L. Narayana (✉) · A. P. Gopi
Vignan's Nirula Institute of Technology & Science for Women, Guntur, Andhra Pradesh, India
e-mail: lakshmanv58@vignannirula.org

R. S. M. Patibandla
Department of IT, Vignan's Foundation for Science, Technology and Research, Guntur, AP, India

T. Choudhury et al. (eds.), *Blockchain Applications in IoT Ecosystem*,
EAI/Springer Innovations in Communication and Computing,
https://doi.org/10.1007/978-3-030-65691-1_16

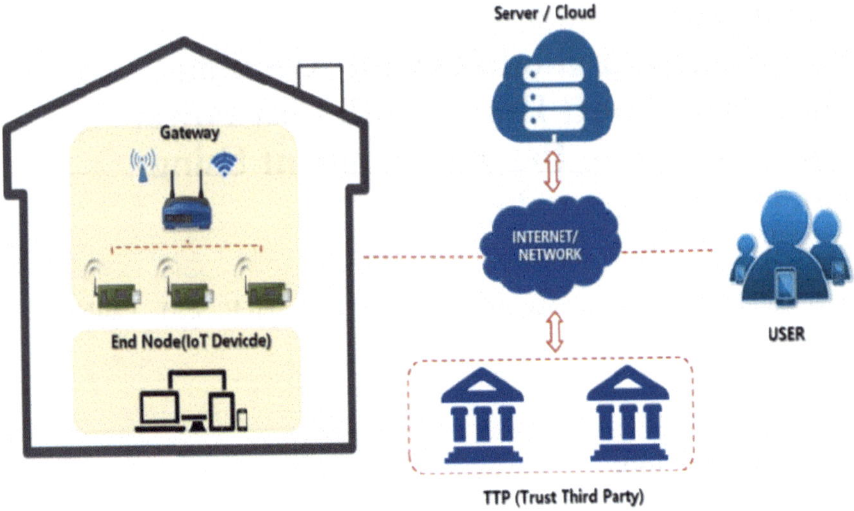

Fig. 16.1 IoT devices and user communication

of these systems should be interconnected through the Internet. Since these tools operate with minimal microprocessor resources, memory, and power specifications, they find applications in any area. According to estimates from Cisco IoT products, $14.4 trillion in value would be produced by all sectors over the next decade. Figure 16.1 depicts the process of communication between users and IoT gadgets.

The Internet of Things (IoT) is a fairly common concept [2] that is only beginning to gain a lot of momentum. Through linking all possible things to the Web, from cameras and tablets to refrigerators and wind turbines, people are easily coming up with innovative ways to utilize the IoT to enhance our quality of life. The challenge with this fast-growing sector is that defense inevitably takes a back seat to flexibility and manufacturing costs. Most devices in IoT systems are very easy, but that doesn't imply they're harmless at all. This is shown, for example, by the recent large-scale DDoS attack, which left many major websites (Twitter, Amazon, PayPal, etc.) unreachable due to the flood of key DNS servers [3]. In this attack, a botnet comprising (among other things) poorly protected IoT devices was used, sometimes with default passwords.

16.1.1 Delivery of Infrastructure to IoT

Internet of Thing (IoT) devices have brought a new wave of interconnected devices with many applications. Such systems can fill the divide between analog hardware and the modern universe as a consequence of increased usage. While such tools may be the key component of data collection and segregation. As such tools are commonly utilized, an outline of some of the most relevant applications offered will be given [4].

Smart homes: With the use of the Internet of Things devices, there is a massive and continuous change in the field of home automation. All of these devices interact with each other through wireless networking or the Internet and create a smart house. Smart home strengthens the protection of others, and the owner of the house may even monitor the atmosphere of the building. This is often energy effective because these machines operate with scarce resources and are committed to their scheduled tasks. Google's smart home, Amazon, Belkin, and Philips are the leading manufacturers of these devices. The Nest smart thermostat is a groundbreaking system that is a self-learning tool.

Wearables The wearables are one of the most commonly used or common items. Many of these smart systems come with built-in IoT applications and are commonly applied to safety monitoring and other entertainment fields. As all the wearables are powered by the batteries, IoT devices may be an acceptable option. Manufacturers like Apple, Samsung, Fitbit, Jawbone, etc. are big manufacturers in the world of wearables.

Environmental Monitoring IoT devices can be deployed as environmental monitoring systems. We may track water quality, atmospheric condition, and soil condition through this method. And thanks to the introduction of these tools, we may track air emissions, as well as early alert measures for natural hazards such as earthquakes, tsunamis, etc.

Agriculture IoT tools lead to the introduction of modern and advanced agricultural techniques. Convergence of physical equipment, cellular or Internet access, and cloud systems can allow farmers to collect data or critical information on the environmental or agricultural conditions of the region. Through the introduction of such machines, farmers can now determine whether the field is dry or whether or not it has been fertilized. IoT apps may also help farmers forecast potential yields.

Medical and Healthcare IoT devices have an important role to play in the Wireless Body Area Network. Through this network, we can remotely monitor the vital patient information about the heart rate, blood pressure, etc. Some specialized sensors may also be deployed to monitor the health of senior citizens or newborn infants. Smart beds have already been installed in several hospitals. Such beds are capable of monitoring the patient's motions.

16.1.2 IoT Security Using Blockchain

When authentication or audit procedures are not given, the issue of trust in information systems is extremely difficult. In particular when concerned with the sensitive details of virtual currencies, as an economic transaction, two controversial theories were brought up by Satoshi Nakamoto in 2008, in that sense [4]. The effect was immense. The first of which is Bitcoin, a machine cryptocurrency, that even requires assistance to keep its worth. Central entity or financial institution. However, Bitcoin is a decentralized P2P network of actors. Collectively and effectively, they form an auditable and verifiable network. The second of the Principles, the popularity of which goes even beyond Cryptocurrency itself, is blockchain.

Blockchain is a system which allows a group of anonymous actors to monitor transactions. It gives an unalterable, open, safe, and auditable booklet. The blockchain can be openly and completely downloaded and used by any entity at any time, providing links to all transactions that have taken place since the first operation of the system. This is the blockchain protocol that constructs information in the blockchain, where every block holds a series of Bitcoin transactions. Blocks are linked to the former pin to build a thread.

Routing, storage, wallet services, and mining [5] should be the following features: network partners to assist and operate with the blockchain. Various node styles will be part of the network based on the resources they provide. The routing feature must be part of the P2P network, which involves transaction and distributing chains. The job of the stock ministry is to hold the chain in the node duplicated. Customer security keys are required for the wallet suppliers. Transactions that work with your Bitcoins (Fig. 16.2).

IoT products, like numerous glitches and security vulnerabilities, frequently arrive with outdated built-in operating systems and applications. Which occurs after 3 years with a cheap computer even if the goods are fitted to the new

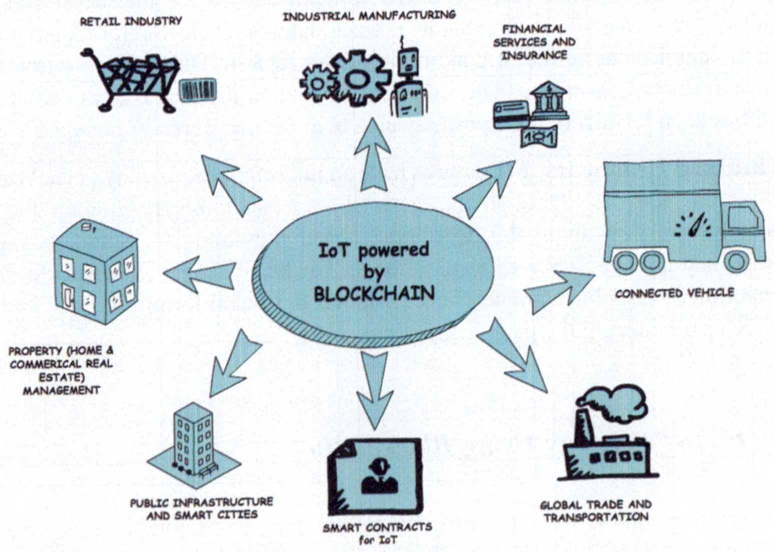

Fig. 16.2 Blockchain in IoT

technology? Sellers do not generally promote technological upgrades, as they are not commercially significant, and therefore an obsolescence of these items is also expected. Safe IoT systems should be design-safe and imperfect from the start or provide critical notifications over their lifetime.

Stable IoT applications will provide specific things such as access management, device security, and encryption in effect. Nevertheless, it is far more challenging to do so given the nature of the programs, their lack of processing power, and their preferably small size. The main problem is that there is no consensus at the present time on direct application of IoT protection in the system. Another challenge is standardizing users' hardware and software 3. As the IoT tackles important tasks and sources of data through applications, this is an enticing target for attackers. The consequences of safety problems become more severe, leading to personal injuries, irreversible property damage, and extended downtime [1].

The study of the IoT honeypot high-interaction packet caps also includes the investigation of this mission. The analysis concludes general network threats as well as active attacks on certain network equipment. Finally, there are several attacks to a specific IoT system which serves as a central control unit to identify vulnerabilities and misdefense in general in IoT devices. The aim of this study is to analyze the Internet of Things and recognize its dangers as a system for its challenges. IoT packet capture analysis has provided valuable insights into the present state of IoT and has shown that devices and the basic principles of communication lack protection. Successful attempts on the D-Link Hub have demonstrated that the systems used have a legitimate safety issue.

IoT reflects a fusion of a number of realms which may be viewed as a central concept, a perception of the future of the Web in which sensor, operating which connectivity capable consumers, computing systems and ordinary artifacts collaborate with unmatched ease and financial advantages. While allowing for seamless computer communication for IoT devices, the present IP-based communication network plays a critical function.

16.2 Literature Survey

Most IoT networks consist of a very small number of (unattended) players with poor computational capabilities. In definition, these are both forms of violence, either as a means to get into the program or as a means to misuse the mechanism to target another DDoS. Due to this aspect, the existing problems of securing any information system will be further aggravated. If only one of these actors is improperly secured, the entire system can become vulnerable. There are also drawbacks with regard to the small amount of electricity these machines are permitted to use, possibly because they cannot absorb additional power (passive systems) or because their batteries are intended to function for a very long period. Such limits will not preclude ways of cryptology [6], but approaches will need to be established with these constraints in mind, which could contribute to higher production costs.

A widely used idiom when thinking about protection is that all bets are off because the intruder has physical access to the system, which ensures that it is virtually difficult to provide security assurances for such products. Physical access attackers have many new ways to attack devices, from directly reading memory to some side-channel attacks. A characteristic of IoT systems is that they, or at least some of their elements, will often be installed in publicly accessible places. This adds to the need to protect the entire program, realizing that many of the members might be affected.

16.2.1 IoT Security Issues

The origination of wireless network sensor systems has provided a major advancement of wireless networking technology. The development has introduced the latest version of such wireless sensor network applications that we call the Internet of Thing apps that are the new phase of wireless sensor network applications. Thanks to its ongoing growth, these products are increasing attention from various industrial producers, such as home automation service providers, etc. When people from all over the world are searching for automation in their day-to-day life, such tools are becoming part of many home appliances such as lamps, refrigerator, etc.

Such IoT systems are linked through the Internet and are low energy. Since these machines are constantly linked to each other through the Internet, they are susceptible to any cyberattacks. The design of IoT devices is not veteran compared to other wireless networking devices; a lot of research is still underway to consider the safety of these devices. These devices also share a nearly similar network protocol to the traditional network protocol.

According to a 2014 survey, 39% of people were more concerned about the safety of these apps by implementing IoT technology [1]. Back in January 2014, one of Forbes' writers has claimed that hackers would spy on people at home via such IoT devices as smart home technology is becoming common all over the place [3]. In 2008, hackers demonstrated a remote insulin pump by cracking the biosensor allowed by these IoT devices [7]. Throughout 2016, hackers disabled DNS servers and main websites, and hackers introduced coordinated denial-of-service attacks by IoT computers operating Mirai malware [4]. Mirai is a malware that transforms every computer connecting to the Internet into a slave or a bot, and the infected computers become part of a wide network of botnets. Researchers at the University of Michigan also launched a successful attack on the Samsung SmartThings platform and were able to control the installed home automation system and also used an eavesdropper to accumulate the PIN code used for the same devices [1].

The IoT is currently a hot subject in both academics and industry. A very helpful survey by Atzori et al. is an illustration of the IoT being discussed. Al, [14] which includes some context, IoT concepts, various functional technology, and examples usage cases. In brief, this is a really useful starting point for someone undertaking

IoT-related work without any previous information. A related article, which also focuses on current and potential trends [8], is a strong compliment to the article described above. Applications for the IoT have already gained a great deal of publicity, and a variety of various publications with suggestions have been produced. See, for starters, Smart Meter [1, 8] or Wind Speed Prediction [9]. Smart healthcare is also a commonly heard topic; see, for example, [4] the proposed infrastructure that integrates the IoT. Two especially important articles on IoT implementations are one that provides a Chinese viewpoint on the potential of the IoT and where it can be used [6] and a proposed work on the development of a smart city system by the IoT [5]. The value of these works stems from the fact that they show a variety of possible cases of combined use.

16.2.2 *Threat Types*

16.2.2.1 Denial in Systems

Denial-of-service attacks, summarized as DoS attacks, are popular to be used as a kind of attack which typically overwhelms traffic machines, systems, or networks to overload target resources and to prevent legitimate users from using them. DoS is three separate assaults on the IoT world's edge computing nodes.

The size limitations of IoT devices contribute to low, very tiny batteries. Battery drainage. This has made the drainage of batteries a powerful instrument that leads indirectly to serious consequences such as node failure and operational failure. For example, the whole fire detection system would be compromised if an intruder is able to deplete a smart smoke detector's battery. If a battery cannot be removed or recharged on a computer, the network node may be damaged by an attack. This assault can be executed by sending several random nodes and pressing them to run search mechanisms including packet authentication and checksum.

Sleep Deprivation This is an attack of a particular type which, by targeting the node's battery, is similar to the vaguely drain attack. Batteries with reduced power capacity are the products that are susceptible to this attack. Attackers submit a variety of lawful requests in time to circumvent the node's sleep or work. It is much hard to detect, as packets are real, rather than a battery-driven attack.

Outage Attacks This is the most popular kind of DoS for stopping information on edge nodes and causing an outage. This stops computers from working properly and slows down their operation in certain situations. This may occur as a consequence of an accidental error in the development cycle, the battery attacks described above, unwanted physical access to the computer, or the insertion in malware.

16.2.2.2 Definite Node Modulation

Access to the network physically is a chance for further attack. Based on the form of behavior taken by the assaulting node, direct network manipulation attacks may be classified into two groups. An uptick in such attacks may inflict severe harm and often contribute to a major decrease in network ability.

Server replication is a type of attack by replicating the server identifiers as an attacker adds a new node to a set node array. New node is always violent, which lets the attacker exploit packets arriving at the replica quickly or misdirect them. Such an attack allows the intruder to obtain the necessary permissions, by executing node cancelation protocols or to recover cryptographic shared keys or even to revoke access to valid and authorized knots.

Camouflage attacks are used for the installation of a false edge node or the covering of an approved node at the bottom. This malicious node operates in standard mode for collecting, storing, sending, or redirecting active packets. On the other hand, passive mode is characterized by node function which is designed only for the analysis and collection of traffic data.

16.2.3 Routing Attempts

Routing attacks challenge the routing of packets through a communication layer. An intruder can spoof the packet, redirect it, misdirect it, or drop it from the network completely. The routing information can be simply altered to create routing loops or incorrect error messages. Subsequently, routing attacks could be listed as a few:

The use of a malicious node to draw all traffic on the network segment through advertising with the shortest path to the intended destination describes black hole attacks. Both packets are sent to a vulnerable node as part of this assault so that they can be simply read or destroyed. This attack variation is called the gray hole attack, in which nodes drop some packets selectively.

Wormhole attacks are far worse than black or gray hole attacks, as they can be carried out and in all interactions with integrity and confidentiality. The aim of this assault is to catch packets at one stage in the network and then at another. The essence of the wormhole nodes is that the route in a network is shorter than the original. The tunnel therefore has to be a fast relation, to establish the illusion of a very resemblance between the two wormhole nodes (Fig. 16.3).

16.3 Proposed Method

In order to evaluate the effect of an algorithm regarding the prevention of cyber threats, we need to find a difference in the power usage of IoT or WSN apps. When we interact with the tools that operate on limited resources, we need to recognize the

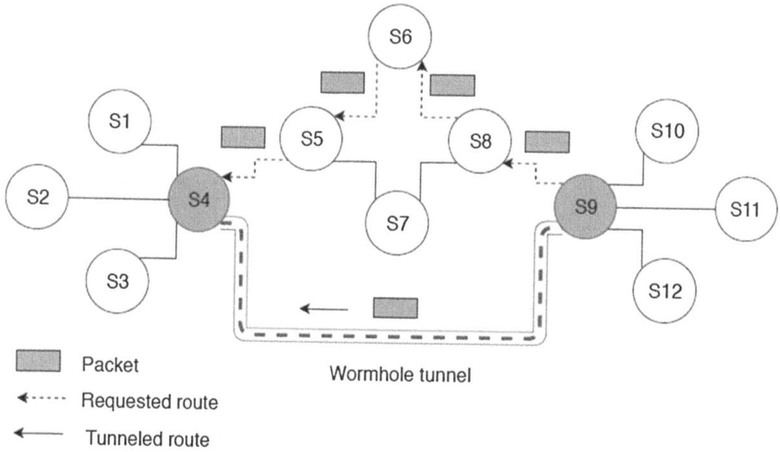

Fig. 16.3 Wormhole attack in the IoT

use of such resources as a significant factor. Blockchain has also created a platform where the principle is focused. You should grasp the smart contract. The smart contract simply applies to code protocols or programs that cause this to happen. Contract to be promptly enforced/coerced, taking into account a range of predefined requirements. For example, smart contracts describe the rationale for the system to be carried out whenever a transaction exists. It is achieved through the trade of cryptocurrencies. In the case of smart contracts, functions and requirements may be specified beyond the scope of business, such as asset validation in a given currency and the number of trades with nonmonetary elements. It is ideal to extend blockchain technology to other areas.

As such instruments have gained prominence among most researchers and manufacturers, they are becoming a vital part of many systems with strong computing capabilities. Higher computing criteria and low resource usage result in decreased power use. Deviation of power usage may be found due to a variety of essential reasons, such as intervention. Intrusion of contact exists due to other causes other than some attacker or cyberattack, such as environmental intrusion, electrical interference or magnetic interference, etc. Such external influences have an effect on the power usage of the system or the whole network, depending on the form of interruption present during ongoing contact. As there is a variation in the power usage of these products, the battery life of these products would also be impacted.

The work discussed in this study was performed step-by-step. Fundamentally, we performed a variety of studies considering the operating environment, the lighting conditions, and the topology of the Zolertia Z1 mote network. Such systems share common design with younger generation devices called Internet of Things (IoT) devices. While smart home automation systems were originally designed to improve energy efficiency, their breadth of effect increased rapidly. Figure 16.4 represents the use of blockchain in the IoT for enhancing security.

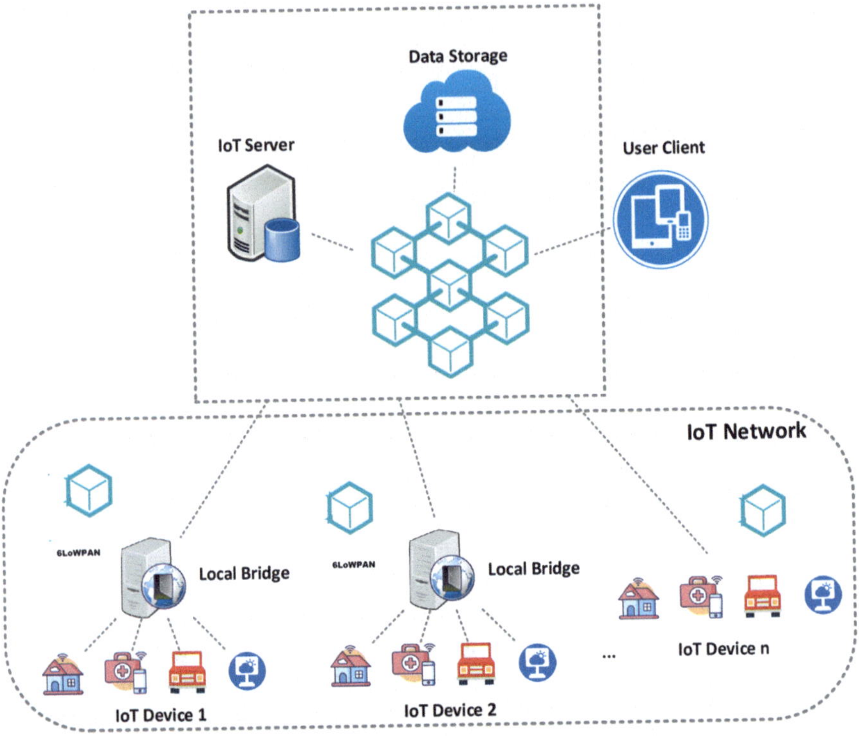

Fig. 16.4 Blockchain in the IoT

The IoT is human operation automation and removal. In order for them to be aware of the digital world, it provides access to data quantities. The concept promotes the development of insightful technologies such as improving community management and quality of life by digitalization of public services. But, in recent years, that has been introduced by cloud storage technology. Provide information and turn it into real-time actions and knowledge [8] to the IoT with the requisite device and measuring capabilities. It has given up exponential development in the IoT. There are different networking tools, such as communication channels and knowledge exchange. The concept of transparent data is a central aspect in these policies. But one of the most critical faults, as has occurred in many instances, Hon. The lack of such initiatives. Confidence. Center architectures such as the one used in cloud computing have played a major role in the IoT growth. Nevertheless, they are serving as black boxes in terms of data transparency so that network members cannot clearly see when and if they should use the details.

It has proved indispensable to integrate emerging technologies such as the IoT and the cloud computer. We do recognize blockchain's enormous ability to revolutionize the IoT. Through providing a secure networking system where information is reliable and traceable, blockchain can boost the IoT.

Data references may be defined at any point, and data stays unchanged, increasing the stability over time. In situations where knowledge on the IoT should be exchanged safely between many participants, such incorporation will reflect a crucial transition. For starters, comprehensive traceability of various food items is a crucial factor in ensuring the health of food. Traceability of food could require the involvement of all of the participants: processing, cooking, care, delivery, and so on.

Smart home automation systems are usually equipped with a large number of surveillance cameras that track the whole internal and external environment or, more often, fully on the external environment and portions of the internal environment. Typically such sensors, which are to be monitored and noticed if there is a possible fire hazard, are placed in critical areas by the doorways or in bedrooms or kitchens.

Many applications, aside from fire protection, may also be rendered in the area of childcare with parents who are, for example, on a lunch break, to be able to track the house to see what their children are doing and whether they are vulnerable to any sort of risk in case that they are home alone. There is still a large usage of water leakage that can be detected on time or verified whether there were any water leakage during family holidays, for example, when there were no people at home. It is also possible to use surveillance cameras along with certain sensors or other tools to build a full image of the danger.

When it comes to smart homes, it is very important to handle the threats associated with them very cautiously and with complete focus. It is attributed to the reality that SH produces a vast volume of highly confidential knowledge regarding the home inhabitants and their activities that may be misused if they are not used for the reason for which they have been gathered. After the inception of IoT apps, there have been several studies that rationalize the variance in power usage, but most of these studies are not unreservedly accurate, because most of these studies are simulation-based. As IoT devices have limited resources, our goal is to enforce the power usage of every reliability. Implementation alone would raise the power consumption, and, in turn, there will be disturbance in the environment which will raise the variance of the power consumption. Most of the previous studies presented are simulated, and very few are real-world analyses. The purpose of this analysis is to predict the battery life of the device prior to the implementation of the proposed algorithm and to compare the results after implementation in order to validate the proposed security implementation.

For research, we found diverse climates and lighting conditions and performed tests in the indoor lab, auditorium, and basketball arena under various lighting conditions. This power trace example provides detailed results for various modes such as Tx, Rx, idle mode, and active mode. With power trace, we often upload multiple scenarios, such as broadcast and unicast or one-to-one correspondence, to verify the effects of the increase in power usage. The precision of the power trace was experimentally confirmed with an accuracy level of 94 percent. This module calculates the power usage at the node level.

After gathering data from the different tests, we measured the IoT system power. The battery life calculation should help to validate the effect of the intrusion on the battery life of the system. We also carried out many studies without any protection application in various settings and lighting situations in the real world using the same topology for all studies. With the same topology, we run simulations with a power trace case, broadcast, and unicast or one-to-one contact. Yet according to our study, we found that the simulation findings and the outcomes of real-world studies were in comparison to each other. In order to verify the efficiency of the implementation of the proposed algorithm in terms of resource usage that we cannot depend on simulation data, a real-world mote study is needed to analyze the behavior of the devices under different conditions.

Within a wormhole intruder, a fake route is generated for a node communication reason. This route or path may view itself as a shorter route or path relative to other routes throughout the network. This causes uncertainty within the node routing system and requires the other nodes to pursue the corrupted path. An attacker may add one or more malicious nodes, and this assault often provides an attacker with the aid of malicious nodes to create a tunnel. During a wormhole attack, the node will catch the transmitted packet and forward it to a malicious node that locates the remote node and transmits all the collected packets to a compromised position or node. It is possible to launch an attack for an attacker on a network that compromises the entire network or any legit node, or it can even help the attacker to break through the cryptographic implementation of the network. Figure 16.5 illustrates the wormhole attack model.

This attack is primarily aimed at dropping packets that were to be forwarded to the node. This assault was known as a denial-of-service operation. If the node of the attacker is located specifically inside the network, it can allow the attacker to separate the goal node from the network. For such assaults, it is possible to manipulate any of the protocols used and maximize the power demand of the whole network. It is also necessary to initiate this assault with a range of denial-of-service assaults. Mitigation: There are many ways to counteract this threat, but the simplest of them is to luxuriate the routes between the source and the destination of the nodes.

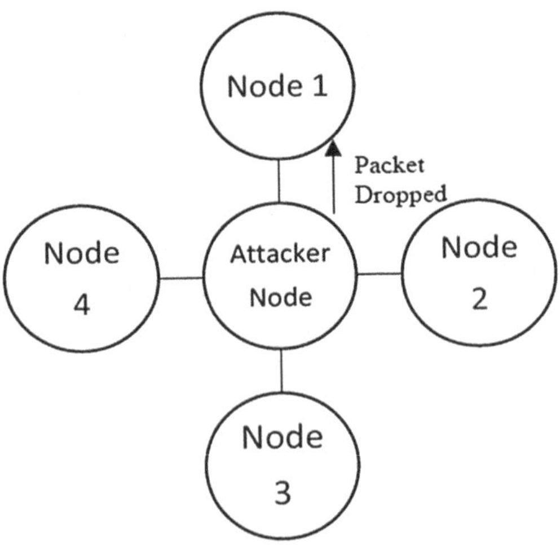

Fig. 16.5 Wormhole attack topology

But it's going to be complicated if we work with a huge number of nodes. Analyzing the traffic flow of the application level between the nodes and implementing security or cryptographic implementations are possible for a small network as well as for a large network consisting of a large number of no devices.

Algorithm Blockchain generation for IoT Devices
Input: Nodes, range, gadgets count
Output: energy balancing with trusted nodes.
for $i = 1$ to n **do**
for $j = 1$ to i
Perform routing Ri= List(Nodes)
Distance dij = distance (i, j)< ϕ
if $D > \phi$ then
N(i) $\leftarrow$ Trust(Node(j))
Attacker Flag=1
N(i) $\leftarrow$ N(j) + N(j+i)
If (Trust(N(i)> λ)
Initiate Communication
Else
Mark as malicious
end for
end for

The suggested algorithm is structured to take into account attacks by RPL. As we are grappling with attacks by RPL that do not have a particular framework to defend it, we've introduced a packet relay wormhole assault on the network, so it's possible to initiate an RPL assault. Wormhole detection mechanism may categorize hardware dependent, RTT based and statistical analysis, clock driven, etc. Since we are working with low-power hardware devices, the hardware design will not be suitable for implementation. The proposed method is focused on software because this methodology offers consistency when optimizing it. We've introduced a wormhole packet relay attack.

16.4 Results

The WSNs that we deployed to conduct experiments were fitted with transmit, unicast, and power trace instances. In the first place, we selected broadcast as an illustration to analyze the power consumption without any threat. The same procedure was followed for a unique example. The testbed that we generated to conduct the experiment consists of eight motes in the grid that we mentioned earlier, and the ninth mote will track the ongoing power consumption through the power trace example. After the attacks were carried out, the same technique was replicated, and the same procedure was repeated for the security implementation of these tools. The findings we have obtained from these many studies have shown a decrease in power consumption. As the increase in power consumption is inversely proportional to the battery life of these devices, calculating the battery life of these devices will be useful for analysis. Contiki OS has a built-in simulator called COOJA simulator for wireless network sensor devices. It is difficult to evaluate the behavior of these instruments in an optimal setting by means of this simulator, because this simulator does not have any facility for the option of operating atmosphere and lighting conditions. We simulated the effects with and without an attack first and then with the IDS implemented or the planned implementation (Fig. 16.6).

The primary aim of this work is to improve the protection of IoT devices, but these devices are low powered and have minimal resources. This is therefore important to examine the effect of intrusion in the environment, every cyberattack, and defense implementation. In order to analyze the impact of these applications, we are performing multiple tests and presenting data or findings on the energy consumption performance and the battery life calculation of these tools. As for the implementation of the cyberattack prevention method, an analysis of the resource use is important because this application itself would be dangerous without any analysis. As described in previous pages, the power consumption is directly proportional to the life of the battery. The higher the power consumption, the longer the battery life, and this will be harmful to the entire network.

The calculation of the energy consumption can be made using the standard equation, and the parameters shown in the figure are taken into account for the accurate calculation (Figs. 16.7, 16.8 and 16.9).

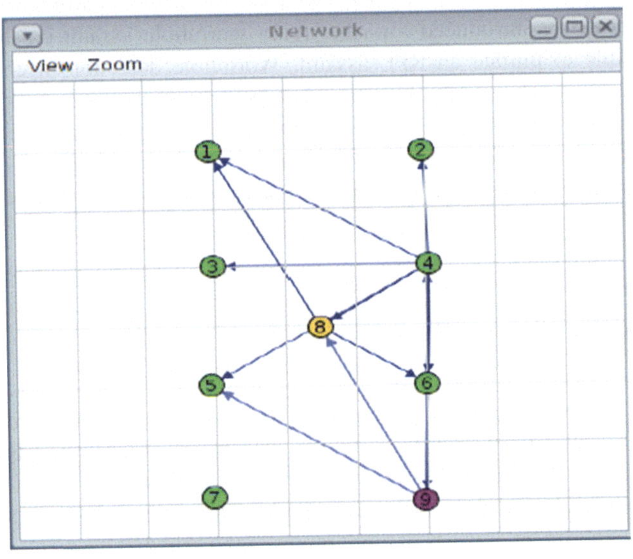

Fig. 16.6 Simulation of IoT gadgets

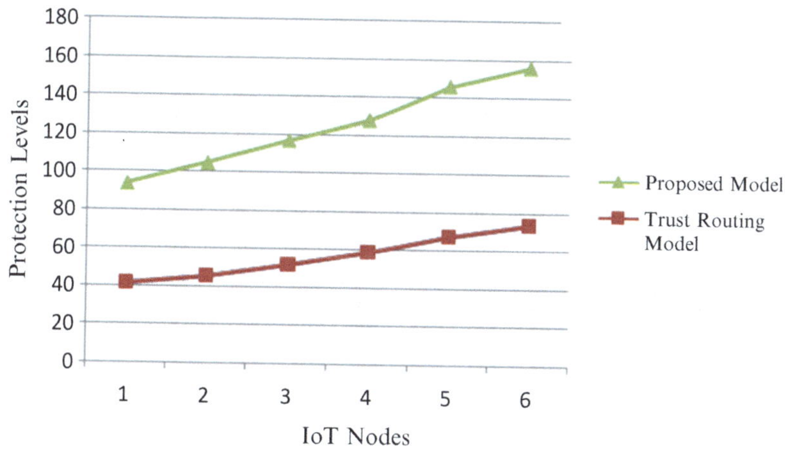

Fig. 16.7 The degree of assault protection

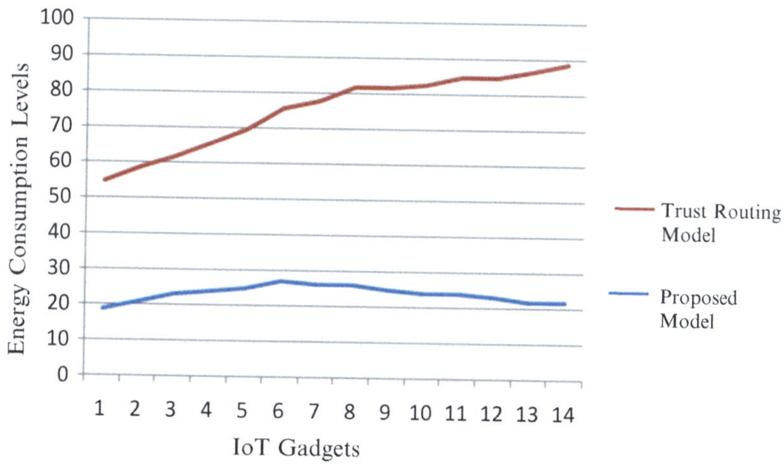

Fig. 16.8 Energy consumption during data communications

$$\text{Energy (mJ)} = \frac{\text{CPU} * 0.5 + \text{LPM} * 0.0005 + \text{Tx} * 17.4 + \text{Rx} * 18.8) * 3}{32768}$$

where

CPU = period when the mote was running
LPM = total time for the low-power mode
Tx = cumulative time of transmission
Rx = full time to respond
Fig: rate of electricity usage

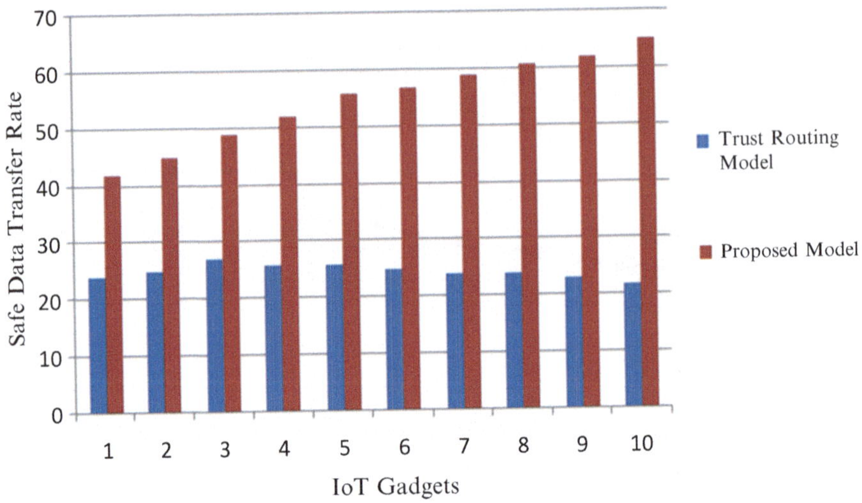

Fig. 16.9 Safe data transfer rate

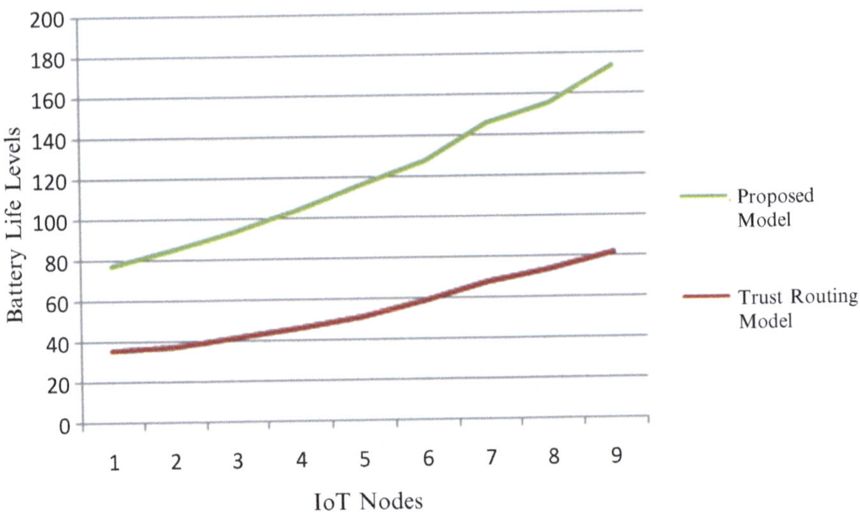

Fig. 16.10 Battery life levels

As far as our research is concerned, we can conclude that interference has an adverse effect on the communication of these low-power wireless network sensors (Fig. 16.10).

We calculate the battery life in hours using the following equation.

$$E = \sum_{i=0}^{n} Vi * Ii * Ti$$

where

E = energy in joules; I = current drawn; T = time; and V = voltage required.

16.5 Conclusion

The Internet of Things (IoT) is an increasingly rising area of technological, social, and economic significance. Objects are being paired with Internet access and strong data processing tools that aim to change the way we function and live. The IoT is a dynamic system of equipment, sensors, software, and apps that need to be able to interact with various geographic locations. Second, data control is and would undoubtedly be a challenging problem for years to come, although it is expected to change to getting exposure to and being willing to use data for research. In fact, the combination between the digital and physical environment would need strong safety levels to deter risks. The Internet of Things applications or IoT apps are being rapidly widespread in several fields. Before using such tools for a variety of specific and critical purposes, such as health tracking, military operation, region surveillance, etc., an effective protection implementation is required. Such applications would be effective in terms of resource use and improve network protection. The proposed model focuses primarily on the study of current risk assessment approaches that can be used for smart home risk assessment relevant to cyber threats from three separate perspectives: person, community, and government. The implications of the measurements to be measured are linked to monetary loss, computer loss, and abuse of computer.

References

1. B. Xiong, K. Yang, J. Zhao, K. Li, Robust dynamic network traffic partitioning against malicious attacks. J. Netw. Comput. Appl. **87**, 20–31 (2017)
2. P.K. Sharma, S.Y. Moon, J.H. Park, Block-VN: A distributed blockchain based vehicular network architecture in smart city. JIPS **13**, 184–195 (2017)
3. T. Ahram, A. Sargolzaei, S. Sargolzaei, J. Daniels, B. Amaba, Blockchain technology innovations, in *2017 IEEE Technology & Engineering Management Conference (TEMSCON)*, (2017)
4. K. Gu, L. Yang, B. Yin, Location data record privacy protection based on differential privacy mechanism. Inf. Technol. Control. **47**(4), 639–654 (2018)
5. P.K. Sharma, S. Rathore, J.H. Park, DistArch-SCNet: Blockchain-based distributed architecture with li-fi communication for a scalable smart city network. IEEE. Consum. Electr. Mag. **7**(4), 55–64 (2018)
6. J.H. Park, M.M. Salim, J.H. Jo, J.C.S. Sicato, S. Rathore, J.H. Park, CIoT-Net: A scalable cognitive IoT based smart city network architecture. Human Comput. Inf. Sci. **9**(1), 1–29 (2019)

7. J. Chandramohan, R. Nagarajan, K. Satheeshkumar, N. Ajithkumar, P.A. Gopinath, S. Ranjithkumar, Smart home automation and security system using arduino and Wi-fi. Int. J. Eng. Comput. Sci. **6**, 20694–20698 (2017)
8. C. Yin, B. Zhou, Z. Yin, J. Wang, Local privacy protection classification based on human-centric computing. Human Comput. Inf. Sci. **9**(1), 33 (2019)
9. J. Wang, Y. Gao, W. Liu, A.K. Sangaiah, H.J. Kim, Energy efficient routing algorithm with mobile sink support for wireless sensor networks. Sensors **19**(7), 1468–1494 (2019)

Chapter 17
Integrating Blockchain with Edge Computing for a Secure and Reliable Data Flow

Aditi Kaushik

17.1 Introduction

Many technologies have emerged in this century. These technologies have made possible many impossible tasks. Blockchain being the latest incredible technology has found its use in different areas like transaction management, cash management, etc. and for managing the diverse data that different devices generate.

"A blockchain is a budding list of blocks, that are chained using cryptography" [3]. Blockchain is said to be resistant to modification of data as each block in a blockchain is divided into parts where transaction data, timestamp, and a cryptographic hash of the previous block are stored.

Edge computing works as a dispersed computation model that improves latency and saves the transmission bandwidth by bringing processing and storing of data near to the location where it is required.

Since, with the increased diversity of the devices, spatially distributed sensors are connected to the Internet, the data size is increasing drastically. Therefore, there is a need to secure this increasing data as well as to perform fast computation so that reliable results can be generated.

For this, the integration of blockchain and edge computing can be applied.

So, what are blockchain and edge computing?

Let us discuss them one by one.

A. Kaushik (✉)
School of Computer Sciences, Starex University, Gurugram, India

T. Choudhury et al. (eds.), *Blockchain Applications in IoT Ecosystem*,
EAI/Springer Innovations in Communication and Computing,
https://doi.org/10.1007/978-3-030-65691-1_17

17.1.1 What Is Blockchain?

Blockchain or we can say "chain of blocks" was discovered by Satoshi Nakamoto. According to many peoples, blockchain and Bitcoin are the same, but this is not the case. Both **blockchain and Bitcoin are different** (Fig. 17.1).

Bitcoin (cryptocurrency) acts as a medium to exchange, create, and store electronically using encryption techniques so that the monetary units and their transfer can be verified and controlled efficiently. It is not redeemable to another commodity and has no physical form [4]. Blockchain works as underlying technology of Bitcoin.

In simple terms, blockchain is a timestamped series of immutable records (blocks), i.e., the records of data that cannot be tampered or altered. A cluster of computers managed that series. Before proceeding further, we should know what a **block** is in this blockchain (Fig. 17.2).

Block
When a new data is stored in a block, that block is inserted to the blockchain. After that, the following things must occur:

(a) A transaction must occur.
(b) Verification of transactions.
(c) Transaction amount and the digital signature of seller and purchaser are stored in the block.
(d) A unique code called "hash" is given to each block. After that, it is added to the blockchain [5].

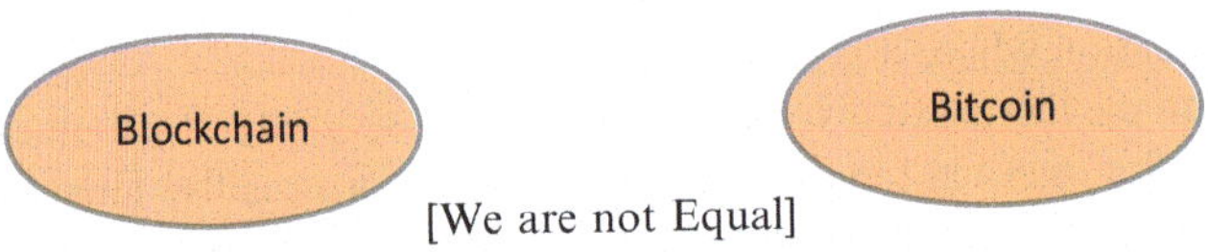

Fig. 17.1 Blockchain and bitcoin

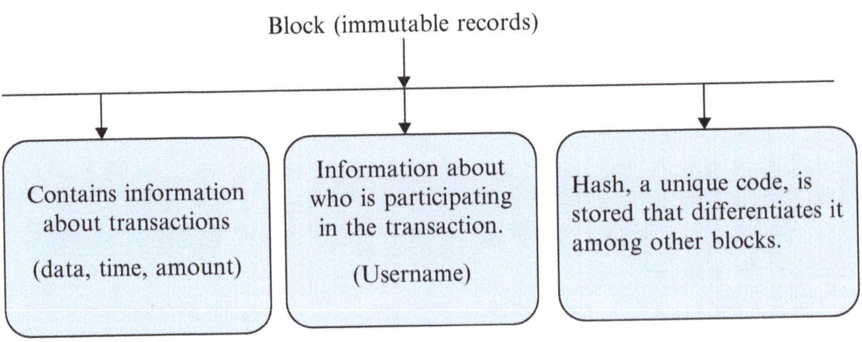

Fig. 17.2 Information in a block

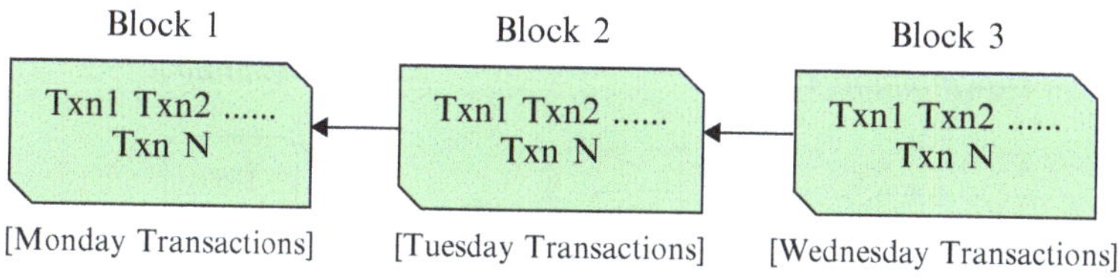

Fig. 17.3 Block example

Each block in a blockchain deals with the valid ledger. A valid ledger is a record sheet that remembers the valid transactions, and valid transactions are a very important part of creating the correct block.

For example, a transaction is valid if *the sender's balance* > =*amount* sender wants to send to the receiver.

Each block in a blockchain has transactions from a specific period. The period could be daily/hourly (Fig. 17.3).

A block contains zero or more transactions with some additional metadata [6].

C++ Code to Demonstrate the Creation of a Block

```
Class transact
{
    acct_t sender;
    acct_t receiver;
    uint64_t amt;
};
Class t_block
{
  std::vector<transact*> txns; // list of pointers to transactions
  t_block* prev_b;
};
std:: list<t_block*> blockchain;
```

Each block is assigned with a valid hash so that it can become part of the blockchain. To identify a block, a cryptographic hash that works like a digital signature is used.

Structure of a Block

The three main components of a block are:

1. Header: It contains additional information about the block. Additional information includes [7]:
   (i) Hash of the previous block as it is used to create the hash of a new block.
   (ii) Mining competition to gain a valid hash.
   (iii) Merkle trees: This is also called a binary hash tree. This tree is a structure that contains all the transactions of a block [8].

So, after briefing "block," we can say that a blockchain is nothing but the linked list of blocks where each block is linked to its previous block as well as to the next block.

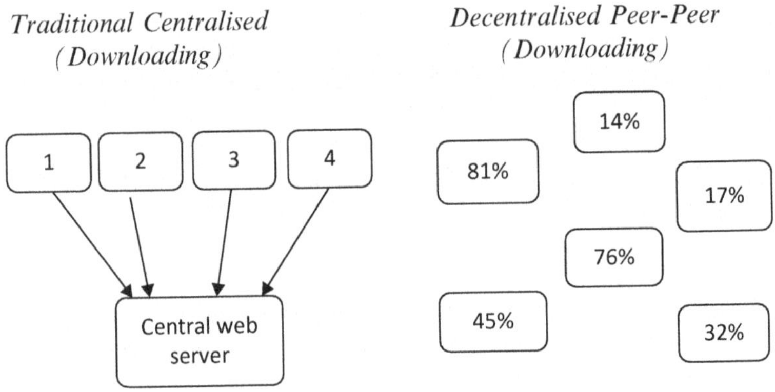

Fig. 17.4 Demonstration of a centralized and decentralized network

Blockchain

Blockchain works on the concept of *distributed database technology*. Each block in the blockchain is immune to any unauthorized changes, and the blocks are bind together through strong hashing.

There is no central authority in the blockchain. The information stored in this chain of a block is open to everyone. So, it can be said that a blockchain is transparent and everyone involved is accountable for their actions.

Maintaining the Blockchain

The peer-to-peer network is used to maintain the blockchain. This network has multiple nodes (device) connected with each other. The benefit this peer-to-peer network gives is torrenting. But in the case of client-server model, the simple task of file downloading becomes slow due to the request-response nature of the model. The client-server model is also not robust in nature as a failure in server can cause the failure in the whole network.

Let us understand it with the help of diagrams (Fig. 17.4).

Consensus Algorithm

A consensus algorithm is a method that establishes reliability and trust between peers in a network by making the peers achieve a common agreement about the present state of the distributed ledger [9]. The consensus algorithm plays a very important role in blockchain as in a decentralized network with no central authority; it handles the work of validation and authentication of transactions before recording them in a block.

There are various consensus algorithms in the blockchain. Let us look at them:

(i) *Proof-of-Work (PoW):* "This algorithm is used by Bitcoin. The main idea of this algorithm is to select a miner for the next block generation" [10].
(ii) *Practical Byzantine Fault Tolerance (PBFT):* Sometimes, some of the nodes are not able to give a response with the correct information or fail to give a

response. This situation is handled by a feature of a distributed network to reach to same value agreement. This feature is Byzantine fault tolerance.

(iii) *Proof-of-Stake (PoS):* In this algorithm, validators validate the blocks by placing the bet on them. The validators get a reward depending on the blocks that are added in the blockchain [11]. The reward is proportionate to the validator's bets. So, this algorithm can be used as an alternative to PoW as it motivates the validators to validate block more accurately and in less time.

(iv) *Proof-of-Burn (PoB):* The principle is to allow miners to "burn" virtual currency tokens. In response to the proportion of coins burnt, a right is granted to them to write blocks.

Why Blockchain Has Become so Popular?

1. **Reduced Time:** The lengthy time to verify and to perform clearance is reduced due to blockchain resulting in faster settlements of trades [12].
2. **Immutable Transactions:** The transactions are stored in such a way in the blockchain that they cannot be altered or changed.
3. **Transparency:** It hides the person's identity through cryptography, and other users can communicate with each other through their public address only.
4. **Decentralized:** Since no central authority is supervising anything, the blockchain is decentralized in nature. It follows standard rules that depict how a node exchanges the blockchain information.

Till now we have overviewed the basics of blockchain with its advantages. Let us proceed to the other popular technology which is somehow related to cloud computing but works faster than that. We are proceeding our discussion to "edge computing."

17.1.2 What Is Edge Computing?

"The cloud computing can be defined as the practice of using a network of remote servers hosted on the Internet to store, manage, and process data, rather than a local server or a personal computer" [13].

Cloud computing is used as a centralized server. This server is placed at location far from its clients. The benefit of using this approach is that it provides better communication between different kinds of mobile devices like smartphones, laptops, etc. Cloud computing follows request-response approach, and the clients are at large distances from each other and from the central controller, i.e., server.

Due to this, sometimes the users face the delay in the computation of their data, and with the advent of IoT, blockchain, and many other real-time applications, it becomes the need to introduce the technology that could reduce this delay in the computation of data. To resolve this challenge, edge computing comes into role.

It is a computing technology where the information is stored and processed close to the device, i.e., at the edge of that device.

Table 17.1 Differences between edge computing and cloud computing

Cloud computing	Edge computing
Advantages 1. Scalable 2. Big data processing 3. Large storage capacity	*Advantages* 1. Reaction in real time 2. Latency rate is low 3. Improved data security as it can work without clouds 4. Charges low bandwidth costs and reduces storage requirement as well as network traffic
Disadvantages 1. Reaction time is low 2. Latency rate is high 3. Cloud does not have an offline mode	*Disadvantages* 1. Limited storage capacity 2. Difficult to maintain security 3. IoT devices consume high power

Table 17.2 Challenges in the IoT [15]

Challenges	Observations
Data inefficiency	Insufficient data to generate an accurate result
Authentication	Absence of a model to prevent cyberattacks
Delay in generating the result	Absence of a model that can handle lightweight data analytics
Insecure transmission	The threat of insecurity in the transmission of information between nodes
Centralized network	Failure at the centralized node can cause failure in the whole network

Since the central location or a server can be thousands of miles away from the device, the edge computing at its initial level computes the data and stores the information closer to the same device where the data is generated.

The main motive of doing so is to produce real-time data so that the performance of the application is not affected by unnecessary delays.

The edge computing has changed the scenario of handling, processing, and delivering of data around the world. There is no need to send data to the centralized data processing warehouse. Edge computing is not just a branch of small clouds. Before proceeding further about edge computing, let us first understand the basic differences between cloud computing and edge computing Table 17.1 (Fig. 17.5).

After defining edge computing and examining the differences between edge and cloud computing, let us see some use cases of edge computing (Fig. 17.6).

Benefits of Edge Computing

1. ***Ultralow Latency:*** The typical latency is in milliseconds, significantly lower compared to centralized data computation.
2. ***High Network Throughput:*** Edge computing provides high network throughput as the content is either generated locally or cached locally.
3. ***Context awareness:*** The edge has access to the radio network. Therefore, the information provided by the radio access network can be used by other applications.

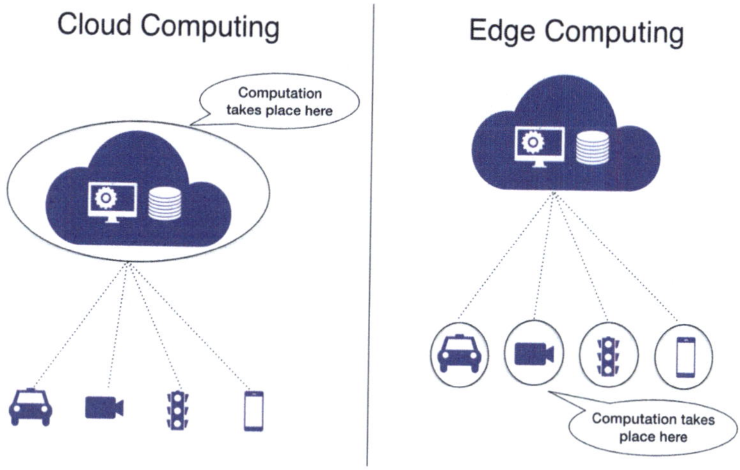

Fig. 17.5 Cloud vs edge. (Image source: towardsdatascience.com)

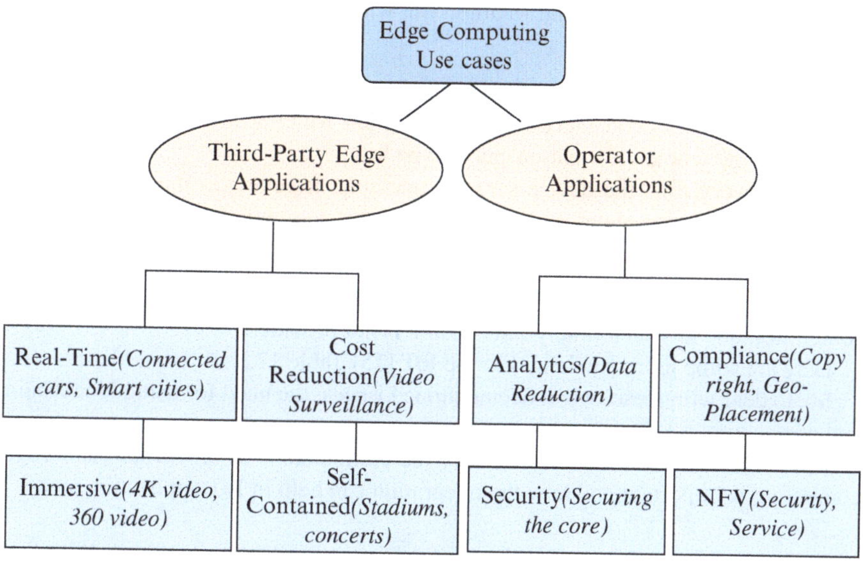

Fig. 17.6 Edge computing use cases

4. ***Data filtering and reduction:*** Data analytics applications at the edge can substantially reduce the amount of data sent upstream.
5. ***Compliance:*** The compliance feature helps the edge in dealing with privacy laws and data location laws [14].
6. ***Security:*** This feature protects users against data theft or unauthorized access to data [14].

Now, let us brief about another emerging technology, the "Internet of Things," i.e., IoT.

17.1.3 IoT

The IoT stands for the "Internet of Things." The technology in which the devices are interrelated through the Internet and each device has its unique identifier. The physical devices can be sensors, vehicles, actuators, etc. that communicate to a distant data center. The data generated in these devices are sent to a centralized data center located in the cloud for the computation purpose.

The computed result is then sent back to the devices. Based on these results, the device generates the signals. The main characteristics of these layer devices are that they require less computational resources, are less powered, and lack sufficient memory for storage.

No doubt, the IoT has emerged as a technology boost to create smart cities, smart healthcare, smart education, and whatnot. But with this achievement, there are some flaws also. The two main flaws are:

(i) Data is sent to distant centralized cloud storage, due to which there are high chances of attacks on data during transfer. Plus, with the down of the centralized server, the whole information can also be lost.
(ii) Since every time the device senses something, it generates the data, and that data is sent to the cloud for the heavy and light computations.

After the computations are performed, the result is sent back to the devices to give a response to the sensed thing. This whole process becomes time-consuming sometimes, due to which timely information is not provided.

Here are some more challenges in the IoT [15] Table 17.2:

So, to deal with these two abovementioned issues, the need for using blockchain and edge computing arose.

Let us discuss further points on why the combination of blockchain and edge computing is important and how the integration can help in IoT application areas.

17.2 Literature Review

The Internet of Things has become an eminent technology in today's world. Smart cities, smart transportation, smart parking, etc. are the products of this technology. As the IoT devices are easy to hack, the need is to pose some security procedures. For this, Khan, M.A. and Salah, K. [16] have analyzed the security issues for the IoT. Furthermore, blockchain is discussed to find a way to solve these issues. To gain security and reliability, it needs the integration of blockchain and edge computing in its framework. A comprehensive study of blockchain by Zibin Zheng,

Shaoan Xie, Hongning Dai, Xiangping Chen, and Huaimin Wang [17] has presented the characteristics and architecture of blockchain along with the discussions of consensus algorithm. Furthermore, the challenges or issues with blockchain are also listed.

To address the needs of IoT framework, Nyamtiga BM, Sicato JCS, Rathore S, Sung Y, and Park JH [18] proposed an architecture demonstrating the idea of integrating blockchain and edge computing in IoT framework for secure storage of IoT data and transactions.

Blockchain provides security, but with edge computing we can get reliability. A design presented by Ines Sitton Candanedo and Juan Manuel Corchado [19] demonstrated the way edge computing can enhance the functionality of the IoT. Furthermore, the features were also discussed by them. To give a better insight on the IoT with blockchain, Upul Jayasinghe, Gyu Myoung Lee, Áine Mac Dermott, and Woo Seop Rhee [20] proposed a network with the name TrustChain. The design has enhanced the privacy as well as the efficiency of the services specifically in accordance with the distributed environment, i.e., the IoT. Yu W., Liang F., He X., Hatcher W.G., Lu C., Lin J., and Yang X [21] conducted the survey and provided some work to demonstrate the integration of these three technologies.

17.3 The Need of Integrating Blockchain with Edge Computing

Undoubtedly, blockchain provides many countermeasures to protect user data by using hashing and by verifying each block before adding it to the chain. But, to do such a tedious task, a large amount of energy and time is wasted since all the computations are to perform at the cloud side. So, edge computing comes into role as it performs all the computations at the edge of the devices and thus the processing becomes efficient.

No doubt, edge computing has come up as a decentralized alternative to work with the Internet of Things. It works collaboratively. But the organizations still have doubts about the security, reliability, and data protection powers of edge computing. So, to deal with these flaws of the edge computing, the use of blockchain works perfectly.

The main thing with edge computing is that after placement it will allow an infinite number of devices to connect and communicate, thus resulting in the formation of a massive system. The only method at least for now to control that massive system is blockchain.

So, both edge computing and blockchain go hand in hand. The integration of both technologies solves the issues and challenges of each other.

Blockchain solves the issue of data security in edge computing by providing hashing schemes.

Now let us see what issues of blockchain does edge computing solves and how..

17.3.1 Issues in Blockchain

(i) *Integrity*: By the construction, blockchain resists the data modification and thus ensures data integrity. But, the issues with integrity arise when the compromise is being done with the correctness and validation of the transactions. If the miner is dishonest, he would pose an attack on the integrity that would result in the downgrading functionality of blockchain.
(ii) *Anonymity*: Blockchain does not provide full anonymity. Consider the case, when a user has transacted a Bitcoin, the wallet address and the transaction details are recorded in the blockchain. The transaction will stay anonymous until there is no link established between the wallet address and the user's identity. As when as the link is created between the wallet and the user's identity, the anonymity of the user will be blown. So blockchain is pseudonymous.
(iii) *Adaptability*: The architecture and operation system of blockchain demands that every participant must check, validate, and permanently store each new block added to the chain. As the number of transactions grows, the chain will become more complex. This results in the requirement of large storage, high power to perform computation, and large bandwidth, which leads to obstruction of blockchain's scalability.

With edge computing, these abovementioned issues can be handled, and thus a smooth integration of both the technologies can be used for making efficient IoT-based applications.

Let us discuss the requirements in the design of integration.

17.3.2 Requirements in the Design of Integration (Fig. 17.7)

The successful integration of edge computing with blockchain ensures fulfillment of the following requirements:

1. *Anonymity*: The integration ensures the anonymity of the user in a way that only transaction details be used for the authentication purpose instead of the identity of the users.
2. *Adaptability*: Since edge computing computes at the edge of the devices, this can efficiently handle the increasing bandwidth requirement and growing transactions.
3. *Data integrity:* To ensure that there are no faults during the modification of data, the actions of consumers and data owners are verified.
4. *Low power*: With the performing computations at the edge, verifying transactions, and hashing of the blocks at the edge by each user, the requirement of high power and energy to do this tedious task automatically lowers down.
5. *Offloaded computation*: To overcome the hardware limitations of a device, such as limited computational power, storage, and energy, offloading computation

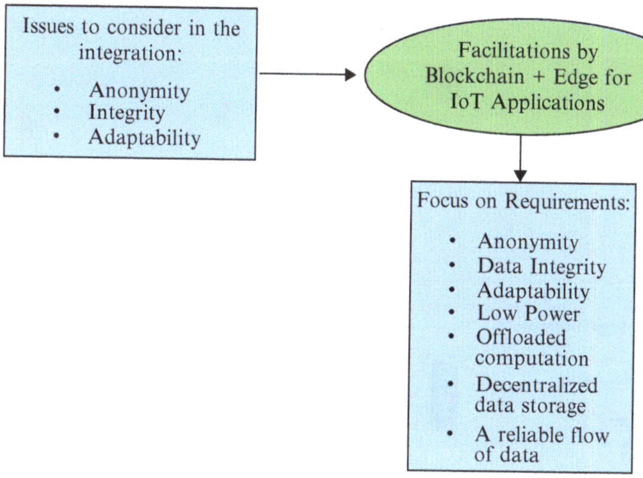

Fig. 17.7 Block diagram to focus on major requirements

plays a very important role [8]. It transfers the transactions to the edge servers for verification and computation purpose.

6. *Decentralized data storage*: To ensure increased storing capacities of IoT devices, the amalgamation of blockchain with edge computing should complement each other by making the participating entities to combine their storage capacities on a P2P basis so that the transaction can be stored and shared. The benefit it will give is the fast computation at the edge of the devices.
7. *A reliable flow of data:* The entities and organizations involved in the IoT naturally lack trusted relationships, which poses significant challenges to the reliable transmission of data. There is a lack of trust between the participating entities and organizations. This lack of trust, somehow buds' challenges to achieve reliable data flow. The securing nature of blockchain removes this lack of trust.

17.4 Integrated Framework of Blockchain and Edge Computing for the IoT (Fig. 17.8)

17.4.1 Terminologies in the Integrated Framework

1. *P2P*: The P2P stands for peer-to-peer. This network is distributed in nature. All the computers and devices that are part of the network are called peers. They all are equal and they share and exchange their workloads.
2. *Private Blockchain*: It is also termed as "permissioned blockchain." Their work is based on access controls that restrict who can take part in the network. Only

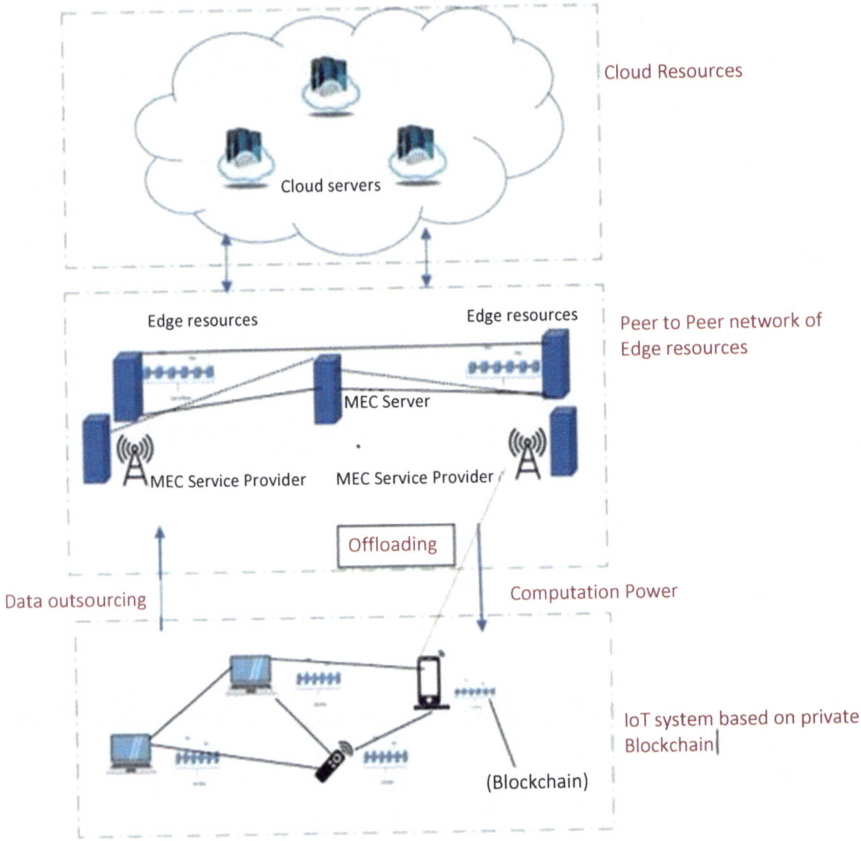

Fig. 17.8 Integrated IoT framework with Blockchain and edge computing

the authorized participants will know the transactions, and other nonparticipating entities will not be able to access it.

3. *Data Outsourcing*: This is the new paradigm of data management. In this, the owner of data is no longer responsible for the management of the data. A portion of data is sent to external providers who offer data function and management functionalities. For example, functionalities provided by the cloud server of Google.
4. *Computation Power*: In simpler terms, the computation power of a device relates the devices' power with the amount of time it takes to solve the instruction. The more the speed, the higher is the computation power of the device.
5. *MEC Servers*: Mobile edge servers or edge servers collect the data from the IoT devices and perform the lightweight analytics on that to give result in less time.
6. *Distributed Cloud Resources*: The **distributed cloud** is the application of cloud computing technologies to connect data and functions which are in different physical locations.

17.4.2 Working of the Integrated Framework

To improve the performance of the IoT network, the IoT framework is divided into three layers and each layer is integrated with blockchain.

1. *Private Blockchain-Based P2P IoT System*: This layer contains peer-to-peer connected IoT devices that are secured by blockchain technology that works privately so that no information or confidential data can be accessed by unauthorized party/person.

When the devices connected through the Internet senses something, the IoT users at this layer will collect the data generated by the devices and offload them to the edge server located at the layer above them.

It is already defined earlier that offloading computation means to transfer data to the edge server for computations and verification to overcome the hardware limitations of the device like high energy consumption, low storage, etc.

2. *P2P Network of Edge Resources:* "Edge computing comes into the concept to bring services closer to the end devices to reduce delays in processing and faster computation" [18]. Furthermore, the edge servers in this layer distribute messages with each other to create replicated storage. To achieve this, blockchain technology is deployed at each edge server so that transmission of data and information can be performed securely between the edge servers. A pool containing mobile resources is created by this layer to achieve faster communication, processing, and analytics. It also provides storage capacity for small amount of time

The P2P network layer of Edge resources utilizes a consensus algorithm availed by blockchain to validate and devise requirements for computation and service claims. In this way, the consensus algorithm establishes a safe communication and trust among unknown peers in the distributed computing environment.

Most of the time, the edge servers are capable to handle and perform computations and lightweight analytics of the data to generate a fast result for the devices. But sometimes, the computations are so heavy that the edge servers are not able to handle them. In that case, edge servers offload their workload to the cloud servers located at the layer above them and request for the cloud services.

3. *Cloud Layer:* Cloud layer in any framework handles all the intensive and heavy computations as well as stores a large amount of data. But in this amalgamation of blockchain and edge computing, the cloud layer functions a little more.

The cloud layer also uses the consensus algorithm of blockchain to maintain the complete replicated records shared between the cloud layer nodes. To achieve safe, low-cost, and real-time access, the blockchain mechanism is used at the cloud layer.

17.4.3 Advantages from the Integration

With integrating blockchain and edge computing for IoT applications, we mainly get the following advantages:

1. *Data security*: The smaller number of information is sent to the cloud, the more secure the data is. Also, we are applying blockchain at each layer to facilitate the secure transmission of data between peer nodes.
2. *Better app performance*: As the computations are being done closer to the devices that generated the data, the result generation and responsiveness of the device have become fast.
3. *Reduced operational cost*: By making the data to store and process closer to its source, the lag time can be reduced and thus will improve the overall app performance. As a result, we can analyze the data in real time, without delays.
4. *Improved reliability:* Since the blockchain consensus algorithm is maintaining the trust between the peers in the network to achieve common agreement about the addition of a block, the reliability and authenticity of the process increase.

With the above advantages, it can be said that combined features of blockchain and edge computing in an IoT framework provide the following:

Secure Data Flow The consensus algorithm applies a procedure through which a common agreement is maintained between the peers. This agreement ensures that all the peers are aware of the present state of the distributed ledger. This results in safe communication among unknown peers in the distributed environment.

Reliable Flow of Data "Reliability is the degree to which the result of a measurement, calculation, or specification can be depended on to be accurate." The consensus algorithm maintains trust among all the peers by making data stored securely using complex cryptography and resistant to modification and unauthorized access. In this way the integrity of data is maintained which results in accurate calculations.

17.4.4 Case Study

Introduction

Does the integration of blockchain and edge computing will work well for emergency warning system?

Disasters are unpredictable as they can happen at any time. We need to be always alerted to handle these unpredictable disasters. For this, there is a need for a secure emergency alert system that can work in real time. The main purpose of this system is (1) to check whether the information obtained from various sources is valid or not, (2) to achieve reliable information, (3) to maintain authenticity and to protect the identity of users involved, and (4) to generate alerts in real time.

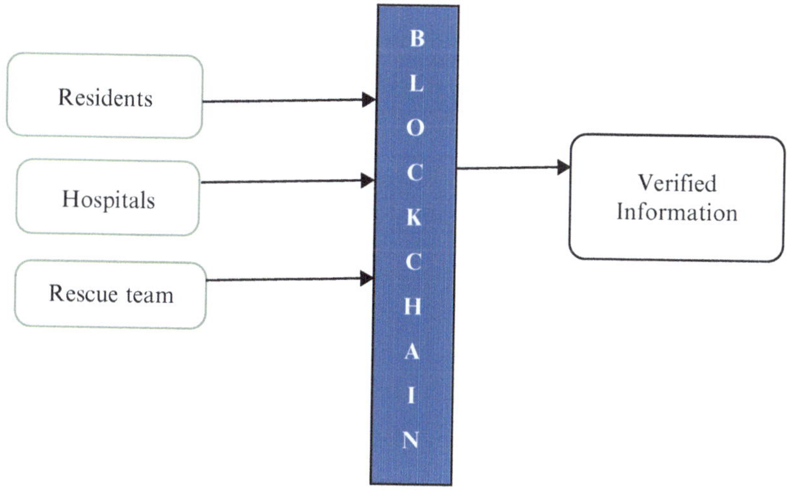

Fig. 17.9 Verification of user's information at IoT layer

In the next section, let us see how the integration can help to build such system.

Background

The study uses blockchain technology with edge computing around the emergency alert system. There are many ways through which the government alerts its citizens about impending disasters. Since so many devices are connected through IoT, there is a need to collaborate with residents, government, and rescue team to one common platform where information is exchanged for timely actions.

The model comprises three layers:

1. IoT layer: 1st layer from the bottom consists of the IoT devices like mobile devices and laptops used by residents, hospitals, and rescue teams. These users must be registered. Each device's information is verified by a consensus algorithm of blockchain so that only authenticated users can do the registration to avoid any kind of misinformation of a disaster. Thus, the integrity and reliability of data are maintained (Fig. 17.9).
2. Edge layer: The next layer consists of P2P edge servers and service providers that store the information generated by the user's devices like location, name, etc. Consensus algorithm is applied here to protect the confidential information of the users from unauthorized access (Fig. 17.10).
3. Cloud layer: The last layer consists of cloud resources that store a large amount of data to perform heavy computations. To achieve secure, inexpensive, and real-time access to best and effective computing services, the blockchain mechanism is used at the cloud layer.

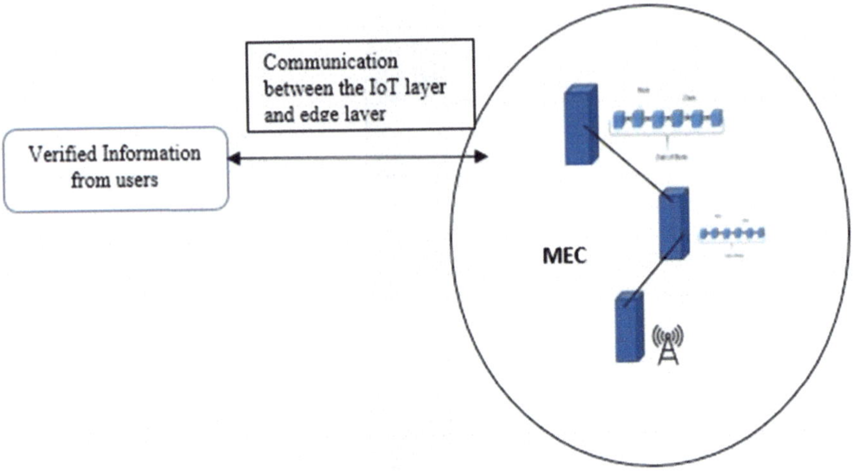

Fig. 17.10 Passing verified information to edge layer

These layers have the potential to solve the issues during and after the disaster like finding the location of the nearest rescue center, locating missing people, to protect people from fraudulent behavior, etc.

Findings
The following are the advantages of the model:

- The collaboration of external databases and different kinds of users to a single platform will ensure transparency.
- With edge computing, it becomes able to handle all requests and to generate responses corresponding to the request in a timely manner. The consensus algorithm of blockchain ensures that the information is verified and immutable before passing on to the network, hence providing a reliable flow of data.
- Since all the users are required to get registered in the network, there is no existence of ambiguity and duplicity. The consensus algorithm of the blockchain network will be able to detect any suspicious code and block that node from sending any information.
- The privacy of the user is maintained by the cryptographic hash function. So secure data flow is guaranteed here. This will help in taking the right decision timely.
- Since the important and required information is stored at edge servers, the speed of computation increases.
- Due to less power needed in computing results at edge servers and high-speed computation, the model shows its cost-saving nature.

Therefore, the integration fulfills our two main objectives:

1. *A reliable flow of data:* As edge devices are capable to store as well as process data, it ensures reliability. The connection loss with the cloud will not affect smart device operations.
2. *The secure flow of data*: Every node or IoT device is registered and then validated by blockchain. The data generated by these nodes are again validated before transferring on to the network to ensure no discrepancy or ambiguity of information as it would lead to wrong decisions. The information stored at edge servers and clouds is again protected by consensus algorithm to avoid any tampering.

17.5 Conclusion

In this chapter, a comprehensive view of the blockchain technology, edge computing, the IoT, and their flaws is presented. A model of blockchain integrated with edge computing in the IoT framework is presented. With this architecture, it is summarized how the integration removes the flaws of blockchain and edge computing. In addition, the benefits we get by their integration are explained. The working of the architecture with its three layers is analyzed, and what advantages this integration will give is studied thereafter. Finally, the chapter is concluded with a case study to show how the integration is to be implemented in real-world application like emergency alert system.

References

1. Ahire. J, *Blockchain: The Future?* in *Lulu.com*, (2018)
2. R. Yang, F. Richard Yu, P. Si, Z. Yang, Y. Zhang, Integrated Blockchain and Edge computing Systems: A survey, Some research issues and challenges. IEEE Commun. Surv. Tutor. **21**(2), 1508 (2019)
3. Blockchain. https://en.wikipedia.org/wiki/Blockchain. Accessed July 2020
4. easternpeak.com
5. Z. Pratap, *What is Blockchain and how does it work.* https://www.freecodecamp.org/news/what-is-blockchain-and-how-does-it-work/. Accessed 2nd July 2020 (2020)
6. J. Gubbi, R. Buyya, S. Marusic, M. Palaniswami, Internet of Things (IoT): A vision, architectural elements, and future directions. Future Generations computer systems. **29**(7), 1–19 (2013)
7. Overview of Blocks. https://subscription.packtpub.com/book/big_data_and_business_ intelligence/9781789139396/1/ch01lvl1sec19/overview-of-blocks. Accessed 20th July 2020
8. https://www.researchgate.net/publication/285388098_Multi-agent_System_for_Controlling_a_Cloud_Computing_Environment
9. Blockchain basic questions. https://www.edureka.co/community/73666/b. Accessed 1st Aug 2020
10. Meet97_patel Consensus Algorithms in Blockchain. https://www.geeksforgeeks.org/consensus-algorithms-in-blockchain/. Accessed 13 Aug 2020 (2020)

11. Ameer Rosic What is Blockchain technology? A step by step guide for beginners. https://blockgeeks.com/guides/what-is-blockchain-technology/. Accessed 2nd Aug 2020
12. https://www.javatpoint.com
13. M. Dryfhout, S. Hewer, *What is Cloud Computing*. https://scouttg.com/blog/articles/what-is-cloud-computing/. Accessed June 2020 (2019)
14. https://www.sciencedirect.com/science/ar
15. https://sciencedirect.com/science/ar
16. M.A. Khan, K. Salah, IoT security: Review, blockchain solutions, and open challenges. Future Gener. Comput. Syst. **82**, 395–411 (2018). https://doi.org/10.1016/j.future.2017.11.022.
17. Z. Zheng, S. Xie, H. Dai, X. Chen, H. Wang, An overview of Blockchain technology: Architecture, consensus, and future trends, in *Proceedings of IEEE 6th International Congress on Big Data*, ed. by G. Karypis, J. Zhang, (Honolulu, CPS, 2017), pp. 557–564. https://doi.org/10.1109/BigDataCongress.2017.85
18. B.M. Nyamtiga, J.C.S. Sicato, S. Rathore, Y. Sung, J.H. Park, Blockchain-based secure storage management with edge computing for IoT. Electronics **8**, 828 (2019)
19. Inés Sittón-Candanedo, Corchado JM (2019): An edge computing tutorial ISSN: 0974-6471, Orient. J. Comp. Sci. Technol 12, No. (2), Pg. 34–38.
20. U. Jayasinghe, G.M. Lee, A.M. Dermott, W.S. Rhee, *TrustChain: A Privacy-Preserving Blockchain with Edge Computing*. (Wireless Communications and Mobile Computing, Hindawi, 2019) pp. 1–17
21. W. Yu, F. Liang, X. He, W.G. Hatcher, C. Lu, J. Lin, X. Yang, A survey on the edge computing for the internet of things. IEEE Access **6**, 6900–6919 (2018)

Chapter 18
Role of Technologies in Revamping the Supply Chain Management of Kirana Stores

Irfat Ahmad and Shailja Dixit

18.1 Kirana Stores and Indian Retail Sector

Indian retail is one of the most vibrant and fast-growing sectors which is experiencing exponential growth and has been termed as the fifth largest preferred retail destination globally now [1]. According to a recent report by RedSeer, the sector is set to grow to $1.3 trillion by the fiscal year 2025 from $1 trillion now. It accounts for over 10% of the gross domestic product (GDP) of India. The retail segment in India is supported by both organized and unorganized players with the later one ruling the market at 84% share. In the FMCG and grocery category though, their contribution will continue to be almost 90%, the report said. Organized retail, both online and offline, is poised to grow from 16% to 22% in the next 5 years. The unorganized segment is largely driven by neighborhood corner stores or grocery stores which in general parlance are called kirana stores.

These kirana stores are the lifeline of Indian consumers and an integral part of the distribution network of food and grocery products. More than 15 M stores are operating in every nook and corner of India serving millions of Indians both from higher-income groups and to lower-income groups. Despite the deeper penetration of organized retail and e-commerce platforms, these stores have not lost their consumer base and their relevance in the life of the consumers. There are millions of Indians from lower-income groups, for whom getting into an organized retail outlet is still a dream but for them, kirana stores have been their saviors. These stores

I. Ahmad (✉)
Research Scholar & Corporate Professional, Amity Business School, Amity University, Lucknow, India

S. Dixit
Amity Business School, Amity University, Lucknow, India
e-mail: sdixit1@amity.edu

T. Choudhury et al. (eds.), *Blockchain Applications in IoT Ecosystem*, EAI/Springer Innovations in Communication and Computing,
https://doi.org/10.1007/978-3-030-65691-1_18

serve the immediate neighborhood and build and keep a close connect with their consumers. They know their consumers by name, their buying pattern, and their shopping preferences and offer credit also to help them maintain their livelihood. There are huge numbers of daily wagers in India who earn their livelihood by doing temporary jobs and paid daily. For these consumers, kirana stores are not only a necessity but a source of credit also. Their families buy essential food items on credit from these stores during the daytime, and they pay back to these stores in the evening when they return home after days' work.

Most of these stores are run by a single individual, from the premise of their home and with limited capital and space. They store and stock a range of products – grains, cereals, packaged goods, beverages, nonfood items from a category of personal care and homecare, fruits, vegetables, and dairy items. These stores score very high on parameters of customer acquaintance, connect and knowing the buying patterns. Without spending any advertising or marketing budgets on branding, these stores develop a strong emotional connect with their consumers which lasts for years. The entry barrier in this business is very low and thus the competition is very high so every store tries to do their best to maintain their committed consumer base.

While these stores have proved themselves the true companion of Indian consumers for a long time, many of the high-end consumers understood their importance during the recent lockdown post-outbreak of Covid-19. When all the established organized players both from brick-and-mortar format to e-commerce and hyper-local format were struggling to operate due to lack of manpower and restriction for movements, these stores served their consumers even while risking their own life. The resilience shown by these stores during that critical time helped them expanding their consumer base to new territories and with a new consumer segment also. During that time when most of the other business was unable to operate, these stores earned laurels and praise from many brands which started wooing them to place their set of brands on these stores. Many hyper-local businesses and established e-commerce companies which were trying to collaborate with these stores to ensure their last-mile deliveries even before lockdown stepped up their efforts to build a partnership with these stores. These partnerships have blurred the gaps between online and offline channels and have provided opportunities for both the players to co-opt with each other rather than compete. While organized players have ensured last-mile deliveries using the vast network of kirana stores, owners of these stores have in return are eying for additional business from these partnerships.

Almost all the leading online and e-commerce players are trying to partner with these kirana stores to expand their reach and ensure last-mile deliveries. Jio Mart, which is promoted by Reliance, is also eying on kirana stores with WhatsApp-based order placement and home deliveries. The objective of these firms is to capitalize on the locational advantage of kirana stores to ensure faster deliveries of their goods while offering additional revenue to these stores.

While this newfound partnership is there to stay as it benefits both the players, at least the time being, it has allowed kirana stores to relook into their business models. While at the front-end operation where consumer experience matters the most, these stores have started working upon by using tech-based solutions. Cashless payment, order taking through messaging apps, listing with local online directories, and partnership with local logistics partners for home deliveries are a few of the initiatives taken by these stores in recent months. But the larger issue is still open for these stores in terms of managing the supply chain. Due to capital issues, they were already struggling to maintain the inventories and range of SKUs, but now with going beyond their traditional consumer base, they have another challenge in terms of managing a wider range and product assortment. This has led to a situation where they need to revamp their entire supply chain.

18.2 Kirana Store's Supply Chain and Its Key Components

A supply chain is termed as the process of making and selling commercial goods, including the procurement of raw material, conversion into finished goods, and its transportation and distribution, for sale to end user. In a more simple term, supply chain can be further defined as an integration of three components buy, make, and move. Here buy refers to sourcing of raw materials, make is to convert this raw material into finished goods, and move is about the transportation of these finished goods to point of sales.

Supply chain management is the management of every step in to supply chain to make it cost-efficient and time-bound to help businesses generate more sales and create more happy customers.

In the context of kirana stores, supply chain starts at the procurement of finished goods from wholesalers or open mandis and ends at a sale to customers at their store. This can be depicted as shown below (Fig. 18.1).

Under newly found convergence between offline and online channels and the emergence of hyper-local channels, there is another leg being operated in the supply chain where logistics partners of hyper-local platforms and e-commerce companies collect the delivery from these stores and deliver to customers directly without the customer making a physical visit to the store. This additional leg in the supply chain can impact inventories and storage part of the supply chain of kirana stores.

Fig. 18.1 Traditional supply chain of kirana stores. (Source author)

18.3 Key Components of SCM and Scope of Technology

The traditional distribution network of kirana stores is highly fragmented, leading to various challenges to operate the stores. Limited storage space and working capital are the given constraint, whereas due to overdependence on wholesalers and company distributors for procurement, their supply chain becomes larger leading to wastages in the system. Various research studies suggest that up to 15% wastage in cereals and 30% in fruits and vegetables are due to storage-related issues and age-old handling methods alone.

Due to the challenge of overdependence on wholesalers and constraints of storage and working capital, supply chain management becomes more critical for kirana stores. Out of the different components of the supply chain, three can be taken as the most prominent one to study further from the perspective of technological intervention. These components are procurement or sourcing, inventory management, and storage and logistics. The component of logistics has been slightly downplayed here as most of the supplies to these stores are inbound and delivered at their stores by suppliers except the loose grains they buy from open mandis. But with the emergence of hyper-local channels and convergence of online and kirana stores, some outbound deliveries are also happening with the help of third-party logistics partners. The key components of SCM have been discussed below with the scope technology can have in their operations.

18.4 Procurement or Sourcing

Sourcing of goods is one of the key components of the supply chain which requires time, money, and understanding of demand projection. Since kirana stores sell both the packaged and loose items, getting estimates on what to buy and from where to buy to meet the financial constraint is a task and where the technologies can help these stores.

18.5 Inventory Management

Inventory management is a critical component for kirana store SCM. For them, it is highly critical to keep the right size of inventory seeing the constraints they have. Inventory management is a concern area that requires attention due to different expiry times of various SKUs in the food and grocery segment. Even to maintain the freshness of dairy items, chocolates, soft drinks, and vegetables for a longer time, these stores would be in need of cold storage with which most of them don't have until and unless provided by suppliers. To get an optimum inventory level under operating condition, these stores need the technological solutions which can suggest them average sales through rate, demand projection basis trends, and accurate estimates of stock lying with them.

18.6 Storage and Logistics

These stores buy mostly on a weekly basis, and quantities depend on credit amount extended by the distributors and suppliers for FMCG products. Due to constraints of capital and no data on how much to buy, frequent stock-outs, especially during the second half of the month, are a common phenomenon for these stores. The reason is that they buy more during the first week of the month seeing the purchase cycle of consumers, especially the salaried one. In the case of loose grains, the buying cycle most of the time happens on a harvest cycle basis, and thus the issue of warehousing is very common. The challenges of storage are directly linked with the inventory planning and procurement cycle and thus also require solutions based on technology to improve the efficiency and operating process.

18.7 Types of Technologies and Their Application in SCM of Kirana Stores

With the growth in the retail sector and the emergence of e-commerce, and the concept of omnichannel presence, supply chain management has become the most critical and talked about function of the retail sector across the globe. E-commerce companies are operating with a global supply chain where some of the goods are procured from the outside country and then shipped and stored in the local country for further distribution. The Amazon effect, a term which was coined after same-day delivery option to its customers, announced by Amazon has further pushed the retail sector to experiment with newer technologies. Digital, RFID, and of late blockchain technologies are some of the most popular technologies being used in supply chain management across the globe. Robotics and the IoT are getting used for supply chain management but are still in a nascent stage from the supply chain perspective. The identified three technologies have immense potential to drive not only cost efficiencies but scores high on parameters of acceptance, usability, and affordability also.

18.8 Blockchain

Blockchain is an Internet-based technology that is prized for its ability to publicly validate, record, and distribute transactions in immutable, encrypted ledgers. It came

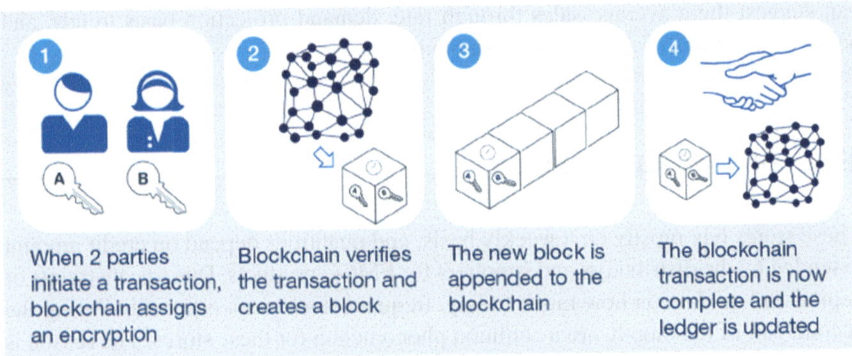

Fig. 18.2 How to create a blockchain transaction. (Source Mckinsey & Company)

into light in 2008 as a tool for securely tracking cryptocurrency transactions. It is a record-keeping technology that is nearly impossible to tamper with.

Blockchain uses mutually distributed ledgers that are built on a series of innovations and used to organize and share data in a digital form. As defined by Seebacher and Schüritz [6] "A Blockchain is a distributed database, which is shared among and agreed upon a peer-to-peer network". It consists of two key elements:

1. Blocks which are a storage unit of transaction
2. Time-stamped transactions that are secured by public-key cryptography (i.e., "hash") and verified by the network community

Once an element is appended to the blockchain, it cannot be altered and it becomes an archive of all the past transactions and activities (Fig. 18.2).

Blockchain as a technology has got immense potential, but its usage is still limited in supply chain management. In today's business set up, most of the firms operate their supply chain operations without using Block Chain Technology. But block chain has immense potential to excite the SCM professionals and they have already started building the initial cases and projects to check the usability and feasibility of this technology in supply chain operations. Walmart is already using it to trace the origin of the products, e.g., mangoes which are shipped from Mexico to the USA. It says that the use of blockchain has shortened the time to track produce from 6 days to 2 seconds, which helped them solve several issues related to food safety, customs and regulatory filings, and automated payments. China-based retailer JD.com has also been using blockchain to track supplies of beef from Australia to China to address the problems of food contamination, misrepresentation, brand erosion, and product theft.

Out of various properties and comfort in which blockchain offers, there are two characteristics – transparency and traceability – which are critical for supply chain management in the context of local market conditions of India.

Transparency Transparency is referred to as the information available to each of the players involved in a supply network. Awaysheh and Klassen [2] define transparency as the extent to which information is readily and easily available to both the counterparties in exchange and also to outside observers who are involved. Transparency is critical as it ensures visibility to all concerned parties about the movement of goods and stages it passes through from procurement till final sale to consumers. It provides a bird's-eye view in the value chain with a guarantee of the proper handover of third-party goods and final product labeling also.

Traceability Skilton and Robinson [7] define traceability as the ability to recognize and substantiate the components and chronology of events in all steps of a process chain. Traceability is related with information to trace the origin of the product in which consumer is buying as most of the consumers now demand information on the source of the product they buy. Blockchain technology can enhance product traceability by reducing counterfeiting and by streamlining product recall.

Besides the above two key properties, blockchain offers other benefits also which can be used in building a more efficient supply chain.

- Reduced errors
- No product delays
- Prevention of fraudulent activities
- Increased consumer/supplier trust

Blockchain can help stores in providing verified information to customers regarding where products are, how it got there, and when it arrived at the shop.

18.9 Role of Blockchain in Supply Chain Management of Kirana Stores

Procurement is the key function of kirana stores. Blockchain can help them manage their procurement as it provides two valuable benefits to them. End-to-end visibility of blockchain promotes fair and complaint practices that can prevent any malfunctioning during the procurement process. With almost no intermediaries in between, the procurement process would become a less costly affair for these stores. Besides that, it would reduce the chances of human errors. For kirana stores, procuring grains is a big challenge as they buy it straight from mandis or farmers, and without any visibility, once the order has been placed, chances of mixing of low-quality grains are very high. The usage of blockchain technology in this procurement process would ensure transparent information at all levels of the supply chain and reducing the possibilities of quality degradation at any level.

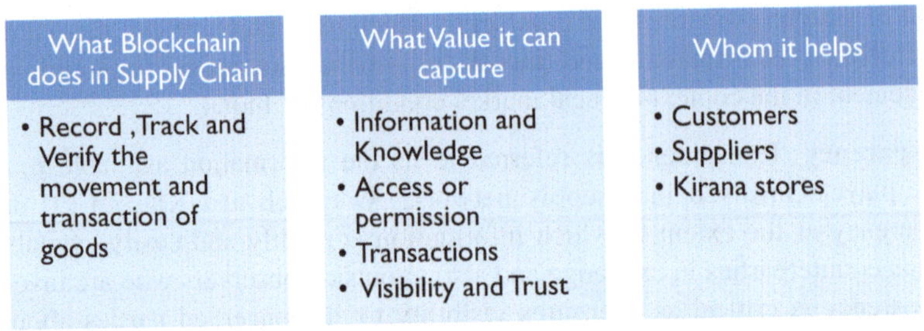

Fig. 18.3 How blockchain can help kirana stores managing supply chain. (Source author)

The benefits of blockchain can be transferred in the delivery cycle of the supply chain also which directly impacts the consumers. With the increased partnership between these kirana stores and hyper-local and e-commerce platforms, the delivery part of goods to consumers has been outsourced to third-party logistics partners in most of the cases. These logistics partners deploy their own vehicles to deliver the goods to the end user. Since three parties are involved in the process from order taking to order delivery (online platform to kirana stores to logistics partners), issues at any level can impact the customer experience. Blockchain can come handy in these situations where delivery vehicles can be integrated with technologies such as GPS, which in turn works as an input source of information for the blockchain. Once this mechanism gets in place, it almost removes all the possibilities of data forging and would track any hampering or mishandling of the goods dispatched for delivery to customers.

The third area where blockchain can be used by kirana stores is to track their goods especially food items where freshness and hygiene are most sought after attributes from a consumer perspective. Blockchain can track food items throughout supply chains to help reduce inefficiencies and give speed to the flow of goods. It can provide visibility to everyone involved in the process of sourcing, storing, distribution, and sale of food items to help them work together. It can solve two purposes – customers would have clear visibility about the place from where their food comes from and would reduce the chances of waste of food items due to spoilage and delay in delivery timing. For example, if a plum is plucked from a plant and transferred into a storage space, the blockchain records the status of the plum. In the absence of a proper record, a supplier might think that the plum is still on the tree.

Diagram given below is showing how blockchain can help kirana stores in managing their supply chain with the help of blockchain (Fig. 18.3).

18.10 RFID

RFID stands for "radio-frequency identification" where digital data is encoded in RFID tags or small labels and captured by a reader via radio waves. Rouse [5] defines RFID as a technology that incorporates the use of electromagnetic or electrostatic coupling in the radio-frequency (RF) portion of the electromagnetic spectrum to uniquely identify an object, animal, or person.

In basic characteristics, RFID is quite similar to barcodes when data from tag or label is taken by a device that stores it in a database. While comparing RFID and barcode, RFID data tags have the functionality to be read beyond the line of sight where the barcode needs to be in the range of optical scanner for retrieval of stored data. RFID systems consist of three components – RFID tag or label, RFID reader, and an antenna. RFID tag contains an integrated circuit, and antenna is used to communicate data to the reader which in turn converts the radio waves into different data formats and transfers this to the host computer via a communication interface.

The criticality of RFID technology lies in its capability to help store owners to trace the location and quantity of inventory without doing manual search and counting [8, 9]. It enables store owners to meet the consumer demand by inventories at the right place, at the right time, and in the right quantity (Fisher et al. 2000). RFID improves the traceability of goods to help retailers track the flow of goods in the physical distribution channel at any given point which leads to reduced inventory levels [4].

Walmart experimented with RFID and successfully placed RFID tags on individual garments that can be read by a handheld scanner [10]. This helped in ensuring optimal stock at shelves and maintaining the required inventory level. In an organized retail scenario, there is no need to scan products with RFID tags separately but can be accounted while being remained in the shopping basket when the customer passes through a doorway equipped with appropriate readers. This leads to faster checkouts, saves labor cost, and adds to enhanced customer experience.

RFID helps in creating delighted customers as it can provide them information about the origin of the product especially in the case of perishable and organic food items. Retailers can use the RFID technology to create differentiation from their competitors by adding and communicating the values hidden in the products itself to the customers.

Like any other technology, RFID also has positives and negatives. While on the positive side, it can revolutionize the supply chain management for both big and small stores, on the negative side it can be misused to track the locations of customers also. Privacy advocates argue that retailers could misuse the technology to track the location of the customers beyond the store also using RFID tags as these tags are traceable even after being removed from the products bought by the customers. Although no such issue has been reported in the recent past, companies have been working on this problem to reduce the privacy concern of the consumers.

18.11 Role of RFID in Supply Chain Management of Kirana Stores

For kirana stores, RFID has its potential to help them in two key areas of supply chain management – storage and inventory management. Managing inventories is a big challenge for kirana stores due to two reasons – space constraints and low working capital. Normally a kirana store stocks and sells 3500–4000 SKUs in limited space, thus making it difficult to locate the low selling SKUs at the right time. Another challenge is in the absence of a proper record of the stock count, and frequent stock-out position is witnessed at stores. RFID helps kirana store not only to locate the products but also to keep a proper track on stock availability of different SKUs so orders can be placed in time to avoid stock-out situations. Proper tracking also helps in saving products from being damaged and dented and thus incurring possibilities of losses for these store owners. Less stock-out situations lead to high customer satisfaction, thus leading to higher sales for these stores.

A core capability of RFID is the ability to have a piece of complete information from the date item is received and data at which it is sold. It helps retailers to know their sale through rate and in turn support them in inventory design and replenishment cycle. It leads to lower wastages and reduces the losses due to spoilage.

With the help of RFID technology, store owners can track the food items all the way from the original location to the consumers it is sold, thus bringing transparency in the entire value chain. Most of the kirana stores sell items that are famous for being produced at one particular location. Like mangoes from Malihabad, Lucknow, in northern India. RFID can help track the origin of these mangoes and would come as handy for consumers to know the source of the product that they are buying.

RFID helps in keeping a complete track of products in the store as retailers want to know the status and condition of goods. RFID tags can update on a real-time basis the information like the temperature of products during transportation. It offers them the opportunity to do the correction and take actions if required to maintain the specific temperature or conditions for the critical products during any stage of the supply chain process.

18.12 Digital Technologies

Deloitte [3] defined digital technology as an as technology-enabled combination of resources (can include instruments, devices, bots, tools, processes, networks, methodologies) which enable the availability of content (can be data, information, expert/social reviews, reports, analysis) for the user to make more productive decisions and satisfying choices. Digital technologies are electronic tools, systems, devices, and resources that generate, store, or process data. Well-known examples include social media, online games, multimedia, and mobile phones.

In the traditional supply chain, information is used to travel in linear order with each step dependent on the one before it leading to cascade the inefficiency at one step in subsequent steps. Digital technologies help in building a coherent supply chain, which is more dynamic and integrated. The digital supply chain employs real-time data and provides greater transparency to help make more informed decisions. It also provides more avenues to collaborate across the entire supply chain network.

Digital technologies are in the process of reshaping the supply chains and are poised to give a complete makeover to the existing process of procurement. Application based on cloud computing and IoT has disrupted the fundamentals of the old-fashioned supply chain. These technologies are helping retailers in strategic sourcing by using predictive analytics and improved transactional relationships between retailers and suppliers. Digital procurement solutions are helping retailers by giving them access to data that was hidden from them to make them use it to drive complex analysis and building more efficient business operations. With the presence of this data, retailers are now ready to partner with brands directly which are already experimented with virtual showrooms and virtual reality.

A digital supply chain (DSC) is a technology-supported smart and value-driven process to produce revenue and business value for the organizations by leveraging new approaches and methods. DSC is not about whether goods and services are digital or physical; it is about how supply chain processes are managed with a wide variety of innovative technologies and solutions.

Digital technologies provide the capabilities that can completely re-engineer the existing supply chain process by collaborating with each player involved in the process and thus focusing more on shoppers alone with the objective of improving sales and shopping experience. The digitally connected supply chain can ascertain the right inventory required to support demand from the market in conjugation with cost to serve the demand.

18.13 Role of Digital Technologies in SCM of Kirana Stores

Since kirana stores operate on low capital and small space to store goods, a digital- based inventory model will help them store the economic quantities and faster replenishment of stock as and when required. Today's consumer is digitally connected and migrates frequently between online platforms and offline channels to ensure the best deal for him. With more retailers getting connected with digital technologies, these consumers place their orders before arriving at the store, thus making it easier for store owners to check their inventories in time. If store owners have clear visibility of collective orders, regardless of their origin especially in the case of the active hyper-local model, stores would have a much better chance to eliminate two things that are the key challenges – out of stock and improper inventories. Another area of application of digital technology at kirana stores is the tracking of food items. Freshness (meat, seafood, deli, bakery, produce) has become the battleground in food. Retailers are focusing on the quality and freshness of the food items to serve their customers and reduce their objection levels. With the help of technology, retailers can weight products randomly, track code dates, enforce safety requirements, and thus ensure food safety as a part of the culture of the store operation.

Kirana stores operate with a space of 200–500 sq ft and always complain about space as in the absence of proper planning the available space even falls short of placing goods in order and sequence, leading to a situation where some of the items get lost in haphazard storing and finally lead to damaged or OOD which again cause loss of cost. Digital technologies deal with macro- and micro-space analysis to improve store layout and store planning as well as optimization of shelf space to assure the most optimal utilization of space and arrangement of goods. Digital displays are now used very frequently for merchandising at kirana stores which have improved store layout and store planning as well as optimization of shelf space to assure the most optimal utilization of space and arrangement of goods.

For FMCG products, procurement is largely dependent on the visit of suppliers at kirana stores. A sensor-based POS terminal keeps track of sales and stock position for various SKUs for packaged goods with barcode if entered properly in the system. This functionality helps small kirana stores managing their replenishment and further procurement.

18.14 Conclusion

Technology has been instrumental in revamping supply chain management across the globe in recent years. Most of these initiatives have been taken largely by organized retailers but later being replicated by unorganized sector also depending upon factors of affordability and ease of usability. In the last few years, kirana stores in India also started experimenting with technological solutions in the supply chain. Eroding market share, changing consumer demographics, increased digitalization, and the onslaught of organized retail and e-commerce were key factors that not only posed challenges before these stores but also provided an opportunity to revamp the existing process both at front-end operations as well back-end operation. Increased affordability of smartphones, mobile data, and availability of the Internet in vernacular languages worked as an enabler in this transformation of kirana stores. In terms of the application of blockchain technology at kirana stores, it looks quite premature seeing the cost and absence of proven examples in the Indian context. While scope for digital technologies and RFID look promising, it would require more sincere efforts to help retailers understand the benefits and operational efficiency these technologies can brought for them. The role of suppliers and distributors would be critical for helping, motivating, and handholding kirana stores to start using these technologies. It can be concluded that technologies like blockchain, RFID, and digital have enough scope to address the issues of traditional supply chain and restructure it according to the needs of today's time. These technologies can help in not only reducing the cost of operation but to improve the operational efficiencies to increase consumer footfall and finally growth in business.

References

1. IBEF (Jan, 2019). Indian Retail Industry Analysis. Retrieved 10 October 2018, from https://www.ibef.org/industry/indian-retail-industry-analysis-presentation
2. A. Awaysheh, R.D. Klassen, The impact of supply chain structure on the use of supplier socially responsible practices. Int. J. Oper. Prod. Manag. **30**, 1246–1268 (2010)
3. Deloitte, Disruption in Retail through Digital Transformation. Retrieved from https://www2.deloitte.com/content/dam/Deloitte/in/Documents/CIP/in-cip-disruptions-in-retail-noexp.pdf (2017)
4. K. Kalyanam, R. Lal, G. Wolfram, Future store technologies and their impact on grocery retailing. in *Retailing in the 21st Century: Current and Future Trends*. 2nd edn. (Springer, 2010)
5. M. Rouse, *RFID (Radio Frequency Identification)*: Definition, Accessed at http://searchmanufacturingerp.techtarget.com/definition/RFID on October 21, 201224–26 May 2017; pp. 12–23 (2007, April)
6. Seebacher, S., & Schüritz, R. (2017, May). Blockchain technology as an enabler of service systems: A structured literature review. In *International Conference on Exploring Services Science* (pp. 12–23). Springer, Cham
7. P.F. Skilton, J.L. Robinson, Traceability and normal accident theory: How does supply network complexity influence the traceability of adverse events? J. Supply Chain Manag. **45**, 40–53 (2009)
8. N. Dehoratius, Inventory record inaccuracy and RFID. Proceedings: *15th North* Am. Res. Sym. Purch. Supply Manag. **15**, 67–76 (2004)
9. P.M. Dunne, R.F. Lusch, Retailing, 5th ed. (Mason, South-Western, 2005)
10. M. Bustillo, Wal-Mart radio tags to track clothing. Wall Street J. **23**, A1–A14 (2005)

Chapter 19
Application Potential of Blockchain Technologies in the Travel and Tourism Industry

Diptiman Banerji, Waleed Rashideh, Bharat Arora, and Aditya Ranjan Pratihari

19.1 Introduction to Blockchain

A blockchain is a distributed ledger made up of a list of transaction bundles of closely linked blocks (or constituents). Under normal circumstances, it is not possible to modify these constituents once they become parts of the entire chain. Blockchain constitutes a peer-to-peer network in which decentralized nodes keep copies of the entire chain, which enables two parties—who do not know each other—to enter a smart contract and facilitates the trusted conclusion of online agreements (e.g., on a platform like Ethereum). Miners, individuals using computer hardware to run algorithms on specific software, validate new transactions on the blockchain and add these transactions to the global ledger (i.e., the blockchain). Blockchain technology creates a decentralized chain and gives access to digital assets in a way that their history remains unalterable and transparent by employing cryptographic hashing. In sum, blockchain is a state-of-the-art technology that promises to reduce risk, minimize chances of fraud, and bring transparency in a scalable way for multiple uses across different sectors and industries.

The banking and financial services industry uses blockchain technology in areas such as cryptocurrency, asset management, and management of insurance claims and cross border remittances. On the other hand, the real estate industry employs blockchain in numerous ways, for example, to manage smart contracts, to

D. Banerji (✉) · B. Arora · A. R. Pratihari
Jindal Global Business School, O.P. Jindal Global University, Sonipat, Haryana, India
e-mail: diptiman@jgu.edu.in

W. Rashideh
College of Computer and Information Sciences, Imam Muhammad Ibn Saud Islamic University, Riyadh, Saudi Arabia
e-mail: wmrashideh@imamu.edu.sa

T. Choudhury et al. (eds.), *Blockchain Applications in IoT Ecosystem*, EAI/Springer Innovations in Communication and Computing,
https://doi.org/10.1007/978-3-030-65691-1_19

increase liquidity in the market through the use of cryptocurrency, and to eliminate intermediaries to achieve cost reduction. A real-life example is "Propy," the global real estate technology company that uses blockchain to enable online real estate transactions through decentralization of its title registry system. Further, there are applications of blockchain technology in healthcare (e.g., for digital IDs, medical recordkeeping, and tracking of drugs all the way through the supply chain to the manufacturer), gaming (e.g., OPSkins, an organization involved in the selling of rare skins, accessories, and emotes for games started using cryptocurrency as a mode of payment[1]), and government services (e.g., digital voting and digital IDs). In sum, blockchain is a disruptive technology that has multiple applications across various industries.

Travel and tourism is another major industry that uses blockchain in its various operations. The adoption of cryptocurrency, whose acceptance is gradually increasing, has directed the industry's attention toward blockchain. As indicated through the website coinmarket.com, there exist 5608 cryptocurrencies with an overall market cap of $267,593,696,533—last updated on June 22, 2020—where the Bitcoin dominance reached 64.7%. For example, the German tourism company TUI is using blockchain technology for reservations and bookings, and companies such as Expedia and Webjet are accepting Bitcoin as a payment method [16]. Other potential applications of blockchain for the tourism sector include maintenance of the authenticity of customer reviews by regulating the identities and eliminating any fraudulent or unverified identities and overseeing innovative customer loyalty programs. This chapter takes a structured look at the various ways in which the travel and tourism industry is employing blockchain technology to date, as well as the potential applications the technology may provide for the industry in the future (Fig. 19.1).

19.2 Methodology

A rigorous, systematic literature review (SLR) [17] of research articles that discuss the concept of blockchain technology published in the leading travel and tourism journals constitutes the first step of the SLR process. Leading travel and tourism journals were identified as those whose names contain either the words "travel," "tourism," or "hospitality" from the ABDC A or A* level publications. A search on the ABDC journal quality list revealed 20 such journals. The next step was the identification of all publications within these 20 journals that contain the word "blockchain." A Google Scholar search identified a total of 30 articles (as of May 2020). Third, a thorough study of these 30 articles identified 10 papers that discuss blockchain-related applications in the travel and tourism industry. In all, 11 themes

[1] As reported on July 3, 2020, by the online publication *The ESports Observer*, the Valve Corporation has shut down OPSkins bots over workaround issues attributed to the company [21].

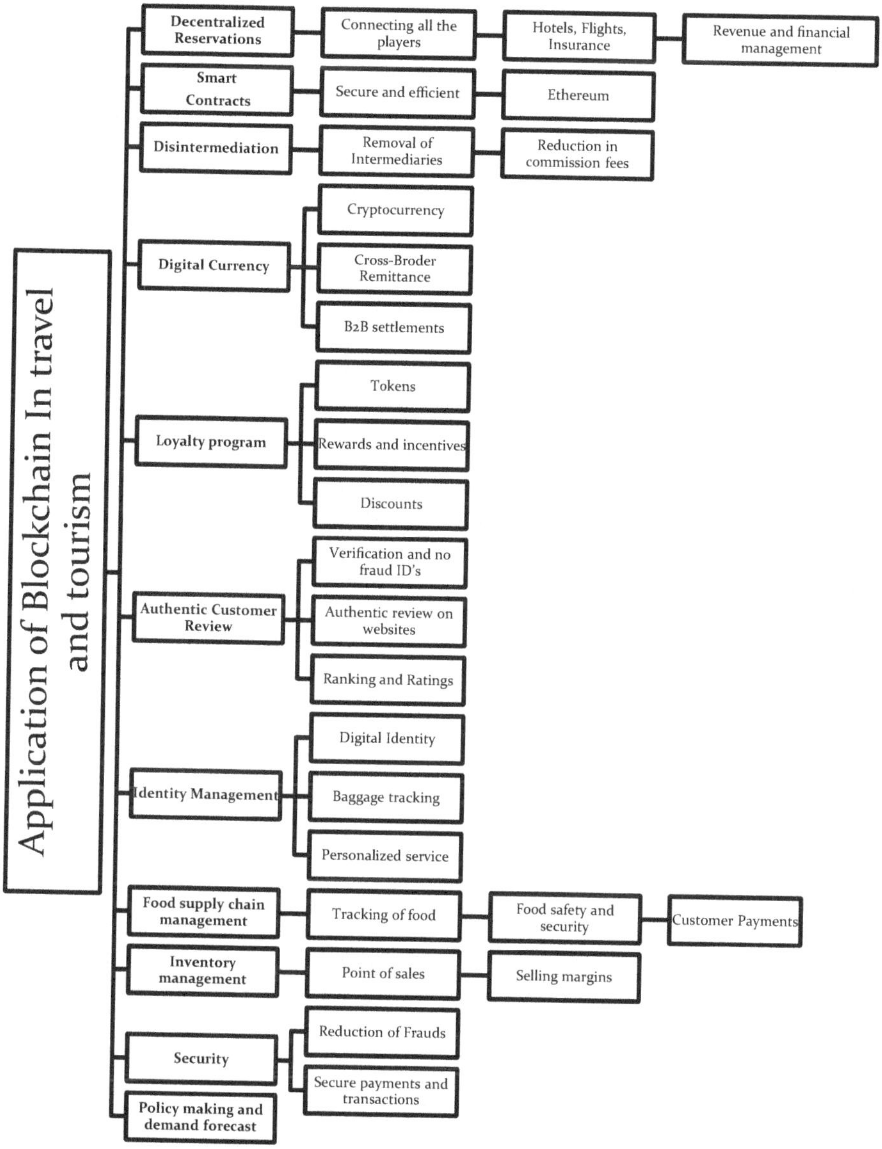

Fig. 19.1 Applications of blockchain technology in travel and tourism

included present and possible applications of blockchain technology discussed in these 10 articles and the reference lists of each. The remainder of this chapter will discuss and build upon these findings and then conclude.

19.3 Applications of Blockchain in Travel and Tourism

In this section, the 11 themes that emerged from the SLR that showcase the areas in which blockchain technology is in use in the travel and tourism industry are discussed.

19.3.1 Decentralization in Booking (Connecting All the Players)

Blockchain technology can connect all hotels through the *One Global Hotel Hub* chain that will offer more choice to customers, accuracy in demand prediction, and direct contact between the two. Similarly, the *Winding Tree* blockchain aims to connect contractors, service providers, and customers of hospitality services, to reduce the barriers to national and international market entry and to eliminate intermediaries [6]. Blockchain helps in the decentralization of bookings. This decentralization connects the players in the marketplace and removes intermediaries, making it possible for customers to choose from more options, make relatable comparisons, pay less for bookings because of the removal of commission fees, and prevent overbooking.

19.3.2 Smart Contracts

A smart contract refers to computer code that (1) helps execute an agreement, (2) is managed by a peer-to-peer chain of computers, and (3) is stored on a blockchain-based system [8]. Smart contracts—likely to underpin the future development of blockchain—play an integral role by allowing parties within the network to create their own agreements without the need for external intermediaries, thus simplifying the processing and management of transactions [6]. Companies will be able to place orders and issue payments without instructions since smart contracts have an autonomous escrow function [15]. Ethereum is the first provider using blockchain for establishing a smart contract that can facilitate more significant disintermediation [16]. In a significant development, the Caribbean island of Aruba has developed an Ethereum blockchain technology platform for travel bookings [10]. Blockchain also ensures that all contracts are stored on a secure and efficient database.

19.3.3 Digital Currency

Blockchain can facilitate the hospitality industry with cryptocurrency, such as Bitcoin. Businesses such as Expedia have also experimented with accepting Bitcoin for travel bookings [8]. Blockchain technology can also safeguard monetary transactions and cashless payments and provide credit facilities to customers and suppliers in the hospitality industry [6].

To boost tourism, Caribbean islands have issued regional governmental cryptocurrency, that is, the Digital Eastern Caribbean Dollar [10]. Cryptocurrency, based on blockchain technology, can provide a secure exchange of money without third parties [16] by making it possible for tourists to pay for travel and reservations through blockchain or cryptocurrency [15]. Blockchain can also help with cross border remittance [10], enables C2C transactions in the primary and secondary markets for tourism products [16], and is not only a tool for cashless transactions but also provides secure payments.

19.3.4 Disintermediation

The most significant impact of blockchain technology on the tourism industry can be the reduction in the level of intermediation with the potential of removing online travel agents from the tourism supply chain [16], thus reducing operational costs for businesses. For example, Lufthansa and Nordic Choice Hotels use blockchain technology for the removal of intermediaries and the reduction of distribution costs, respectively [8]. Further, for the aggregators operating individually in the hospitality segment of the sharing economy such as Airbnb, blockchain can tackle the malicious behaviors by either host or guest [2], thus making the transactions more equitable in the sharing economy segment of hospitality [6, 10]. In fact, blockchain-based systems will not only eliminate existing intermediaries but will also block the path for new intermediaries for the tourism sector [18]. Further disruption through blockchain technology is expected in the sharing economy and reservation platforms [11] and in improvement in the sharing economy overall [7].

19.3.5 Authenticity in the Realm of Customer Reviews

With the implementation of systems that can identify and remove false and unfair reviews or comments, blockchain can help the hospitality industry through the reshaping of traditional communication channels in digital marketing [6]. Blockchain is capable of building a unique identity for each entity by embedding verification processes into the review and rating platforms [8]. Currently, because the industry can manipulate the centralized system (which is not based on blockchain), the trustworthiness of the review system is questionable. Blockchain can provide fair online reviews using a unique private key, which will minimize duplicate reviews [16].

19.3.6 Innovative Loyalty Programs and Loyalty Program Management

Loyyal is a blockchain technology-based platform that improves the interoperability of airline reward programs by allowing for the collection of earned points irrespective of the airline the customer uses and whereby the inter-airline transfer of points is possible. The platform is also suited for hotels and cabs [6, 14]. Adoption of blockchain technology will enhance public relations, provide discounted travel and rewards or incentives such as points to customers, and even attract travelers to new and underserved tourist destinations [10]. Blockchain can solve malfunctions in and improve the competitiveness and efficiency of loyalty programs through automation [8]. Using blockchain technology, frequent flyers will be able to avail themselves of tokens for use at restaurants or for booking flights.

19.3.7 Tracking, Identity Management, and Service Customization

With the use of blockchain technology-based digital identity, tracking and customized customer services are available to guests without the worry of leaks of personal data [8]. Referencing the literature of small island economies by Kwok and Koh [10], blockchain can provide tourists and tourism operators with technology-based, smart, personalized solutions, unique digital identity, and baggage tracking, which will enhance the experience of tourists. Blockchain technology can provide a solution to the critical issue of baggage management problems in the aviation sector through continuous monitoring of AI-based sensors and contribute to identity management in the immigration stage where reduced multiple checks at different stages will decrease wasted time.

19.3.8 Food Chain Supply Management

The possibilities of scope for the use of blockchain technology for supply chain and logistics management in food procurement are many [6]. Blockchain technology can assist from the point of tracking food procurement, to delivery to the customer, and through the customer's payment transactions to the supplier. Blockchain can track food deliveries—including quantity and type—and also store details of the restaurant owner or food supplier. This application is best suited to airlines, large hotel chains, and franchises.

19.3.9 Information to Policymakers and Demand Forecast

Through its use, blockchain businesses can build the exchange of information from industry to policymakers and vice versa. Through their usage of blockchain, businesses can track critical information about consumer statistics such as the number of tourists and preferred hotels. Policymakers can make use of this information in audits, and businesses can use the information for customer retention and reduction of customer dissatisfaction [6]. Blockchain-based systems also provide timely and precise data on business performance, which makes it possible to perform more accurate forecasting of demand, budgeting, and decision-making [6].

19.3.10 Security Reduction of Fraud

Due to being a peer-to-peer network, the process of "decryption key" in cryptography makes it challenging to hack or fake a transaction within the network, thereby reducing the risk of fraud and providing accuracy in billing [6]. This ultimately translates to achieving a higher level of consumer trust and to increasing business outcomes such as revenues.

19.3.11 Data Sharing

Through the distributed ledger, that is, blockchain technology, consumers can share their data with companies in exchange for monetary value such as tokens or cryptocurrency [12]. Smart contracts collect the data and pay in cryptocurrency. The decentralized platform prevents leaks of the data to third parties and provides control of the data to the customers [12]. The blockchain-based company, TravelChain, is making use of blockchain technology for data collection and sharing [22].

19.4 Examples of Present Blockchain Usage in the Travel and Tourism Sector

In this section, some real-life business applications of the concept of blockchain not covered above are discussed.

19.4.1 Loyalty Programs

Singapore Airlines (SIA), KPMG, and Microsoft have jointly developed KrisPay, the first airline loyalty program in the form of a digital wallet based on blockchain technology. The program allows customers to convert their air miles into a digital currency available for future flight bookings or making purchases from SIA's partner merchants [9, 20]. Similarly, another blockchain-based loyalty management program called Trippki provides tokens for hotel reservations for approximately 1.6 million hotels and accepts cryptocurrency as one of the modes of payment [23]. On a similar note, a program called Sandblock works on customer retention. It makes customized blockchain-based token programs that provide more than just reward points for companies to issue to repeat customers. They are crypto assets that the customers can either use for brand-specific rewards or trade as money [19].

19.4.2 Booking and Payments

LockTrip is a property-based rental platform that uses a decentralized system allowing end customers and property owners to deal with each other without any fees or commission charges. It operates as a direct marketplace for hotels and companies to rent out property and as a source of booking for customers, thus lowering the amount of commission paid to zero. LockTrip covers global booking and property management without intermediaries and commissions and accepts all modes of payment from cryptocurrency to credit cards. Through this method, LockTrip can provide its customers a reduction in booking costs of approximately 19–20% for about 530,000 different hotels and properties [13]. A similar application—Winding Tree—uses the decentralized ecosystem for the main objective of connecting all players, airlines, hotels, and travel guides, to make them all available for the customer at a single platform without intermediaries, which can save the customer up to 20% for online bookings. Ethereum and smart contracts regulate payments for better deals. Winding Tree has partnered with airlines such as Etihad and Lufthansa [25]. Further, Beenest, a sharing economy like Airbnb, dealt with lots of intermediaries, which increased the cost for both hosts and guests. Using the decentralized booking systems, Beenest is now able to directly connect hosts with guests—which reduces the integration costs—and provides users with

Bee tokens or cryptocurrency for payments. Beenest uses blockchain technology for disintermediation through which it also verifies hosts and guests for security and for elimination of unverified or fraudulent listings [3].

19.4.3 Identity Management

A few blockchain-based identity management solutions are being increasingly adopted by businesses. For example, a similar application, Known Traveller Digital Identity System, developed by Accenture and the World Economic Forum, is a blockchain technology-based digital identity system used to store travelers' information on a distributed ledger [1]. Airlines and airports use the system to enhance security and to shorten waiting lines. Marriott Hotels is adopting the technology for verifying and securing digital identities to reduce paper documentation.

19.4.4 Other Applications

The application DApps (decentralized applications) allows individuals to interact more closely and efficiently with smartphones or browsers through blockchain technology. Companies can utilize DApps for increased customer satisfaction and retention [15]. Companies can also use DApps to connect and interact with customers and individuals directly without intermediaries [18]. The application Cool Cousin uses a blockchain-based ecosystem to connect local guides with travelers and to provide personalized recommendations of guides to travelers through blockchain using cryptocurrency as a mode of payment available for both guests and travelers. Its rating is trustworthy as blockchain regulates it [5].

The application Webjet is a travel booking website used by Australia and New Zealand with the main purpose of reducing the stress of travel bookings and mistakes. The distributed ledger records all entries in blockchain and provides customers with a less stressful travel booking experience, avoids inaccuracies, and ensures the safety and security of the data. The testing of the technology took place with partners such as Thomas Cook and DidaTravel [24].

Further, the application—TravelChain—rewards users with travel tokens for sharing their information such as preferences, location, food, entertainment options, and the way of living and traveling [4]. Such information is likely to be useful for travelers, who currently have to browse multiple websites, blogs, forums, reviews, and watch videos to make their travel-related decisions. All this content is created for free by individuals because they want to share their experience with those they care about or even the whole world. With TravelChain, on the one hand, users receive a monetized token for data that acts as money, and on the other hand, firms can cut their costs by 50% and formulate specifically targeted offers by using the information that the users are willing to put on the market [22].

19.5 Conclusion

Blockchain is a disruptive technology that offers a variety of advantages that could dramatically transform business as it relates to the travel and tourism industry. Business models, payment systems, security and trust, and traceability and sharing information are all business factors considered core areas in the tourism industry. The positive impact of blockchain technology presents a golden opportunity for companies to obtain benefits in these areas. This chapter seeks to provide stakeholders in the tourism industry with basic facts regarding blockchain technology applications. The benefits offered to businesses through the use of blockchain technology makes it a top priority for adoption. Its use can transform the tourism industry through innovative strategies aimed at improving the efficiency of the business process.

Blockchain technology reduces risk, minimizes chances of fraud, and brings transparency to many industries. Cryptocurrencies—increasingly accepted all over the world as a payment method—increase the advantages that members of the travel and tourism industry can gain by using the blockchain technology. This review sets forth a variety of blockchain applications tailored for the travel and tourism industry and outlines their various uses. An SLR is the method used in this research to gather information on a variety of blockchain technology applications that serve the travel and tourism industry. Most of the applications aim to connect businesses and users without the interference of a third party. Smart contracts play a powerful role in connecting these parties to accelerate the business process and reduce costs. This platform provides trusted monetary transactions by using cryptocurrencies between suppliers and travelers without the need for intermediaries. In fact, this connection leads to more efficient business processes. In addition, travelers feel more confident making travel decisions based on the trusted reviews that employ blockchain technology. Blockchain technology can enhance a supply chain from farm to fork by accurately tracking and reducing fraud. Overall, this relatively new technology enhances and secures loyalty programs, booking, and decentralized applications and simplifies business processes in the tourism industry.

References

1. Accenture, *The Known Traveller: Unlocking the Potential of Digital Identity for Secure and Seamless Travel.*https://www.accenture.com/_acnmedia/PDF-70/Accenture-WEF-The-Known-Traveller-Digital-Identity.pdf (2018)
2. L. Altinay, B. Taheri, Emerging themes and theories in the sharing economy: A critical note for hospitality and tourism. Int. J. Contemp. Hosp. Manag. **31**(1), 180–193 (2019). https://doi.org/10.1108/IJCHM-02-2018-0171
3. Beenest, *What is Beenest.* https://medium.com/@thebeetoken/what-is-beenest-how-the-bee-token-is-revolutionizing-the-home-sharing-market-8da32d79bbbb (n.d.)
4. R. Bova, *How Could Blockchain Transform the Way We Travel? TravelChain CEO Explains.* https://cointelegraph.com/news/how-could-blockchain-transform-the-way-we-travel-travelchain-ceo-explains (2018)

5. Cool Cousins, *Travel Perfection Made Simple-Get Insider Tips Straight from Vetted Locals in Over 100 Cities*. https://www.coolcousin.com/stories/about-us (n.d.)
6. V. Filimonau, E. Naumova, The blockchain technology and scope of its application in hospitality operations. Int. J. Hosp. Manag. **87** (2019). https://doi.org/10.1016/j.ijhm.2019.102383
7. M. Hossain, Sharing economy: A comprehensive literature review. Int. J. Hosp. Manag. **87** (2020). https://doi.org/10.1016/j.ijhm.2020.102470
8. M. Kizildag, T. Dogru, T.C. Zhang, M.A. Mody, M. Altin, A.B. Ozturk, O. Ozdemir, Blockchain: A paradigm shift in business practices. Int. J. Contemp. Hosp. Manag. **32**(3), 953–975 (2019). https://doi.org/10.1108/IJCHM-12-2018-0958.
9. KPMG, *KPMG helps to develop first blockchain-based, airline loyalty program 'digital wallet*. https://home.kpmg/xx/en/home/media/press-releases/2018/09/kpmg-helps-develop-airline-loyalty-digital-wallet-fs.html (2018)
10. A.O. Kwok, S.G. Koh, Is blockchain technology a watershed for tourism development. Curr. Issue Tour. **22**(20), 2447–2452 (2018). https://doi.org/10.1080/13683500.2018.1513460
11. C. Lam, R. Law, Readiness of upscale and luxury-branded hotels for digital transformation. Int. J. Hosp. Manag. **79**, 60–69 (2019). https://doi.org/10.1016/j.ijhm.2018.12.015
12. N.D. Line, T. Dogru, D. El-Manstrly, A. Buoye, E. Malthouse, J. Kandampully, Control, use and ownership of big data: A reciprocal view of customer big data value in the hospitality and tourism industry. Tour. Manag. **80** (2020). https://doi.org/10.1016/j.tourman.2020.104106
13. LockTrip, *Book Hotels and Flights with Bitcoin and ETH, Travel with Crypto*. https://locktrip.com (n.d.)
14. Loyyal, *Reinventing Loyalty through Blockchain*. https://loyyal.com/partner/airline-program/ (n.d.)
15. K. Nam, C.S. Dutt, P. Chathoth, M.S. Khan, Blockchain technology for smart city and smart tourism: latest trends and challenges. Asia Pac J. Tour. Res. *27*, 1–15 (2019). https://doi.org/10.1080/10941665.2019.1585376
16. I. Önder, H. Treiblmaier, Blockchain and tourism: Three research propositions. Ann. Tour. Res. **72**(C), 180–182 (2018). https://doi.org/10.1016/j.annals.2018.03.005
17. J. Paul, A.R. Criado, The art of writing literature review: What do we know and what do we need to know? Int. Bus. Rev. **29**(4), 101717 (2020). https://doi.org/10.1016/j.ibusrev.2020.101717
18. W. Rashideh, Blockchain technology framework: Current and future perspective for the tourism industry. Tour. Manag. **80**, 104–125 (2020). https://doi.org/10.1016/j.tourman.2020.104125
19. SandBlock, *About Sandblock Chain*.https://sandblock.io/#chain (n.d.)
20. Singapore Airlines, *KrisPay*.https://www.singaporeair.com/en_UK/sg/ppsclub-krisflyer/use-miles/krispay (n.d.)
21. The E-Sports Observer, *Valve Shuts Down OPSkins Bots, Users Lose Skins Valued at Approx. $2M Despite Warning*.https://esportsobserver.com/valve-shuts-down-opskin-bots/ (2020)
22. Travelchain, *Connect the travel chain for customers*. https://amadeus.com/en/portfolio.retail-travel-agencies.content.travel-chain (n.d.)
23. Trippki.com, *Blockchain Events: Browse Blockchain events & book hotels with Trippki*https://trippki.com/blockchain-events (n.d.)
24. Webjet, *Booking Flights, Cheap Hotels, Car Hire and Insurance and Holiday Packages*. https://www.webjet.com.au (n.d.).
25. Winding Tree, *Blockchain powered decentralized travel ecosystem*. https://windingtree.com (n.d.)

Index

T. Choudhury et al. (eds.), *Blockchain Applications in IoT Ecosystem*,
EAI/Springer Innovations in Communication and Computing,
https://doi.org/10.1007/978-3-030-65691-1

The manufacturer's authorised representative in the EU is Springer Nature Customer Service Centre GmbH, Europaplatz 3, 69115 Heidelberg, Germany. If you have any concerns regarding our products, please contact ProductSafety@springernature.com

Printed and bound by CPI Group (UK) Ltd, Croydon, CR0 4YY
22/09/2026
02221062-0015